Newnes Electronic Circuits Pocket Book

Volume 2:

Passive and Discrete Circuits

Other books by R. M. Marston

Audio IC Circuits Manual
CMOS Circuits Manual
Diode, Transistor & FET Circuits Manual
Electronic Alarm Circuits Manual
Instrumentation and Test Gear Circuits Manual
Op-amp Circuits Manual
Optoelectronics Circuits Manual
Power Control Circuits Manual

Newnes

Electronic Circuits Pocket Book

Volume 2: Passive and Discrete Circuits

R. M. Marston

Newnes
An imprint of Butterworth-Heinemann Ltd
Linacre House, Jordan Hill, Oxford OX2 8DP

A member of the Reed Elsevier group

OXFORD LONDON BOSTON
MUNICH NEW DELHI SINGAPORE SYDNEY
TOKYO TORONTO WELLINGTON

First published 1993

British Library Cataloguing in Publication Data
A catalogue record for this book is available from the British Library

ISBN 0 7506 0857 9

Produced by Co-publications, Loughborough
Typeset by Sylvester Publications, Loughborough
Printed in England by Clays Ltd, St Ives plc

Contents

To Esther, with love.

Preface

Passive electrical components (resistors, capacitors, inductors, and switches, etc.) and transducers (relays, loudspeakers, 'phones, and thermistors, etc.) and simple 'discrete' semiconductor devices such as diodes, transistors, SCRs, and TRIACs, etc., form the very bedrock on which the whole field of modern electronics is built, and from which all modern electronic circuits (including those using linear or digital ICs) have evolved. This information-packed book is a single-volume applications guide to the most popular and useful of these devices, and presents a total of 670 diagrams, tables, and carefully selected practical circuits, backed up by over 65,000 words of highly informative text. It explains the basic features and important details of modern passive and active discrete components, and shows how to use them in a wide range of practical applications.

The book is aimed directly at those engineers, technicians, students and competent experimenters who can build a design directly from a circuit diagram, and if necessary modify it to suit individual needs. It deals with its subjects in an easy-to-read, concise, and highly practical and mainly non-mathematical manner. Each chapter deals with a specific type or class of device, and starts off by explaining the basic principles of its subject and then goes on to present the reader with a wide spectrum of data, tables and (where relevant) practical applications circuits.

The book is split into twenty distinct chapters. The first three explain important practical features of the available ranges of modern passive electrical components, including relays, meters, motors, sensors and transducers. Chapters 4 to 6 deal with the design of practical attenuators, filters, and 'bridge' circuits. The remaining fourteen chapters deal with specific types of 'discrete' semiconductor device, including various types of diode, transistors, JFETs, MOSFETs, VMOS devices, UJTs, SCRs, TRIACs, and various optoelectronic devices.

Throughout the volume, great emphasis is placed on practical 'user' information and circuitry; all of the active devices used in the practical circuits are modestly priced and readily available types, with universally recognised type numbers.

R. M. Marston

1993

1 Passive electrical components guide

Modern electronic circuit design is based on the interaction between passive electrical components or transducers and various types of active rectifying, amplifying, or switching devices. The practical electronics design engineer needs a good understanding of all these elements in order to generate truly cost-effective and reliable designs that will continue to function correctly under hostile operating conditions.

This opening chapter takes an in-depth look at the five major types of passive component, i.e., resistors, capacitors, inductors, transformers, and switches, and provides the reader with a concise but comprehensive guide to their symbology, pertinent formulae, basic data, major features, and identification codes, etc.

Guide to modern resistors

Either of two basic symbols can be used to represent a resistor, and *Figures 1.1 and 1.2* show their major family 'sets'. Internationally, the most widely acceptable of these is the 'zig-zag' family of *Figure 1.1*; these symbols may be subjected to some artistic variation, with the number of zig-zag arms varying from two to five. The alternative 'box' symbols of *Figure 1.2* are rarely used outside of Western Europe.

The most widely used resistance formulae are the simple 'ohms' and 'power' ones listed in *Figure 1.3*, and the series and parallel 'equivalents' ones shown in *Figures 1.4 and 1.5.* These formulae are valid under d.c. and low-frequency a.c. conditions only; all practical resistors exhibit a certain amount of inductance and capacitance, which may significantly influence the component's high frequency impedance.

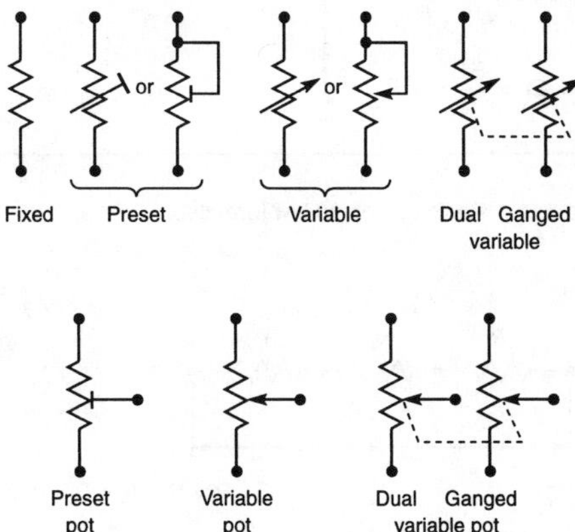

Figure 1.1. Internationally-accepted symbols for various types of resistor and variable potentiometer (pot).

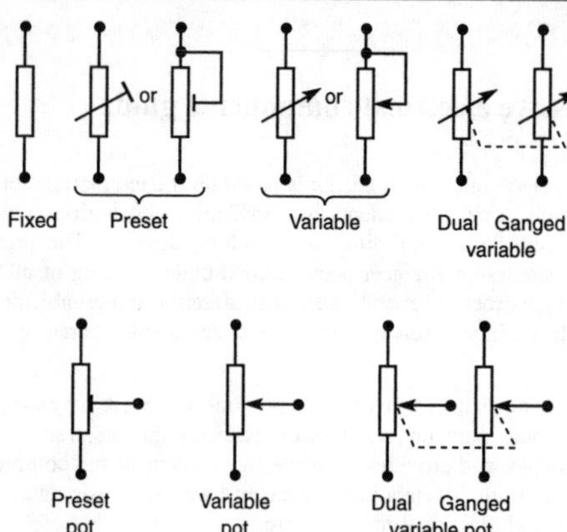

Figure 1.2. Alternative resistor and pot symbols, popular in many West European countries.

$$I = \frac{E}{R} = \frac{W}{E} = \sqrt{\frac{W}{R}}$$

$$E = IR = \frac{W}{I} = \sqrt{WR}$$

$$R = \frac{E}{I} = \frac{E^2}{W} = \frac{W}{I^2}$$

$$W = EI = \frac{E^2}{R} = I^2R$$

E = Voltage (volts)
I = Current (amps)
R = Resistance (ohms)
W = Power (watts)

Figure 1.3. Basic d.c. ohms law formulae.

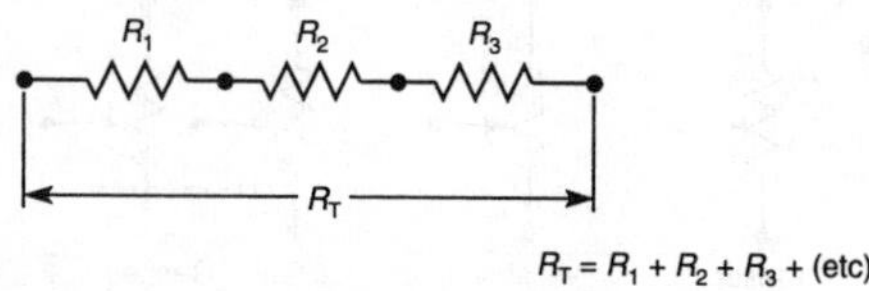

Figure 1.4. Method of calculating combined value of resistors in series.

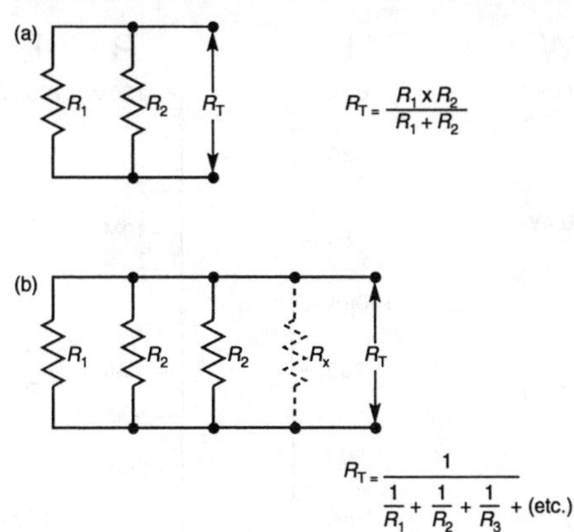

Figure 1.5. Method of calculating combined value of *(a)* two or *(b)* more resistors in parallel.

Precise parameter values can easily be derived from the above formula with the help of a calculator. Alternatively, a whole range of values can be quickly found — with a precision better than ten percent — with the aid of the nomograph of *Figure 1.6* and a straight edge or ruler, which are used in the manner described in the caption; thus, if a 1k0 resistor is used with a 10V supply, a line projected through these two values on the appropriate vertical columns shows that the resistor passes 10mA and dissipates 100mW.

Three basic types of fixed-value resistor are in general use or are likely to be met by practical engineers; these are carbon composition or 'rod' types, 'film' types, and 'wire wound' types.

Carbon composition resistors are now obsolete but are often found in pre-1985 equipment. They consist of a resistive rod (made from a blend of finely ground carbon and insulating/binding resin filler, the 'mix' of which determines the resistance value), to the ends of which are attached a pair of wire leads; the assembly is protected against moisture penetration, etc., by an insulating coating. *Figure 1.7* shows the three most widely used constructional variations of these resistors. In *(a)* and *(b)* the leads are axial (are in line with the component's axis); in *(c)* they are radial.

The most widely used modern resistors are 'film' types. These use the basic construction shown in *Figure 1.8(a)*; a resistive film of carbon compound (in 'carbon film' resistors) or metallic oxide (in 'metal film' or 'oxide film' resistors) is deposited on a high-grade non-porous ceramic rod and electrically connected to a pair of axial leads. The film is then machine-cut to form a helical spiral who's length/breadth sets the actual resistance value; the final assembly is given a glazed insulating coating.

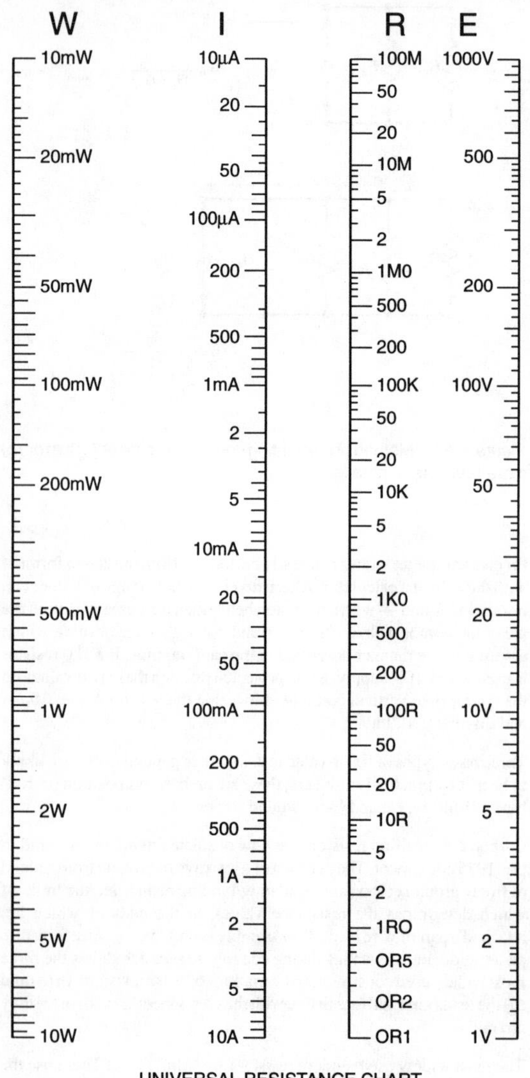

UNIVERSAL RESISTANCE CHART

Figure 1.6. To use this nomograph, lay a straight edge so that it cuts any two vertical columns at known reference points, then read off the remaining two unknown parameter values at points where the edge cuts the other two vertical columns.

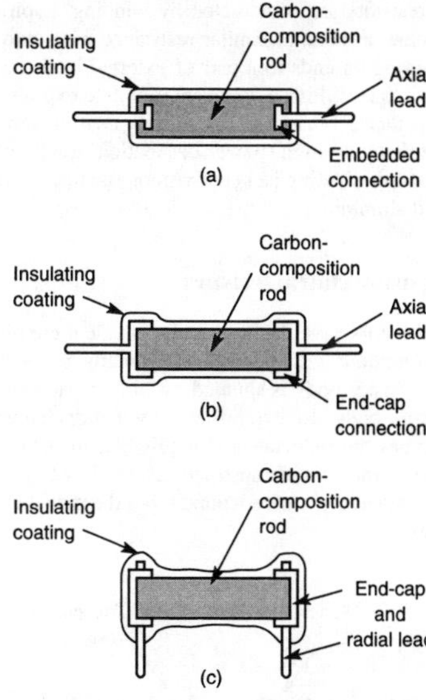

Figure 1.7. Construction of various types of carbon composition (rod) resistor.

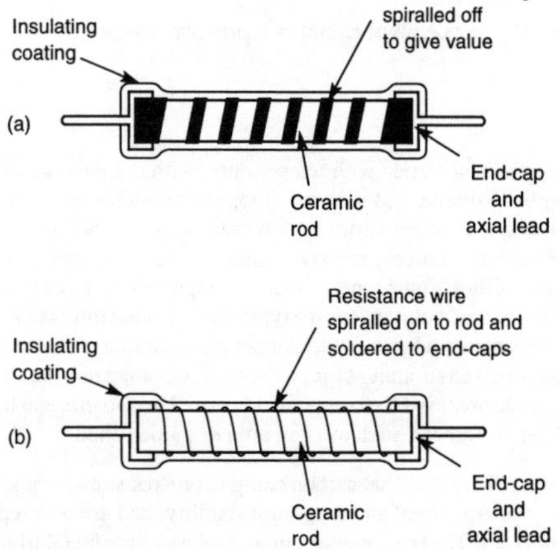

Figure 1.8. Typical construction of *(a)* film-type and *(b)* wire-wound resistors.

Wire-wound resistors are constructed by winding a spiral of high-quality Nichrome, Eureka, or similar resistance wire onto a ceramic former and joining its ends to a pair of external leads, as shown in *Figure 1.8(b)*; if the resulting component is built to exploit its inherent 'precision', it is then given a normal insulating coating, but if it is built to exploit its inherently high 'power dissipation' qualities it may be fitted with a heat-dissipating jacket of vitreous enamel, ceramic compound, or solid aluminium.

Practical resistor characteristics

All practical resistors have the electrical equivalent circuit of *Figure 1.9*. Each lead exhibits self-inductance (typically 8nH/cm or 20nH/inch), and the resistor body is shunted by stray capacitance (roughly 1pF). The resistor body also exhibits some self-inductance; in carbon composition types the inductance is negligible; in 'film' types it is (because of the spiral form of construction) small, but may be significant at high-frequencies; in wire-wound types the inductance is inherently quite high.

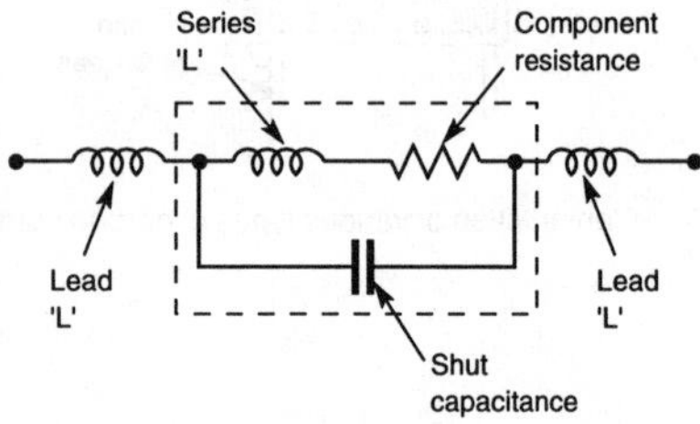

Figure 1.9. Equivalent circuit of a practical resistor.

A resistor's value varies with temperature, with the passage of time, with applied voltage, and with conditions of humidity, etc. The magnitude of these changes varies with resistor type, as shown in *Figure 1.10*. Here, the 'temperature coefficient' is listed in terms of both 'parts-per-million' (ppm) and 'percent' change per degree Celsius. The 'shelf life change' column lists the typical percentage limits of component value change when the component is stored in a stable temperature-controlled environment for one year; this change may be an order of magnitude greater if the component is stored in a hostile and highly-variable environment, such as in an attic or garden shed.

Note from *Figure 1.10* that carbon composition resistors are noisy and have very poor thermal and long-term stability, and are best replaced by 'film' types. The best general-purpose resistors are metal film ones, which offer excellent thermal and long-term stability and low noise. Wire wound vitreous (etc.) types offer very high power dissipation, and wire wound precision types offer superb overall stability.

Resistor type	Resistance range	Power ratings	Tolerances (±)	Ambient temp range (°C)	Temperature coefficient		Noise level	Shelf life change in 1 year
					±ppm/°C	±%/°C		
Carbon composition	2*R*2–22M	1/8W to 1W	10%, 20%	–40 to +105	–1200	–0.12	2µV/V	5%
Carbon film	10*R*–10M	1/4W to 2W	5%	–45 to +125	300–1000	.03–0.1	0.5µV/V	2%
Metal thick film	33*R*–100M	1/2W to 1 1/2W	5%	–45 to +130	100–300	.01–.03	0.25µV/V	1%
Metal film	10*R*–1M0	1/8W to 1/2W	2%, 5%	–55 to +125	50–100	.005–.01	0.1µV/V	0.1%
Precision metal film	1*R*0–10M	0.6W	0.1% to 1%	–55 to +155	15–50	.0015–.005	.01µV/V	.02%
Wire wound, vitreous	0*R*1–22k	2W to 25W	5%	–55 to +200	75	.0075	–	.01%
Wire wound, precision	10*R*–47k	1/3W	0.1%	–	5–15	.0005–.0015	–	.003%

Figure 1.10. Typical parameter values of various types of fixed-value resistor.

Resistance value notation

Resistance is measured and denoted in Ohms, which may be abbreviated to Ohm, R, or Ω. When values of thousands or millions of Ohms are denoted the abbreviations K or k (for Kilohms) or M (for Megohms) may be used; when a denoted resistance value requires the use of a decimal point indicator, the 'point' may be replaced by the appropriate R, K, k, or M sign. The following examples illustrate these facts:

0.1 Ohms	=	0.1Ω	or	0R1		
6.8 Ohms	=	6.8Ω	or	6R8		
120 Ohms	=	120Ω	or	120R		
4700 Ohms	=	4.7K	or	4K7	or	4k7
12,000 Ohms	=	12K	or	12k		
47,500 Ohms	=	47.5k	or	47k5		
1,200,000 Ohms	=	1.2M	or	1M2		
22,000,000 Ohms	=	22M				

Preferred resistance values

By international agreement, general-purpose resistors are made in a number of 'preferred' nominal values, and the number of values per decade is related to the desired resistance precision. Thus, if a precision of ±20% is specified, the entire spectrum of possible resistance values in the 80R to 800R decade can be adequately spanned by just six 'preferred' resistors with the following nominal values and tolerance 'spreads':

100R nominal;	±20% spread	=	80R	to	120R
150R nominal;	±20% spread	=	120R	to	180R
220R nominal;	±20% spread	=	176R	to	264R
330R nominal;	±20% spread	=	264R	to	396R
470R nominal;	±20% spread	=	376R	to	564R
680R nominal'	±20% spread	=	544R	to	816R

Note that these values increase logarithmically, in increments of about 50%, and that the range of values can be expanded in decade multiples and sub-multiples to span all possible component values. Throughout most of the world this particular set of values is known as the '20% tolerance' series, but in most of Europe is known (because it uses six values per decade) as the 'E6' series. Several other series are also available in the 'preferred' family of resistor values; *Figure 1.11* gives details of the most widely used members of the family; this table also applies to the preferred range values of capacitors and so on.

Resistor coding

The value and tolerance of a resistor may be marked on the component's body either directly (i.e., as 4k7, 5%) or by using some form of easily recognised colour or alpha-numeric code system. The most widely used system is based on the standard and well known 'black, brown, red, orange, etc.' colour code system, as shown in *Figure 1.12*.

Any component value in the E6, E12, and E24 preferred values series can be represented by just two digits and a decimal 'multiplier'. Consequently, the value and tolerance of any resistor in these series can

20% (E6) series	10% (E12) series	5% (E24) series		1% (E96) series								
100	100	100	110	100	102	105	107	110	113	115	118	
	120	120	130	121	124	127	130	133	137	140	143	147
150	150	150	160	150	154	158	162	165	169	174	178	
	180	180	200	182	187	191	196	200	205	210	215	
220	220	220	240	221	226	232	237	243	249	255	261	267
	270	270	300	274	280	287	294	301	309	316	324	
330	330	330	360	332	340	348	357	365	374	383		
	390	390	430	392	402	412	422	432	442	453	464	
470	470	470	510	475	487	499	511	523	536	549		
	560	560	620	562	576	590	604	619	634	649	665	
680	680	680	750	681	698	715	732	750	768	787	806	
	820	820	910	825	845	866	887	909	931	953	976	

Figure 1.11. Preferred value series of resistors and capacitors in a single decade.

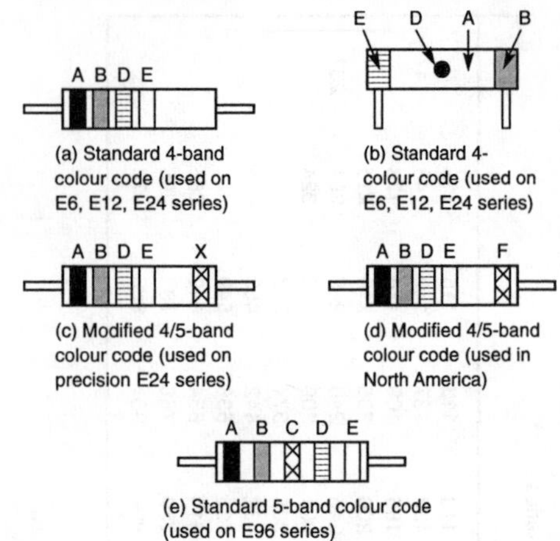

(a) Standard 4-band colour code (used on E6, E12, E24 series)

(b) Standard 4-colour code (used on E6, E12, E24 series)

(c) Modified 4/5-band colour code (used on precision E24 series)

(d) Modified 4/5-band colour code (used in North America)

(e) Standard 5-band colour code (used on E96 series)

Colour	A	B	C	D	E	F	X
	1st digit	2nd digit	3rd digit	Multiplier	Tolerance	Stability	Temp coefficient
Black	–	0	0	1	–	–	200 ppm/°C
Brown	1	1	1	10	±1%	1%	100 ppm/°C
Red	2	2	2	100	±2%	0.1%	50 ppm/°C
Orange	3	3	3	1000	–	0.01%	15 ppm/°C
Yellow	4	4	4	10,000	–	.001%	25 ppm/°C
Green	5	5	5	100,000	±0.5%	–	–
Blue	6	6	6	1,000,000	±0.25%	–	10 ppm/°C
Violet	7	7	7	10^7	±0.1%	–	5 ppm/°C
Grey	8	8	8	10^8	–	–	1 ppm/°C
White	9	9	9	10^9	–	–	–
Gold	–	–	–	0.1	±5%	–	–
Silver	–	–	–	0.01	±10%	–	–
Blank	–	–	–	–	±20%	–	–

Notes:

(1). Stability = relative percentage change in value per 1000 hours of operation.

(2). Grade-1 Hi-stab resistors may be distinguished by a salmon-pink fifth band or body colour.

Figure 1.12. Standard resistor colour code systems and notations.

be indicated by a four-colour code, with the first three colours indicating the resistance value and the fourth indicating the tolerance. On resistors with axial leads this colour code is set in the form of four bands, which are read from left to right as shown in *Figure 1.12(a)*; on resistors with radial leads the code is arranged as a set of body colours, as shown in *(b)*. *Figure 1.13* shows examples of the use of this code.

The basic 4-band colour code system is simple and unambiguous. Note, however, that some resistors use a modified version of this code, with a fifth well separated band added and used to convey additional information, as shown in *(c)* and *(d)*. Usually, in Europe, this fifth band is carried on precision Hi-Stab E24-series resistors only and denotes the component's temperature coefficient, as shown in column 'X' of *Figure 1.12*, but in the U.S.A. it denotes the component's specified stability in terms of percentage change in value per 1000 hours of operation, as in column 'F'.

In the E96 series, any value can be represented by three digits and a decimal multiplier. Consequently, the value and tolerance of any resistor in this series can be indicated by a five-band colour code, in which the fifth band indicates the component's tolerance, as shown in *Figure 1.12(e)*. Thus, the sequence orange-black-white-brown-brown indicates a 3k09, ±1% resistor value. Superficially, this 5-band system may seem unambiguous, but in practice great confusion can result if a resistor using a true 5-band system is mistaken for one using a modified 4/5-band colour code; thus, a 2% component marked red-red-orange-gold-red with a true value of 22R3, may be mistaken for a 22k, 5% resistor with a 50ppm/°C temperature coefficient or 0.1% stability factor if read as a 4/5-band component.

A	B	C	D	Value
Orange	Orange	Silver	Gold	0.33Ω, 5%
Red	Red	Gold	Silver	2R2, 10%
Yellow	Violet	Orange	Gold	47k, 5%
Brown	Black	Yellow	Brown	100k, 1%
Red	Red	Blue	Blank	22M, 20%

Figure 1.13. Examples of resistor values, using four-band or four-colour coding.

Regarding alpha-numeric resistance-coding, the two most widely used of these systems are shown in *Figures 1.14 and 1.15*; each of these systems may be subject to some detail variations. In the *Figure 1.14* system the component value is printed in normal alpha-numeric terms (e.g., as 3k3 or 3.3kΩ), and the tolerance is denoted by a direct indication (such as 5%) or by one of the five code letters listed. Thus, a 47Ω 5% resistor may be notated 47RJ, for example.

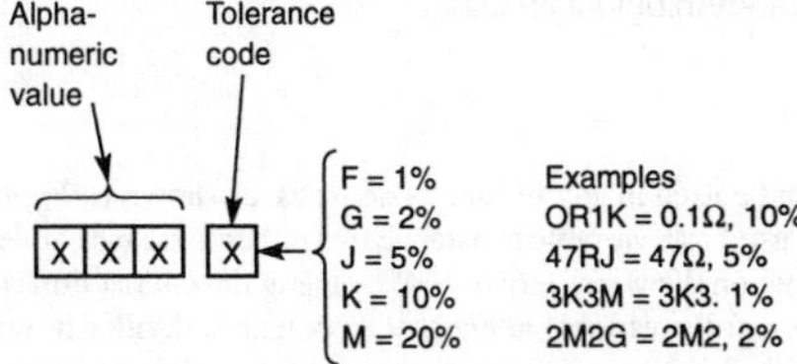

Figure 1.14. Typical alpha-numeric resistor-coding system.

The alternative system of *Figure 1.15* uses a purely numeric three-digit coding system in which the first two numbers equal the first two digits of the resistor value, and the third gives the number of zeros to be added to those two digits. In some variations of this system the component's tolerance may also be indicated, in the same ways as in *Figure 1.14*; thus, the notation 100J indicates a 10R, 5% resistor.

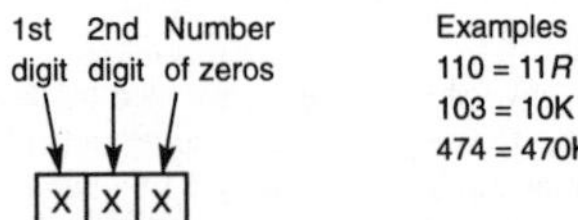

Figure 1.15. Typical numeric resistor-coding system.

Variable & preset resistors ('pots')

Nowadays, these components are universally known as potentiometers or 'pots', although the name 'potentiometer' was originally coined to describe an early ratiometric type of voltage-measuring instrument. A modern pot is the equivalent of the 3-terminal resistive unit shown in *Figure 1.16(a)*, in which a resistive track is formed between pins 1 and 3, and electrical access to any point on the track can be made via a movable wiper that is in contact with pin 2. Practical pots come in a variety of shapes and sizes, but the best known is the 'rotary' type shown in *Figure 1.16(b)*, in which the wiper (pin-2) moves towards pin-3 when the spindle is rotated clock-wise (CW) and towards pin-1 when it is moved counter clock-wise (CCW).

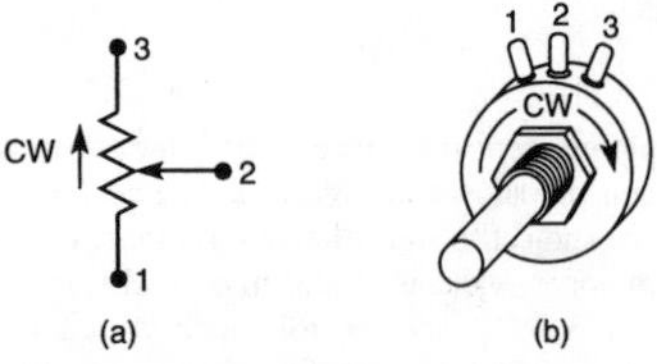

Figure 1.16. Symbolic representation *(a)* and typical appearance *(b)* of a modern potentiometer.

A pot can be used in any of four basic ways, as shown in *Figure 1.17*. It can be used as a variable resistance that either increases or decreases in value when the wiper moves CW by using the connections of *(a)* or *(b)*, or as a fully-variable attenuator or potential divider in which the output either increases or decreases when the slider moves CW by using the connections of *(c)* or *(d)*.

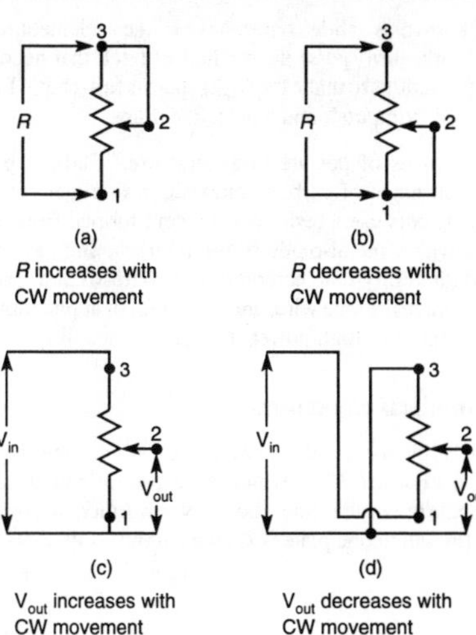

Figure 1.17. A potentiometer can be used as a variable resistor [*(a)* or *(b)*] or as a fully-variable attenuator [*(c)* or *(d)*].

When the design of a resistance-connected rotary pot is such that its resistance value varies in direct proportion to the angular movement of its control shaft it is said to have a 'linear law'. Rotary pots can be designed with a variety of different 'laws', and the four most popular of these are depicted in *Figure 1.18*; the 'S'-law type gives a semi-linear response. Rotary pots are often produced in 'ganged' form, with a single shaft controlling several pot wipers.

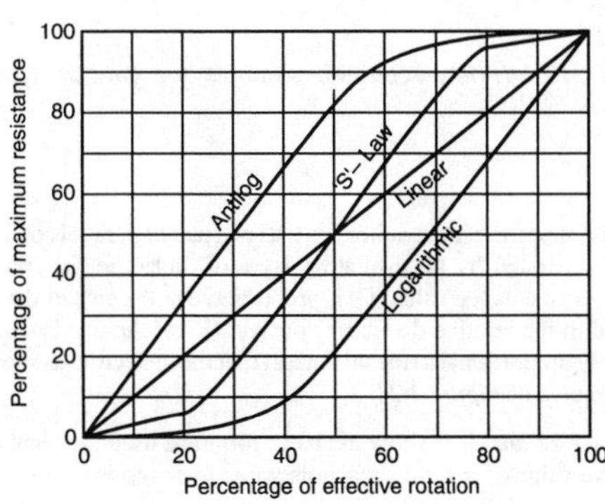

Figure 1.18. Resistance/rotation laws for four types of potentiometer.

Some pots, known as 'slide' types, have a linear element, and some (known as 'multi-turn' pots) have a helical track that needs several control-shaft rotations to make the wiper span its full length. Preset pots are often called 'trimpots', and have a linear law.

Three basic classes of pot are in general use. 'Carbon' pots use a resistance track made of carbon composition in either solid or film form. 'Cermet' pots use a resistance element formed from a ceramic base, coated with a metal-oxide film-type track, and have far better stability and durability than carbon types. 'Wire-wound' pots have a track made from resistance wire, and are useful in applications calling for a low resistance or high power-dissipation capability.

Guide to modern capacitors

Internationally, the most widely used set of basic capacitor symbols is that shown in *Figure 1.19*, in which the capacitor's plates are represented by a pair of parallel lines, but in North America a different set of symbols (in which one plate is shown curved) is used, as shown in *Figure 1.20.*

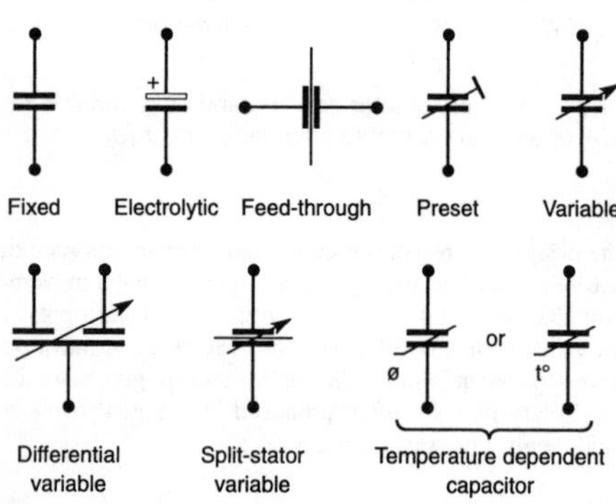

Figure 1.19. Widely-accepted symbols for various types of capacitor.

In its simplest form, a capacitor consists of a pair of parallel conductive plates, separated by an insulating layer of dielectric material. The device's capacitance value (C) is proportional to the area of the plates (A) and to the relative dielectric constant (K) of the insulator, but is inversely proportional to the thickness (t) of the dielectric, as shown by the formulae of *Figure 1.21.*

Figures 1.22 and 1.23 show the basic formulae used to calculate the effective value of several capacitors wired in series or parallel.

The reactance (X) of a capacitor is inversely proportional to frequency (f) and to capacitance value. *Figure 1.24* shows the formula that relates these parameters to one another. Precise parameter values can easily be

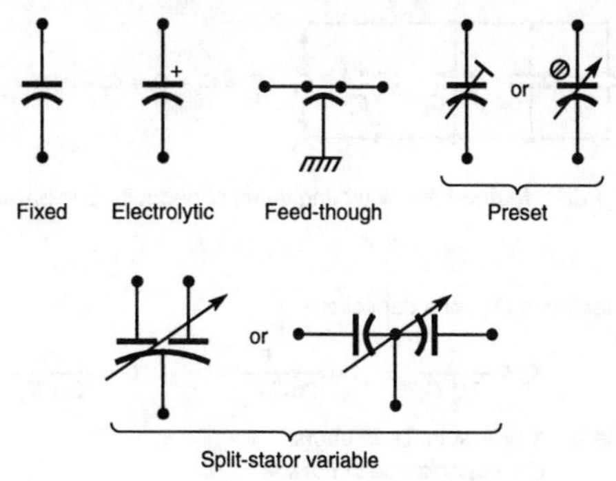

Figure 1.20. North-American symbols for various types of capacitor.

The capacitance value of a parallel-plate capacitor is given by:

$$C = \frac{0.0885\ K.A.}{t}\ (n-1)$$

where C = capacitance in picofarads
K = dielectric constant (air = 1.0)
A = area of plates, in square cm
t = thickness of dielectric, in cm
n = number of plates

Note:

$C = \frac{0.224\ K.A.}{t}\ (n-1)$ if A is in square inches and t is in inches.

Figure 1.21. Basic capacitance formulae.

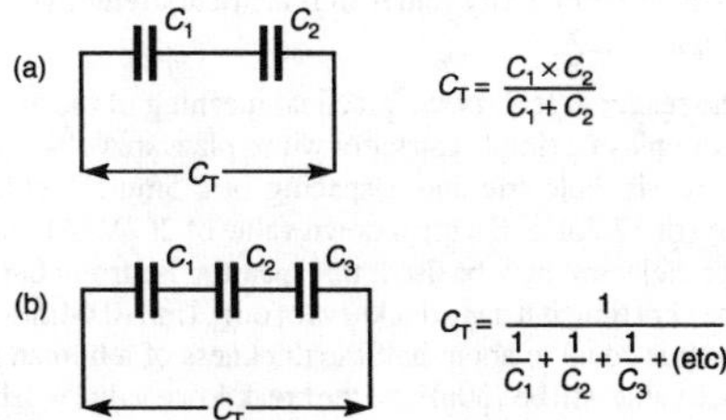

Figure 1.22. Method of calculating combined value of *(a)* two or *(b)* more capacitors in series.

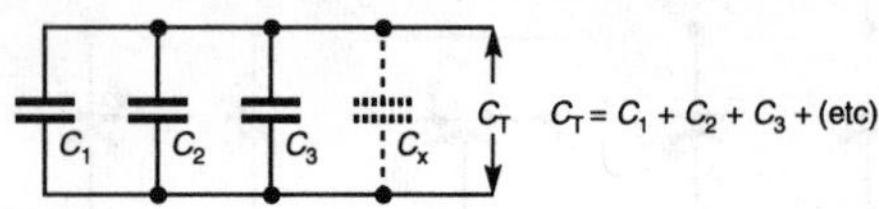

Figure 1.23. Method of calculating value of capacitors in parallel.

Reactance (X_c) of a capacitor:-

$$X_c = \frac{1}{2\pi f.C} \qquad f = \frac{1}{2\pi C.X_c} \qquad C = \frac{1}{2\pi f.X_c}$$

where X_c = reactance in ohms.
C = capacitance in Farads
f = frequency in Hz

Figure 1.24. Capacitive reactance formulae.

derived from these formulae with the help of a calculator. Alternatively, a whole range of values can be rapidly found — with a precision better than twenty percent — with the aid of the nomograph of *Figure 1.25* and a straight edge or ruler; to use this chart in this way, simply lay a straight edge so that it cuts two of the C, f, or X columns at known reference points, then read off the remaining unknown parameter value at the point where the straight edge cuts the third column. This chart can also be used to find the approximate resonance frequency (f) of an L-C network, by using the straight edge to connect the L and C columns at the appropriate 'value' points and then reading off the 'f' value at the point where the edge crosses the f column.

The simplest type of capacitor consists of a pair of conductive plates separated by a dielectric material, as shown in *Figure 1.26*; if the plates are rectangular the capacitor is known as a plate type, as in *(a)*, and if they are circular it is known as a disc type, as in *(b)*. The capacitance value is proportional to the area of the plates and to the 'permittivity' (the dielectric constant) of the dielectric, but inversely proportional to the dielectric thickness, as shown by the formulae of *Figure 1.21*. A dozen or so different basic types of dielectric are in common use, and the approximate permittivity values and electrical strengths of these are listed in *Figure 1.27*.

To give the reader a 'feel' of the practical meaning of the above data, take the example of a simple capacitor with a plate area of 1 square cm. If it uses an air dielectric and a spacing of 0.5mm, it will have a capacitance of 17.7pF and a breakdown value of 262V. Alternatively, a polyester dielectric may be used; this material is strong but flexible and may be very thin; if it has a thickness of only 1mil (0.04mm or 0.001 inch), which is equal to about half the thickness of a human hair, the capacitance value will be 730pF and the breakdown value will be 250V. Again, a ceramic dielectric may be used; this material is rigid and brittle, and thus needs to be relatively thick; for example, if a 1mm high-K ceramic dielectric with a permittivity of 10,000 is used the capacitance will be 88.5nF and the breakdown value 5000V.

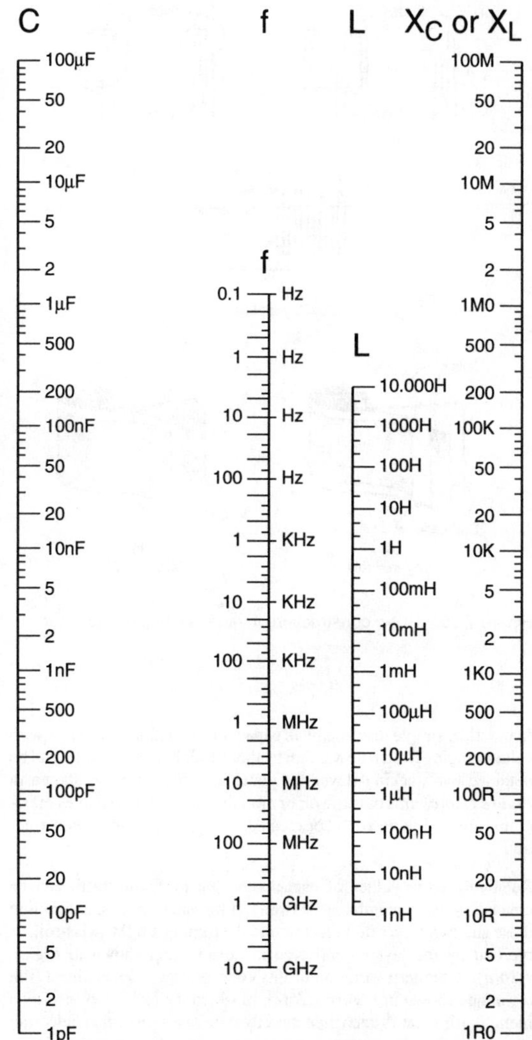

UNIVERSAL REACTANCE AND RESONANCE CHART

Figure 1.25. This chart can, with the aid of a straight edge, be used to find the reactance of a capacitor or inductor, or the resonant frequency of an L–C network; see text for explanations.

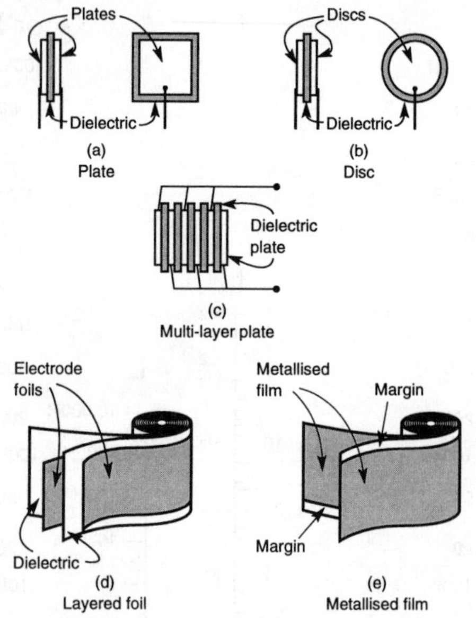

Figure 1.26. Basic construction of various types of capacitor.

In practice, simple disc and plate capacitors are readily available with values ranging from below 2.2pF to about 100nF. Larger values can be obtained by 'stacking' layers of plates and dielectrics, as shown in *Figure 1.26(c)*; this multi-layer or monolithic technique enables maximum values of about 1μF to be created, using an eleven-plate construction.

Relatively large values of capacitance (up to about 10μF) can be created by using lengths of flexible foil instead of plates, and rolling these and two layers of dielectric into the form of a tight swiss-roll, as typified by the layered foil type of construction shown in *Figure 1.26(d)*. A modern variation of this concept uses the metallised film technique shown in *Figure 1.26(e)*, in which the foil is replaced by a metallic film that is deposited directly onto one side of the dielectric; a non-metallised margin is left at one edge of each dielectric, to facilitate the making of external connections during the manufacturing stage.

Irrespective of the basic type of construction used, the completed fixed-value capacitor is always given some form of protection against environmental contamination, either by coating it with lacquer, dipping it in resin, or encapsulating it in some type of plastic or metal jacket.

Dielectric type	Permittivity	Electrical strength in volts per mil (1 mil = 0.001 inch, = 0.04 mm)
Air	1	21
Paper	2.5	200 to 1200
Mica	3 to 8	600 to 5000
Polycarbonate	2.8	200
Polyester	3.3	250
Polypropylene	2.2	600
Polystyrene	2.4	400 to 700
Low-loss ceramic	7	300
Medium-K ceramic	90	250
High-K ceramic	1000 to 40,000	200
Aluminium oxide electrolytic	7 to 9	–
Tantalum oxide electrolytic	27	–

Figure 1.27. Dielectric constants and breakdown voltages.

Electrolytic capacitors

Very large values of capacitance (up to about 100,000μF) can be created by using an 'electrolytic' technique in association with the *Figure 1.26(d)* layered foil form of construction. In aluminium oxide electrolytics the two foils are made of high-purity aluminium; one of these (the cathode or 'negative' foil) is used directly, but the other one (the anode or 'positive' foil) has its surface covered with a thin insulating film of aluminium oxide; this film, which may be only 0.01mil thick, forms the dielectric. The two foils are separated by a layer of tissue that is soaked with an electrolyte which, being an excellent conductor, operates in conjunction with the plain foil to form the true cathode of the final capacitor. Note that, thanks to this conductive electrolyte, the *effective* thickness of the capacitor's dielectric is equal to the thickness of the aluminium oxide film on the anode foil, and is independent of the actual thickness of the electrolytic layer.

Because of the thinness of their oxide film, electrolytics can provide very high capacitance density. An aluminium oxide type with a 2cm by 100cm foil size and a 0.01mil oxide thickness will, for example, have a capacitance of about 35μF. A tantalum oxide type using a similar form of construction and the same dimensions would have a capacitance of about 120μF. These values can be increased by a factor as high as four by chemically etching the aluminium foil prior to the forming process, thus greatly increasing the foil's effective surface area.

The anode foil's oxide film is formed by using the foil as one of the electrodes in an electrolysis bath; when a d.c. voltage is first applied to this electrode it conducts readily, and as it passes current the film of highly resistive oxide begins to form on its surface; as the film builds up its resistance progressively increases and thus reduces the current and the rate of deposition at a proportionate rate, until the film reaches 'adequate' thickness. Consequently, this 'forming' process results in a capacitor that inherently passes a significant leakage current when in use. Typically, in aluminium oxide electrolytic types, this leakage value is roughly equal to:

$$I_L = 0.006CV\ \mu A$$

where C is in μF and V is in volts. Tantalum oxide electrolytic capacitors have far lower leakage currents than aluminium oxide types.

Practical capacitors

All practical capacitors have the electrical equivalent circuit of *Figure 1.28(a)*, in which C represents pure capacitance, R_S represents dielectric losses, R_P represents parallel leakage resistance, L_S the inductance of electrode foils, etc., and L_L the self-inductance of the component's leads (about 8nH/cm or 20nH/inch). At high frequencies this circuit simplifies into the series resonant form shown in *(b)*. Note that the resonant frequency (f_R) of a large electrolytic may be only a few kHz, while that of a small ceramic type may reach hundreds of MHz (*Figure 1.29* shows typical f_R values of several ceramic capacitors with specific total lead lengths). Also note that a capacitor's impedance is capacitive below f_R, resistive at f_R, and inductive above f_R, and that an electrolytic's impedance may thus be quite large at high frequencies, preventing it from removing spikes or transients from supply lines, etc.

At low frequencies a capacitor's self-inductance has little practical effect, but R_S and R_P cause a finite shift in the capacitor's voltage/current phase relationship; this same phase shift can, at any given

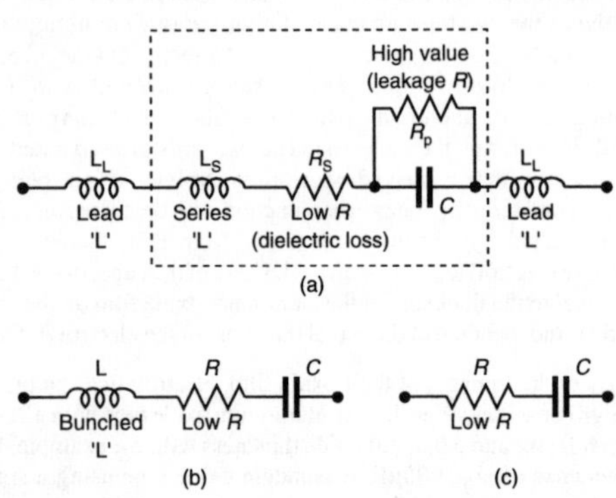

Figure 1.28. The full equivalent circuit *(a)* of a capacitor can be simplified to *(b)* at high frequencies, or to *(c)* at low-value fixed frequencies.

Capacitor value	Self-resonant frequency (MHz)	
	1 cm lead length	1 inch lead length
10n	13	11
3n3	25	20
1n0	46	38
330pF	80	62
100pF	145	120
33pF	240	205
10pF	440	380

Figure 1.29. Typical self-resonant frequency of ceramic (disc or plate) capacitors.

frequency, be emulated by a single 'lumped' resistor in series with a pure capacitor, as in *Figure 1.28(c)*. The ratio between R_S and X (the pure reactance) is known as the 'D' or 'loss factor' of a capacitor, and indicates the component's purity factor; the lower the 'D' value, the better the purity.

A capacitor's value varies with temperature, with the passage of time, and with frequency and applied voltage. The magnitude of these changes varies with capacitor type, as shown in *Figure 1.30*. The following notes explain the most important features of the various capacitor types.

Silver mica

These capacitors have excellent stability and a low temperature coefficient, and are widely used in precision RF 'tuning' applications.

Ceramic types

These low-cost capacitors offer relatively large values of capacitance in a small low-inductance package. They often have a very large and non-linear temperature coefficient; in high-K types this is often so massive that it is specified in terms of percentage change in capacitance value over the device's full spread of thermal working limits (typically -55 to +85°C). They are best used in applications such as RF and HF coupling or decoupling, or spike suppression in digital circuits, in which large variations of value are of little importance.

'Poly' types

Of the four main 'poly' types of capacitor, polystyrene gives the best performance in terms of overall precision and stability. Each of the others (polyester, polycarbonate, and polypropylene) gives a roughly similar performance, and is suitable for general-purpose use. 'Poly' capacitors usually use a layered 'swiss-roll' form of construction; metallised film types are more compact that layered film-foil types, but have poorer tolerances and pulse ratings than film-foil types. Metallised polyester types are sometimes known as 'greencaps'.

Capacitor type	Range	Tolerance	Temperature coefficient	f_R	D	Leakage resistance	Stability
Silver mica	2.2pF–10nF	±1%	+35ppm/°C	1–10MHz	0.002	Very high	Excellent
Ceramic, low-K	2.2pF–330pF	±2%	Variable*		0.001	High	Good
" , medium-K	390pF–4.7nF	±10%	±10%*	5–100MHz	0.03	High	Fair
" , high-K	1nF–100nF	–20% to +80%	+22% to –82%*		0.2	High	Fair
" , monolithic	10pF–0.47μF	±10%	Variable*	10MHz	0.02	Very high	Good
Polystyrene	22pF–0.1μF	±1%	–150ppm/°C	10MHz	0.0005	Very high	Excellent
Polyester	1nF–2.2.μF	±10%	+200ppm/°C	1MHz	0.01	Very high	Fair
Polycarbonate	100pF–10μF	±10%	±60ppm/°C	0.1–1MHz	0.005	Very high	Fair
Polypropylene	100pF–4.7μF	±10%	–200ppm/°C	0.1–1MHz	0.0005	Very high	Fair
Electrolytic (aluminium foil)	0.1μF–47,000μF	–10% to +50%	±1500ppm/°C	50KHz	0.2	Very low	Fair
Electrolytic (reversible)	1.5μF–100μF	±20%	±1000ppm/°C	500KHz	0.1	Very low	Fair
Electrolytic (tantalum)	0.1μF–100μF	±20%	±500ppm/°C	1MHz	0.1	Low	Good

Note: * = see text

Figure 1.30. Capacitor comparison chart.

Electrolytic types

These offer large values at high capacitance density; they are usually polarised and must be installed the correct way round. Aluminium foil types have poor tolerances and stability, and are best used in low-precision applications such as smoothing, filtering, energy storage in PSU's, and coupling and decoupling in audio circuits. Tantalum types offer good tolerance, excellent stability, low leakage, low inductance, and a very small physical size, and should be used in all applications where these features are a positive advantage.

Capacitance value notation

The basic electrical unit of capacitance is the Farad, but in electronics the microfarad (μF) is used as the basic practical unit of measurement and equals 0.000 001 Farad. The μF is divisible into multi-decade sub-units of nanofarads (nF) and picofarads (pF), in which 1pF equals 0.000 001μF, and 1nF equals 0.001μF (equals 1000pF).

Capacitors are readily available in the standard E6, 20% (100, 150, 220, etc.) and E12, 10% (100, 120, 150, 180, 220, etc.) range of preferred values. The component's value may be denoted in a variety of ways, as illustrated by the following examples;

0.1μF	=	100nF				
6.8nF	=	6n8	=	0.0068μF	=	6800pF
220pF	=	0.22nF				
4.7μF	=	4μ7	=	4700nF		

Capacitance coding

On most modern capacitors the component's nominal value and tolerance is, where space permits, printed in clear and reasonably unambiguous terms such as 4n7, 10%, or 68nF, 5%, etc; if the value is given in purely digital form, such as 120 or 4700, this indicates the value in pF. Often, the capacitor is also marked with its specified working voltage, such as 63V (or 63U), 100V (or 100U), etc.

Up until the late 1980s the values (in pF) of tubular-ceramic and some polyester capacitors were often marked by a colour coding system, as shown in *Figure 1.31*. Thus, a ceramic dot-coded 680pF, 5% capacitor would be coded blue-grey-brown-green, and a polyester 470nF, 10%, 400V capacitor would be coded yellow-violet-yellow-white-yellow. This system of coding is now rarely used on new components.

Some tantalum electrolytic capacitors have their value (in μF) and their working voltages marked by the simple colour coding system shown in *Figure 1.32(a)*, but others use a plain numeric system, as in *(b)*.

The values of modern ceramic disc and plate capacitors may be indicated by a variety of code systems; the most popular of these are illustrated in *Figure 1.33*. The simplest of these is that shown in *(a)*, in which the three-figure code contains two digits plus the symbol n, p, or μ, and gives a self-evident indication of the capacitance value, as shown by the examples. In some cases this code may be followed by a capital letter, which indicates the component's tolerance value, using the Electrical Institute of America (EIA) code system of *Figure 1.34*.

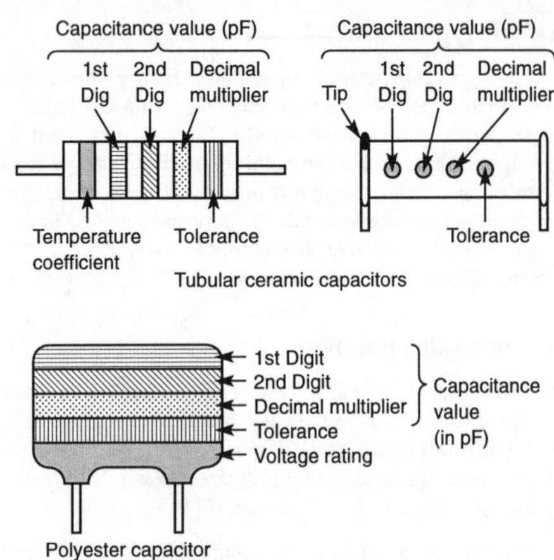

Colour	1st and 2nd digits	Decimal multiplier	Tolerance (per cent)	Temperature coefficient (ppm/°C)	Voltage rating
Black	0	1	–	0	–
Brown	1	10	±1	–30	100 V
Red	2	100	±2	–80	250V
Orange	3	1k	±3	–150	–
Yellow	4	10k	–0, +100	–220	400V
Green	5	100k	±5	–330	–
Blue	6	1M	±6	–470	–
Violet	7	10M	–	–750	–
Grey	8	100M	–	+30	–
White	9	1000M	±10	+100 to –750	–
Gold	–	–	±5	–	–
Silver	–	–	±10	–	–
No colour	–	–	±20	–	–

Figure 1.31. Colour coded system once used to indicate value of tubular-ceramic and polyester capacitors (now obsolete).

The most widely used coding system is that shown in *Figure 1.33(b)*, in which the capacitance value is indicated (in pF) by a three-digit code in which the first two figures give the first two digits of the value, and the third figure gives the number of zeros to be added to give the full value; the three-digit code is followed by a capital letter, which indicates the component's tolerance (see *Figure 1.34*). Sometimes, the capacitor's temperature coefficient may also be indicated, using either a colour code as in *(c)* or an industrial or EIA alpha-numeric code, as shown in *Figure 1.35*. Thus, a capacitor with a temperature coefficient of -220ppm/°C may have a yellow tip or carry the code N220 or R2G.

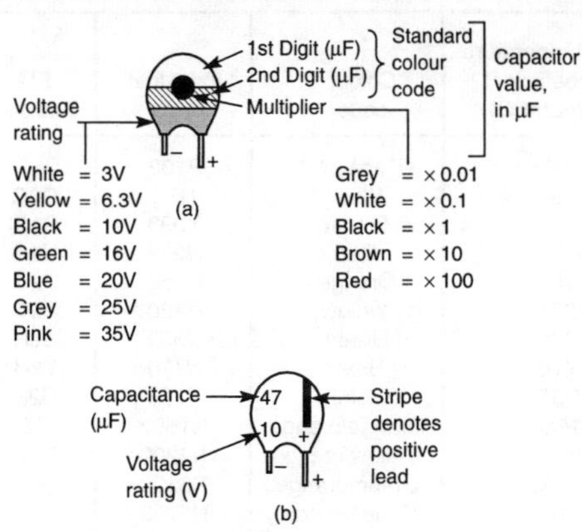

Figure 1.32. Alternative notations used on Tantalum electrolytic capacitors.

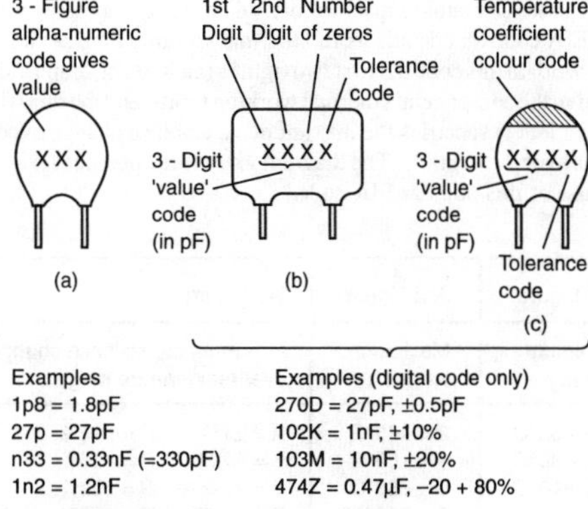

Figure 1.33. Coding system used on modern disc and plate ceramic capacitors.

C = ±0.25pF	M = ±20%
D = ±0.5pF	Q = -10, +30%
F = ±1pF or ±1%	S = -20, +50%
G = ±2pF or ±2%	T = -10, +50%
J = ±5%	Z = -20, +80%
K = ±10%	

Figure 1.34. Popular EIA capacitance tolerance codes.

Temperature coefficient (ppm/°C)	Colour code	Industrial code	EIA code
+100	Red/violet	P100	–
±0	Black	NP0	C0G
–33	Brown	N033	S1G
–75	Red	N075	U1G
–150	Orange	N150	P2G
–220	Yellow	N220	R2G
–330	Green	N330	S2H
–470	Blue	N470	T2H
–750	Violet	N750	U2J
–1500	Orange/orange	N1500	P3K
–2200	Yellow/orange	N2200	R3L
–3300	Green/orange	N3300	–
–4700	Blue/orange	N4700	–

Figure 1.35. Code systems used to indicate capacitor temperature coefficients.

In some cases a ceramic capacitor (or its data sheet) may carry a three-figure EIA code that details its full thermal specification, as shown in *Figure 1.36*. In this code the first two figures (an X, Y, or Z, and a single digit) give the component's thermal working limits, and the third figure (a capital letter) specifies the maximum capacitance change expected to occur over those limits. The diagram gives three practical examples of the use of this popular EIA code.

1st Figure	2nd Figure	3rd Figure	
Minimum temp.	Maximum temp.	Maximum capacitance change over temperature range	
X = –55°C	2 = +45°C	A = ±1%	P = ±10%
Y = –30°C	4 = +65°C	B = ±1.5%	R = ±15%
Z = +10°C	5 = +85°C	C = ±2.2%	S = ±22%
	6 = +105°C	D = ±3.3%	T = –33 to +22%
	7 = +125°C	E = ±4.7%	U = –56 to +22%
		F = ±7.5%	V = –82 to +22%

Examples of use of the code:-

Code		Temperature range		Max. Capacitance change over full temp. range
X7R	=	–55°C to +125°C	=	±15%
Y5V	=	–30°C to +85°C	=	–82 to +22%
Z5U	=	+10°C to +85°C	=	–56 to +22%

Figure 1.36. EIA temperature characteristic codes for ceramic capacitors.

Some film-type capacitors carry a code that indicates their dielectric type. *Figure 1.37* gives details of the most popular of these code systems, which is used by Philips and some other European manufacturers.

Code		Dielectric type
MKT	=	Metallised polyester film (PETP).
MKC	=	Metallised polycarbonate film.
KS	=	Polystyrene film/foil.
KP	=	Polypropylene film/foil.
MK	=	Metallised polypropylene film.
KP/MKP	=	Dual dielectric: metallised polypropylene and polypropylene film.
KP/MMKP	=	Dual dielectric: polypropylene film, foil and double metallised polypropylene film.
MKT-P	=	Dual dielectric: metallised polyester film and paper.

Figure 1.37. Dielectric material codes, used by Philips and some other European manufacturers on film capacitors.

Variable capacitors

Variable capacitors come in two basic forms; fully variable ones are known as 'tuning' capacitors, and preset ones are known as 'trimmers'. 'Tuning' types have a semi-circular set of rotatable plates that intermesh with a set of fixed plates, so that capacitance is greatest when the two sets of plates are fully intermeshed, and least when the movable plates are rotated out of mesh; the precise shape of the rotatable plates may be designed to produce a linear, semi-linear (S), logarithmic, or square-law variation of capacitance as the plates are rotated; the dielectric used between the two sets of plates may be air, plastic film, or some other insulating material. Often, two or more tuning capacitors are ganged together, so that they vary in unison, but such units are rather expensive and have — in recent times — largely been superseded by matched sets of varicap diodes.

Most trimmer capacitors are meant to be adjusted (varied) only occasionally; several basic types are available. 'Vane' types are similar to conventional tuning capacitors and may use an air or plastic film dielectric; plastic film types are now very popular. Ceramic 'disc' types can be regarded as a simple 'vane' types, using only one fixed and one rotatable half-vane; they offer an excellent high-frequency performance (with a very high Q) but may have a very large temperature coefficient. 'Compression' and 'piston' type trimmers give a fairly limited range of capacitance adjustment.

Guide to modern inductors

Figure 1.38 shows the basic set of symbols used to represent various types of inductor. These may be subjected to some artistic variation; the basic 'inductor' symbol may, for example, consist of a set of interconnected loops or of joined-up arcs, as shown, and the number of loops/

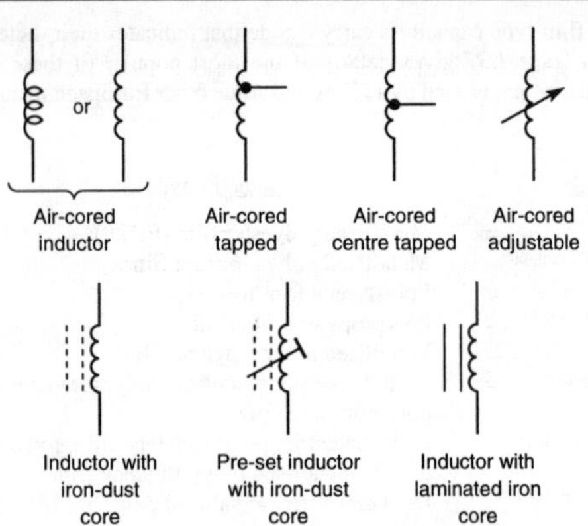

Figure 1.38. Internationally-accepted symbols for various types of inductor.

arcs may vary from three to ten or more. Similarly, in inductors with iron-dust or laminated cores, the number of vertical columns used to represent the cores may vary from one to three.

The basic unit of inductance (L) is the Henry, but sub-multiples of this unit, such as mH (millihenry), μH (microhenry), and nH (nanohenry) are widely used in electronics. An inductor's reactance (X) is proportional to frequency (f) and to its L value; *Figure 1.39* shows the formulae that relates these parameters to one another. Precise parameter values can easily be derived from these formulae with the help of a calculator. Alternatively, a whole range of values can be rapidly found — with a precision better than twenty percent — with the aid of the nomograph of *Figure 1.25* and a ruler; to use the chart in this way, simply lay a straight edge so that it cuts two of the L, f, or X columns at known reference points, then read off the remaining unknown parameter value at the points where the straight edge cuts the third column.

Figure 1.40 shows the basic formulae used to calculate the effective values of several inductors wired in series or in parallel, assuming that there is no inductive mutual coupling between the components.

Reactance (X_L) of an indicator:-

$$X_L = 2\pi f.L \qquad f = \frac{X_L}{2\pi L} \qquad L = \frac{X_L}{2\pi f}$$

where X_L = reactance, in ohms.
L = inductance, in microhenrys (μH)
f = frequency, in MHz

Figure 1.39. Inductive reactance formulae.

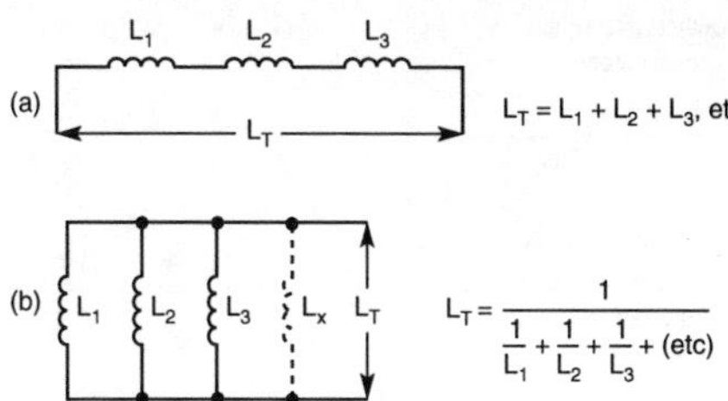

Figure 1.40. Method of calculating combined value of inductors in *(a)* series or *(b)* parallel.

One of the most important parameters of an inductor is its Q value, or reactance-to-resistance ratio; *Figure 1.41* shows the formula for 'Q'.

All practical inductors have the simplified electrical equivalent circuit of *Figure 1.42(a)*, in which L represents pure inductance, R is the component's resistive element, and C_D represents its inherent 'distributed' self-capacitance. All practical inductors thus have a natural self-resonant frequency, dictated by their L and C_D values; *Figure 1.42(b)* shows the characteristic response curve that results; the coil's reactance is inductive below resonance, resistive at resonance, and capacitive above resonance.

The Q of an inductor varies with frequency, and is given by

$$Q = \frac{2\pi f.\ L}{R}$$

where L = inductance, in µH
f = frequency, in MHz
R = effective series resistance, in ohms.

Figure 1.41. Inductive Q formula.

Practical inductors are usually classed as either 'coils' or 'chokes'. Coils are characterised by having high Q, low self-capacitance, and excellent inductive stability, and are used as inductive elements in tuned circuits, etc. Chokes are used as simple signal-blocking or filtering elements, in which Q, self-capacitance, and precise L values are non-critical.

The simplest inductor is a straight length of round-section copper wire. This has an inductance roughly proportional to the wire's length and inversely proportional to its diameter, as shown by the formula and worked examples of *Figure 1.43*; the 0.4mm wire size is roughly equal to the American 26 'A.W.G.' or the British 27 'S.W.G.' gauges, and the 4mm size to 6 A.W.G. or 8 S.W.G. Most wires used in electronics have

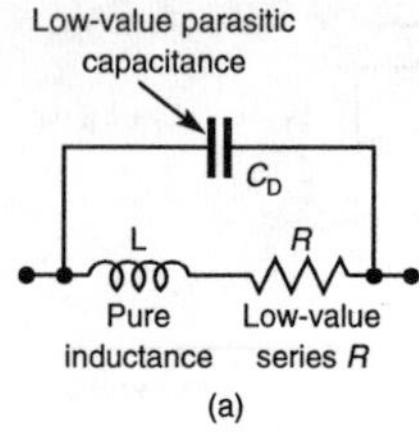

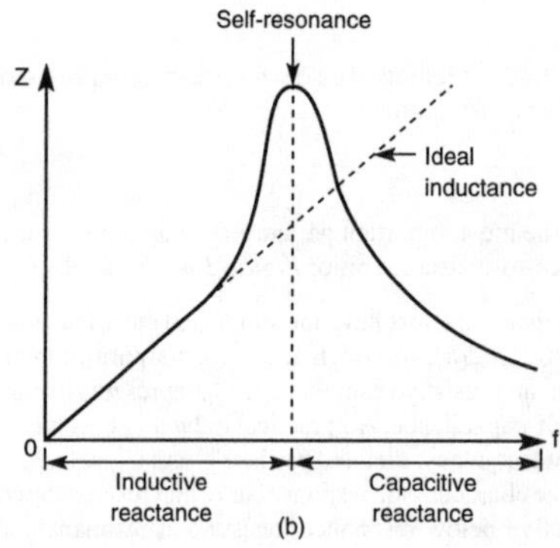

Figure 1.42. All practical inductors have the simplified equivalent circuit shown in *(a)* and exhibit the self-resonance shown in *(b).*

The inductance of a straight length of round-section wire is given by:-

$$L = 0.0002b\ [(2.303 \times \log_{10} \frac{4b}{d}) - 0.8]$$

where L = inductance, in µH
d = wire diameter, in mm
b = wire length, in mm.

Wire diameter	Wire length			
	25 mm (1 inch)	100 mm (4 inch)	250 mm (10 inch)	1000 mm (40 inch)
0.4 mm	0.03µH	0.122µH	0.411µH	1.68µH
4 mm	0.018µH	0.076µH	0.296µH	1.222µH

Figure 1.43. Formula and examples related to self-inductance of a straight wire, in free air.

diameters in the range 0.21 to 2.64mm, and the table of *Figure 1.44* gives details of the standard set of wire gauges within this range; the 'turns per linear inch' winding figures are derived by averaging data from several different sources.

A wire's inductance can be greatly increased by forming it into a coil or solenoid; the resulting inductance depends on the coil's dimensions and on the thickness and the spacing of the wire. A reasonably simple and practical set of formulae that relate these parameters is shown in

Wire diameter (bare)		Wire gauge		Turns per linear inch		
mm	inches	A.W.G.	S.W.G	Enamel	Enamel and single cotton	Double cotton
2.64	0.104	–	12	9.09	8.8	8.4
2.59	0.1019	10	–	9.6	9.1	8.9
2.34	0.092	–	13	10.2	9.9	9.3
2.305	0.097	11	–	10.7	–	9.8
2.04	0.080	12	14	11.9	11.3	10.7
1.82	0.072	13	15	13.17	12.5	11.8
1.63	0.064	14	16	15.0	14.0	13.0
1.45	0.0571	15	–	16.8	–	14.7
1.42	0.056	–	17	16.9	15.7	14.5
1.29	0.0508	16	–	18.9	17.3	16.4
1.22	0.048	–	18	19.7	18.0	16.8
1.15	0.0453	17	–	21.2	–	18.1
1.02	0.040	18	19	23.5	21.0	19.5
0.91	0.036	19	20	26.2	24.0	21.2
0.812	0.032	20	21	29.2	26.1	23.0
0.72	0.028	21	22	33.0	29.1	25.5
0.644	0.0253	22	–	37.0	31.3	27.2
0.61	0.024	–	23	38.3	34.2	28.4
0.573	0.0226	23	–	41.3	37.0	31.0
0.56	0.022	–	24	41.6	37.2	31.6
0.51	0.020	24	25	45.5	39.0	34.0
0.46	0.018	25	26	51.0	43.0	37.0
0.42	0.0164	–	27	55.1	46.2	38.0
0.405	0.0159	26	–	58.0	47.0	40.0
0.38	0.0148	–	28	61.0	50.2	41.0
0.361	0.0142	27	–	64.9	52.0	42.0
0.35	0.0136	–	29	66.0	52.2	43.0
0.32	0.0125	28	30	72.5	55.0	46.0
0.29	0.0116	–	31	77.5	59.8	48.0
0.286	0.0113	29	–	81.6	61.0	51.8
0.27	0.0108	–	32	82.7	62.0	48.0
0.255	0.0100	30	33	88.0	64.0	49.7
0.23	0.0092	–	34	98.3	65.5	52.9
0.227	0.0089	31	–	101.0	69.0	57.2
0.21	0.0084	–	35	105.8	73.0	58.0

Figure 1.44. International copper-wire table.

Figure 1.45; these assume that the wire diameter is small (less than 10%) relative to the coil diameter, and that the wire is close-wound in a single layer, and that the coil's length is within the range 0.4 to 3 times the coil diameter. These formulae can be used to determine the inductance of a coil or to find the number of turns of wire needed to create a specific inductance.

Home-made coils can easily be wound on a variety of commercially available air-cored formers, and *Figure 1.46* list some practical inductance values obtained on these, using formers of various diameters. These values are derived from manufacturer's data, and are similar to those indicated by the *Figure 1.45* formulae. Note that many small (up to about 0.5inch or 12mm diameter) formers have a provision for fitting an adjustable core or 'slug', which enables the coil inductance to be varied over a limited range; if coil length is within the limits 0.3 to 0.8 times diameter, iron-dust, ferrite, or brass cores will give maximum inductance increases of x2, x5, or x0.8 respectively.

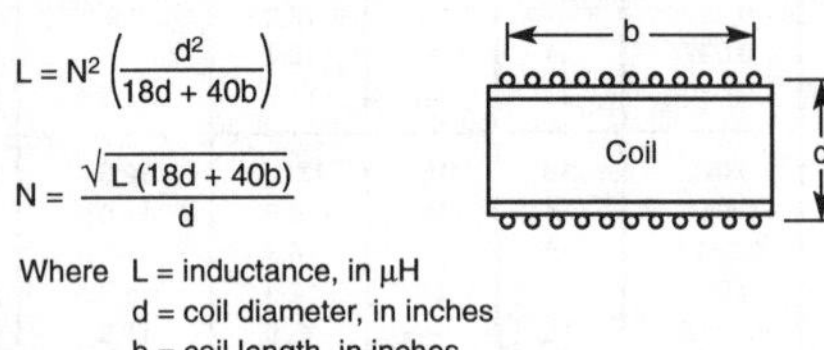

$$L = N^2\left(\frac{d^2}{18d + 40b}\right)$$
$$N = \frac{\sqrt{L\,(18d + 40b)}}{d}$$
b
Coil
d

Where L = inductance, in µH
d = coil diameter, in inches
b = coil length, in inches
N = number of turns

Figure 1.45. Formulae concerning the inductance of single-layer close-wound air-cored coils.

Coil inside diameter and length	Coil turns per inch					
	4	6	8	10	16	32
	Coil inductance, in microhenrys (µH)					
1/2"	0.036	0.08	0.144	0.224	0.58	2.4
5/8"	0.084	0.136	0.33	0.51	1.32	5.4
3/4"	0.108	0.243	0.432	0.684	1.78	7.2
1"	0.28	0.644	1.18	1.85	4.73	19.04
1 1/4"	0.55	1.26	2.24	3.5	8.5	–
1 1/2"	0.975	2.2	3.9	6.125	15.75	–
1 3/4"	1.56	3.54	6.3	9.9	25.5	–
2"	2.38	5.4	9.54	15.1	38.9	–
2 1/2"	4.79	10.8	19.3	30.1	–	–
3"	8.12	18.3	32.5	51.6	–	–

Figure 1.46. Examples of inductance obtained on a selection of commercial air-cored coil formers, using 1/1 coil diameter/length ratios, at various turns-per-inch values.

High values of inductance can be obtained by using various multi-layered forms of coil construction. Alternatively, a coil's inductance can be greatly increased by winding it on a ferrite rod core, or can be increased even more by winding it on a ferrite ring, so that the coil forms a toroid; all of these techniques are widely used on commercially manufactured inductors.

Another way of getting high inductance is to wind the coil on a bobbin that is then shrouded in a high permeability material; in low-frequency applications this may take the form of iron laminations, but at higher frequencies it is usually a ferrite pot core, as shown in *Figure 1.47*; these ferrite cores have a typical permeability or 'μ' value of between 50 and 200, and a typical specific inductance or A_L value in the range 150 to 400nH/turn. The diagram shows the basic formulae used with these pot cores; note that L and A_L must use like units (such as μH) in these formulae. Thus, to find the number of turns needed to make a 10mH

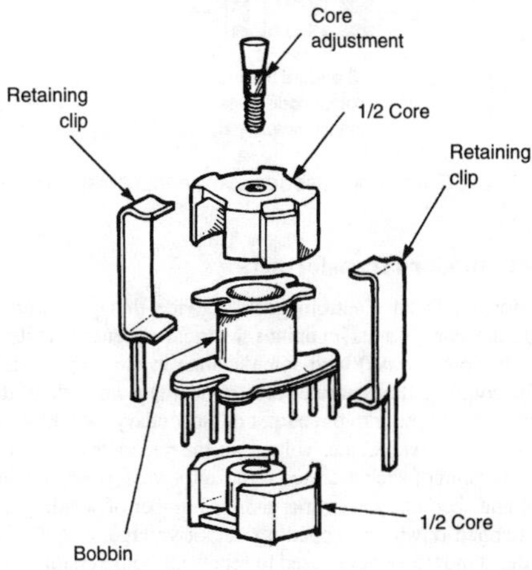

$$L = N^2 A_L$$

$$N = \sqrt{\frac{L}{A_L}}$$

Where L = inductance of assembly.
A_L = specific inductance of pot core.
N = number of turns of coil

Note:
L and A_L are in like units, i.e. μH or mH, etc.

Figure 1.47. Exploded view of pot core assembly, together with basic design formulae.

inductance, using a core with an A_L value of 400nH, simply divide 10,000μH by 0.4μH (= 25,000) and take the square root of the result (= 158) to find the number of turns needed; the core's data sheet then tells what gauge of wire to use.

A vast range of ready-built coils and chokes is available from specialist component suppliers, and the values of these are usually clearly marked in plain language. Some RF chokes, however, have their values indicated by a colour-code system, and this usually takes the form shown in *Figure 1.48*, with a broad silver band to the left, followed by a standard three-band colour code that gives the inductance value (in μH), followed by a narrow band that indicates the component's tolerance (gold = 5%, silver = 10%).

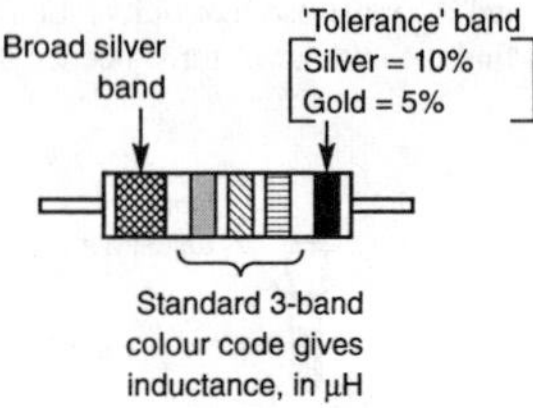

Figure 1.48. Colour code used on some encapsulated RF chokes.

Guide to modern transformers

A transformer is an electromagnetic device that 'transforms' the electrical a.c. energy at its input into electrical a.c. energy at its output, usually (but not always) with some change in the output voltage. It works by coupling the generated electromagnetic a.c. field of its input or 'primary' winding into the output or 'secondary' winding, thereby inducing a turns-related a.c. voltage in the secondary winding. The simplest version of such a device is the autotransformer, in which the primary and secondary networks share a number of windings; it may give a stepped-down or stepped-up output voltage, and *Figure 1.49* shows the standard symbols used to represent both versions, together with the basic formula that defines the relationship between their input and output voltages. Most transformers are of the 'isolation' type, which uses two or more sets of windings that are electromagnetically coupled but are electrically isolated from one another. *Figure 1.50* shows a selection of basic symbols used to represent various types of isolation transformer; these symbols may be subjected to some artistic variation.

The diagrams of *Figures 1.49 and 1.50* show only the *basic* symbols used to represent various types of transformer. In actual circuit diagrams these symbols may be subjected to some degree of elaboration, as shown in the examples of *Figure 1.51*. Thus, an IF transformer may be shown complete with a tuned and slugged primary and a screened can, as in *(a)*, or a pulse transformer may be shown with an in-phase or an anti-phase output, as indicated by the large dots in *(b)* and *(c)*; *(d)* shows a pair of tapped secondary windings connected in-phase, to give an output of 15 volts.

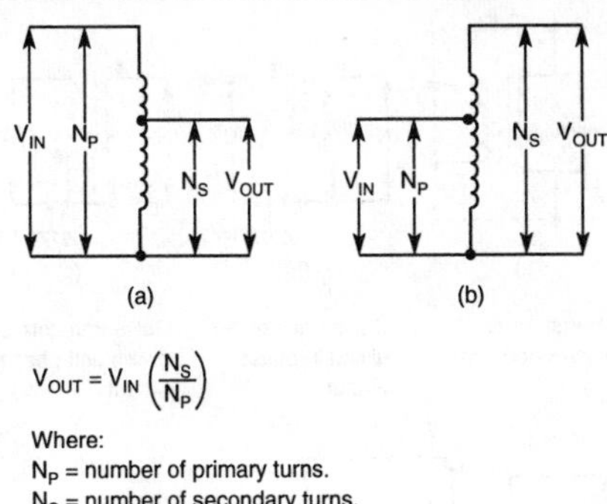

$$V_{OUT} = V_{IN}\left(\frac{N_S}{N_P}\right)$$

Where:
N_P = number of primary turns.
N_S = number of secondary turns.

(c)

Figure 1.49. An autotransformer can be used to provide *(a)* a stepped-down or *(b)* a stepped-up output voltage; in either case, the formula of *(c)* applies.

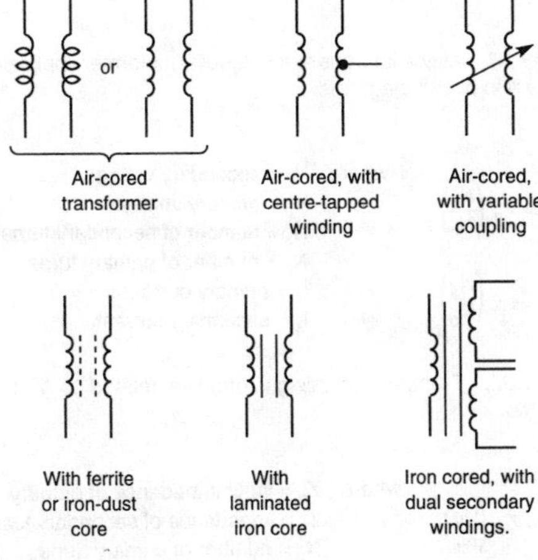

Figure 1.50. Standard symbols for various types of isolation transformer.

A transformer's secondary voltage is proportional to its secondary-to-primary turns ratio and to its input (primary) voltage, and its input current is proportional to its secondary current and its turns ratio, as shown by the formulae of *Figure 1.52.* Its input impedance is proportional to its secondary load impedance and to the square of its primary-to-secondary turns ratio, as in *Figure 1.53.*

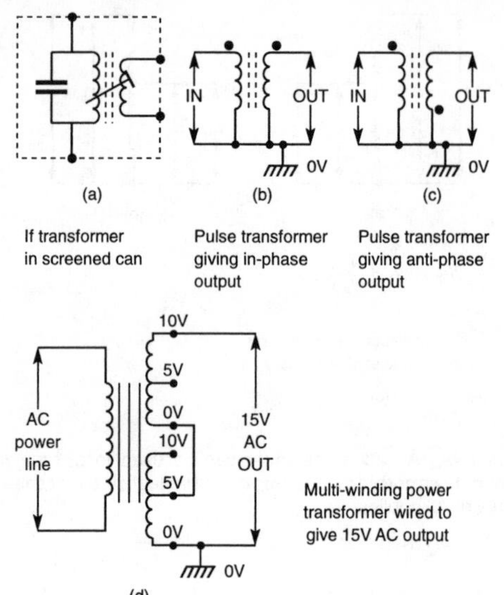

Figure 1.51. Miscellaneous examples of transformer applications, as shown in circuit diagrams.

$$V_s = V_p \left(\frac{N_s}{N_p}\right)$$

$$I_p = I_s \left(\frac{N_s}{N_p}\right)$$

where V_s = secondary voltage
V_p = primary voltage
N_s = number of secondary turns
N_p = number of primary turns.
I_p = primary current
I_s = secondary current.

Figure 1.52. Transformer design formulae related to V, I, and turns ratio.

$$Z_p = Z_s \left(\frac{N_p}{N_s}\right)^2$$

where Z_p = input impedance of primary
Z_s = impedance of secondary load
N_p = number of primary turns
N_s = number of secondary turns.

Figure 1.53. Transformer impedance ratio formula.

Since its input impedance is proportional to its load impedance, one obvious transformer use is as an impedance-matching device, and *Figure 1.54* shows the basic formula that relates to this application, together with a worked example that shows that a 7.07 primary-to-secondary turns ratio can be used to match a 4Ω load to a 200Ω input impedance.

$$\frac{N_p}{N_s} = \sqrt{\frac{Z_p}{Z_s}}$$

where Z_p = required primary impedance
Z_s = impedance of secondary load
N_p/N_s = necessary primary-to-secondary turns ratio

Example:

To match a 4 ohm load to a 200 ohm input impedance:

$$\text{Turns ratio } \frac{N_p}{N_s} = \sqrt{\frac{200}{4}} = \sqrt{50} = 7.07$$

∴ Primary must have 7.07 times as many turns as the secondary.

Figure 1.54. Transformer impedance-matching formula, with worked example.

Transformers are made in much the same way as ordinary inductors; RF types are usually air cored, HF types may be ferrite or iron-dust cored, and low-frequency types are usually potted in ferrite or use a laminated-iron core. A vast range of commercial transformers of all types is readily available. It is usually a fairly simple matter to build one to suit a personal specification.

In the case of a.c. power line transformers, kits of parts, together with adequate instructions and a ready-built primary winding, are available from specialist suppliers. As a hypothetical example of how to use one of these kits, suppose that a specification calls for an 11V, 4A secondary winding. This calls for a minimum transformer power rating of 44VA; the nearest standard size to this is 50VA, and its instruction sheet shows (perhaps) that the required number of secondary turns can by found by multiplying the 11V by 4.8 then adding 1% for each 10VA of loading, thus calling for a total of 55 secondary turns. A further look at the instruction sheet shows that the best wire size in this case needs a diameter of about 1.25mm, and this corresponds to 18 S.W.G or 16 A.W.G. and that completes the design procedure.

Guide to modern switches

To complete this 'passive components' survey, this final section looks at switches and basic switch circuits. Switches come in several basic types, and a selection of these is shown in *Figure 1.55*. The simplest is the push-button type, in which a spring-loaded conductor can be moved so that it does or does not bridge (short) a pair of fixed contacts. These switches come in either normally-open (NO) form *(a)*, in which the button is pressed to short the contacts, or in normally-closed (NC) form *(b)*, in which the button is pressed to open the contacts.

The most widely used switch is the moving arm type, which is shown in its simplest form in *(c)* and has a single spring-loaded (biased) metal arm or 'pole' that has permanent electrical contact with terminal 'A' but either has or has-not got contact with terminal 'B', thus giving an on/off switching action between these terminals. This type of switch is known as a single-pole single-throw, or SPST, switch; *(d)* shows two SPST switches mounted in a single case with their poles ganged together so that they move in unison, to make a double-pole single-throw, or DPST, switch.

Figure 1.55(e) shows a single-pole double-throw (SPDT) switch in which the pole can be 'thrown' so that it connects terminal 'A' to either terminal 'BA' or 'BB', thus enabling the 'A' terminal to be coupled in either of two different directions or 'ways'. *Figure 1.55(f)* shows a ganged double-pole or DPDT version of this switch.

Figure 1.55(g) shows a switch in which the 'A' terminal can be coupled to any of four others, thus giving a '1-pole, 4-way' action. *Figure 1.55(h)* shows a ganged 2-pole version of the same switch. In practice, switches can be designed to give any desired number of poles and 'ways'.

Two other widely used electric switches are the pressure-pad type, which takes the form of a thin pad easily hidden under a carpet or mat and which is activated by body weight, and the microswitch, which is a toggle switch activated via slight pressure on a button or lever on its side, thus enabling the switch to be activated by the action of opening or closing a door or window or moving a piece of machinery, etc.

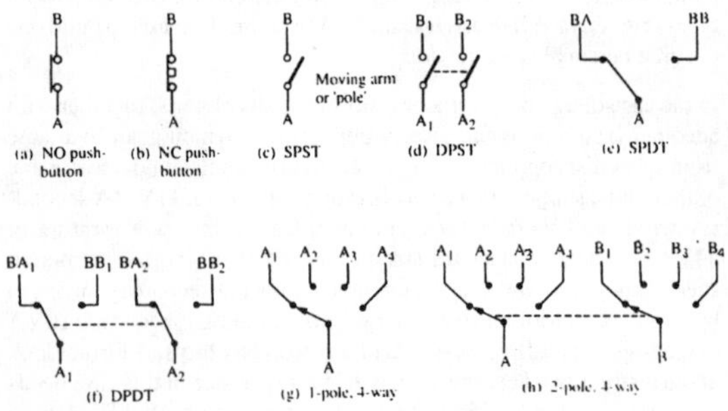

Figure 1.55. Some basic switch configurations.

Basic A.C. power switch circuits

One simple application of a toggle switch is an on/off lamp control. In A.C. line-powered applications this circuit must take the form shown in *Figure 1.56*, with SW_1 connected to the *live, phase* or *'hot'* power line and the lamp wired to the *neutral* or *safe* line, to minimise the user's chances of getting a shock when changing lamps.

Figure 1.57 shows how to switch a lamp from either of two points, by using a two-way switch at each point, with two wires (known as strapping wires) connected to each switch so that one or other wire carries the current when the lamp is turned on.

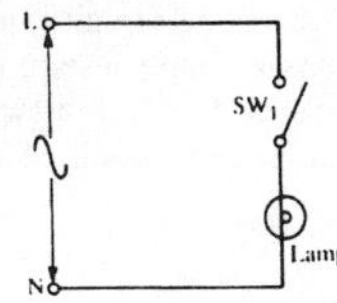

Figure 1.56. Single-switch on/off A.C. lamp control circuit.

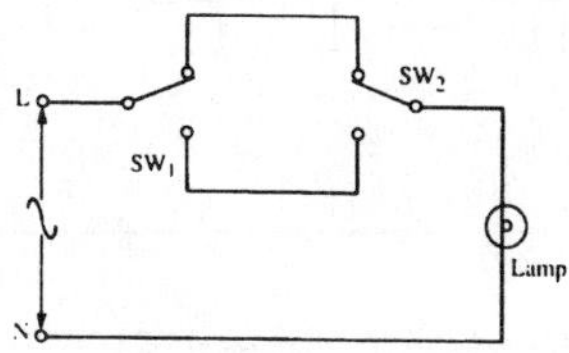

Figure 1.57. Two-switch on/off A.C. lamp control circuit.

Figure 1.58 shows the above circuit modified to give lamp switching from any of three points. Here, a ganged pair of 2-way switches (SW_3) are inserted in series with the two strapping wires, so that the SW_1–SW_2 lamp current flows directly along one strapping wire path when SW_3 is in one position, but crosses from one strapped wire path to the other when SW_3 is in the alternative position.

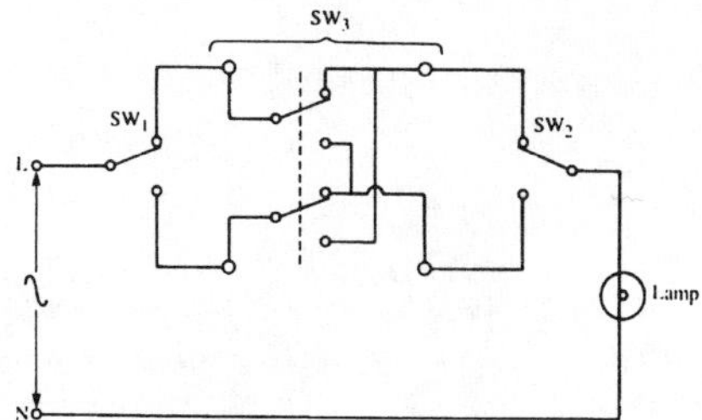

Figure 1.58. Three-switch on/off A.C. lamp control circuit.

Note that SW_3 has opposing pairs of output terminals shorted together. In the electric wiring industry such switches are available with these terminals shorted internally, and with only four terminals externally available (as indicated by the small white circles in the diagram); these switches are known in the trade as 'intermediate' switches.

In practice, the basic *Figure 1.58* circuit can be switched from any desired number of positions by simply inserting an intermediate switch into the strapping wires at each desired new switching position. *Figure 1.59*, for example, shows the circuit modified for four-position switching.

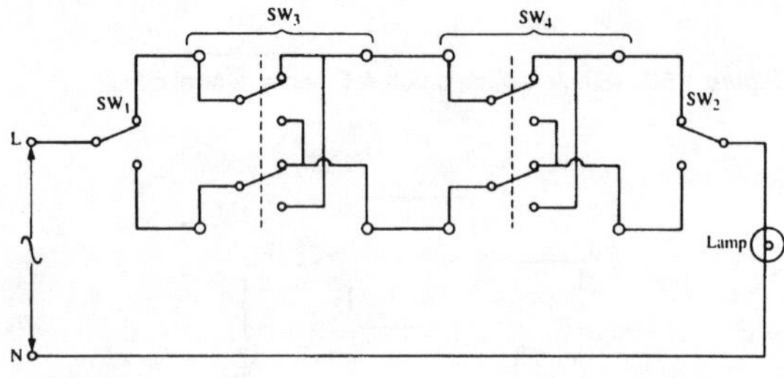

Figure 1.59. Four-switch on/off A.C. lamp control circuit.

2 Relays, meters, & motors

Modern electronic systems make considerable use of simple electro-mechanical devices that convert an electrical or magnetic force into a useful mechanical movement. The best known of such devices are relays, reed-relays, moving-coil meters, solenoids, and electric motors; all these devices are described in this chapter.

Relays

A conventional electromagnetic relay is really an electrically operated switch. *Figure 2.1* illustrates its operating principle; a multi-turn coil is wound on an iron core, to form an electromagnet that can move an iron lever or armature which in turn can close or open one or more sets of switch contacts. The operating coil and the switch contacts are electrically fully isolated from one another, and can be shown as separate elements in circuit diagrams.

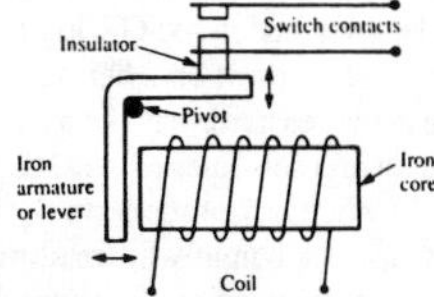

Figure 2.1. Basic design of standard electromagnetic relay.

The main characteristics of the relay coil are its operating voltage and resistance values, and *Figure 2.2* shows alternative ways of representing a 12V, 120Ω coil; the symbol of *(c)* is the easiest to draw, and carries all vital information. Practical relays may have coils designed to operate from a mere few volts d.c., up to the full a.c. power line voltage, etc.

There are three basic types of relay contact arrangement, these being normally closed (NC), normally open (NO), and change-over (CO), as shown in *Figure 2.3*. Practical relays often carry more than one set of contacts, with all sets ganged; thus, the term 'DPCO' simply means that the relay is 'Double Poled' and carries two sets of change-over (CO) contacts. Actual contacts may have electrical ratings up to several hundred volts, or up to tens of amps.

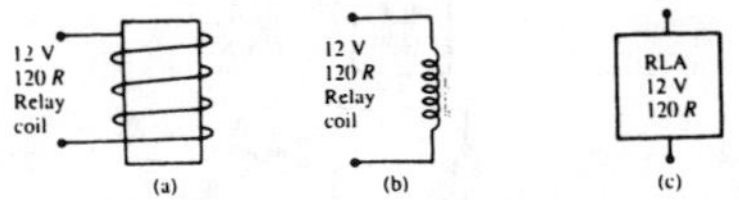

Figure 2.2. Alternative ways of representing a 12V, 120R relay coil.

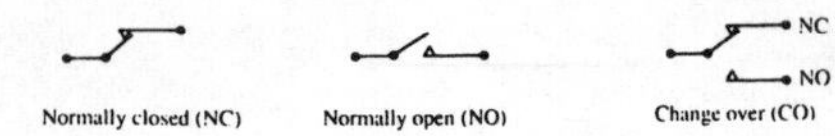

Figure 2.3. The three basic types of contact arrangement.

Relay configurations

Figures 2.4 to *2.12* show basic ways of using relays. In *Figure 2.4*, the relay is wired in the non-latching mode, in which push-button switch S_1 is wired in series with the relay coil and its supply rails, and the relay closes only while S_1 is closed.

Figure 2.5 shows the relay wired to give self-latching operation. Here, NO relay contacts RLA/1 are wired in parallel with activating switch S_1. RLA is normally off, but turns on as soon as S_1 is closed, making contacts RLA/1 close and lock RLA ON even if S_1 is subsequently re-opened. Once the relay has locked on it can be turned off again by briefly breaking the supply connections to the relay coil.

A relay can be switched in the AND logic mode by connecting it as in *Figure 2.6*, so that the relay turns on only when all switches are closed. Alternatively, it can be switched in the OR logic mode by wiring it as in *Figure 2.7*, so that the relay turns on when any switch closes. *Figure 2.8* shows both of these modes used to make a simple burglar alarm, in which the relay turns on and self-latches (via RLA/1) and activates an alarm bell (via RLA/2) when any of the S_1 to S_3 'switches' are briefly closed (by opening a door or window or treading on a mat, etc). The alarm can be enabled or turned off via key switch S_4.

Relay coils are highly inductive and may generate back-emfs of hundreds of volts if their coil currents are suddenly broken. These back-emfs can easily damage switch contacts or solid-state devices connected to the coil, and it is thus often necessary to 'damp' them via protective diodes. *Figures 2.9* and *2.10* show examples of such circuits.

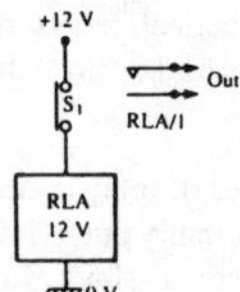

Figure 2.4. Non-latching relay switch.

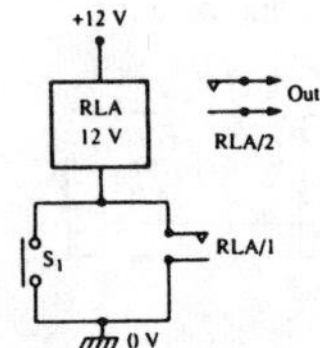

Figure 2.5. Self-latching relay switch.

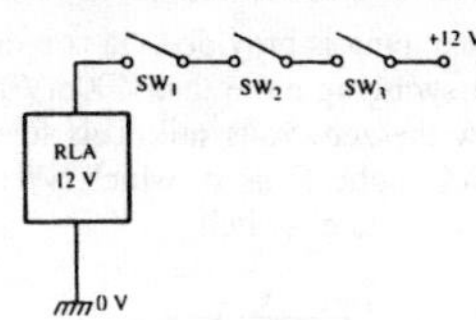

Figure 2.6. AND logic switching.

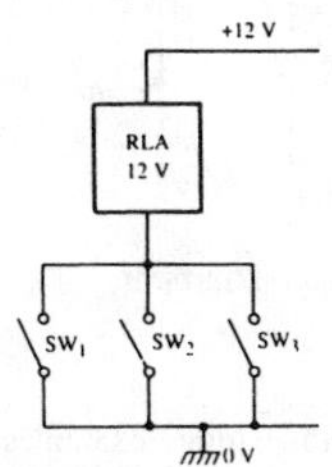

Figure 2.7. OR logic switching.

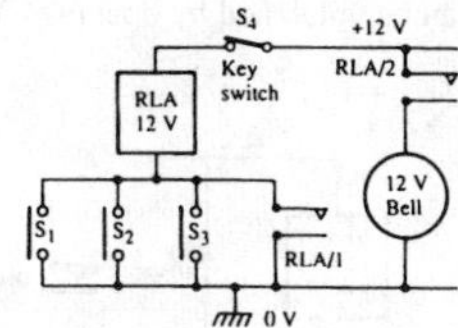

Figure 2.8. Simple burglar alarm.

In *Figure 2.9* the coil damping is provided via D_1, which prevents switch-off back-emfs from driving the RLA–SW_1 junction more than 600mV above the positive supply rail value. This form of protection is adequate for many practical applications.

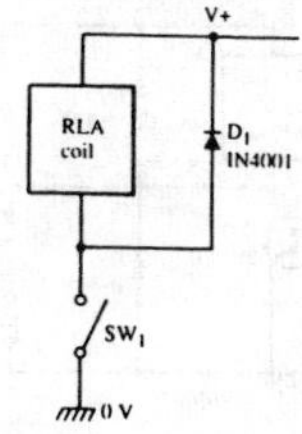

Figure 2.9. Single-diode coil damper.

In *Figure 2.10* the damping is provided via two diodes that stop the RLA–SW_1 junction swinging more than 600mV above the positive supply rail or below the zero-volts rail. This form of protection is recommended for all applications in which SW_1 is replaced by a transistor or other 'solid-state' switch.

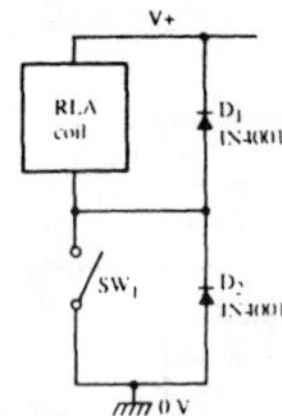

Figure 2.10. Two-diode coil damper.

Figures 2.11 and *2.12* show simple examples of transistors used to increase the effective sensitivity of a relay. The relay is off when the input is low, and on when the input is high; an input current of 1mA or so is enough to turn the relay on. Both circuits are provided with relay coil damping, and the *Figure 2.12* version gives self-latching relay operation (the relay can be unlatched by opening SW_1).

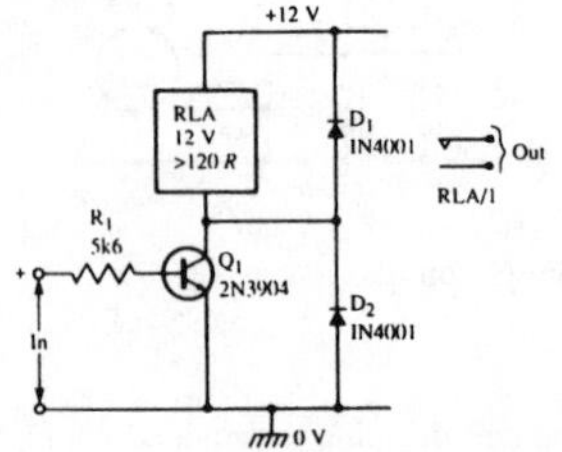

Figure 2.11. Non-latching transistor-driven switch.

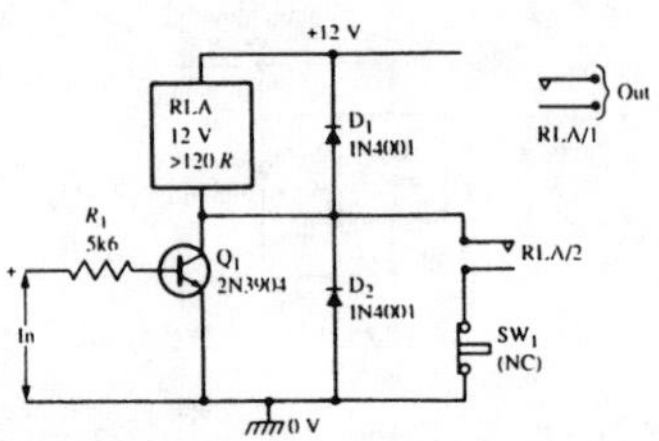

Figure 2.12. Self-latching transistor-driven switch.

Reed relays

A 'reed' relay consists of a springy pair of opposite-polarity magnetic reeds with gold- or silver-plated contacts, sealed into a glass tube filled with protective gasses, as shown in *Figure 2.13*. The opposing magnetic fields of the reeds normally hold their contacts apart, so they act as an NO switch, but these fields can be effectively cancelled or reversed (so that the switch closes) by placing the reeds within an externally-generated magnetic field, which can be derived from either an electric coil that surrounds the glass tube, as shown in *Figure 2.14*, or by a permanent magnet placed within a few millimetres of the tube, as shown in *Figure 2.15*.

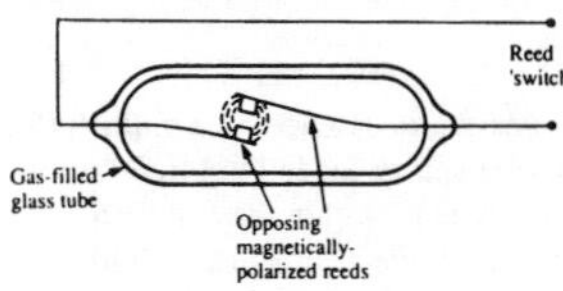

Figure 2.13. Basic structure of reed relay.

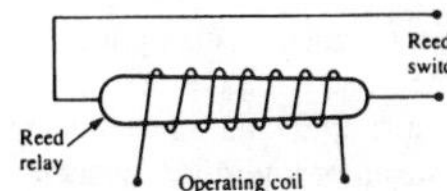

Figure 2.14. Reed relay operated by coil.

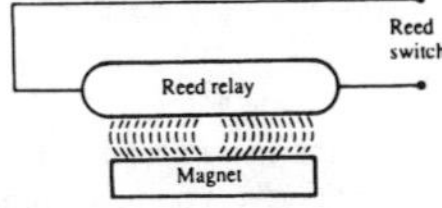

Figure 2.15. Reed relay operated by magnet.

Practical reed relays are available in both NO and CO versions, and their contacts can usually handle maximum currents of only a few hundred mA. Coil-driven types can be used in the same way as normal relays, but typically have a drive-current sensitivity ten times better than a standard relay.

A major advantage of the reed relay is that it can be 'remote activated' at a range of several millimetres via an external magnet, enabling it to be used in many home-security applications; *Figure 2.16* illustrates the basic principle. The reed relay is embedded in a door or window frame, and the activating magnet is embedded adjacent to it in the actual door or window so that the relay changes state whenever the door/window is opened or closed. Several of these relays can be inter connected and used to activate a suitable alarm circuit, if desired.

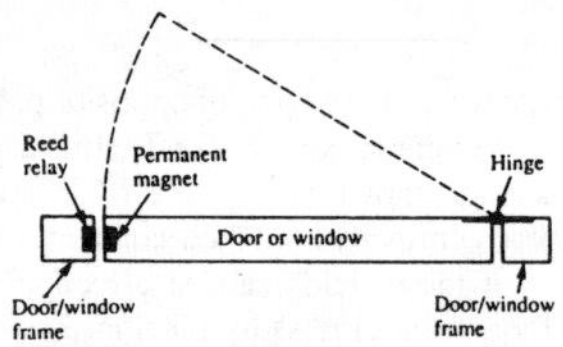

Figure 2.16. Method of using a reed relay/magnet combination to give burglar protection to a door or window.

Solenoid devices

A basic solenoid consists, in essence, of a multi-turn coil wound on a former that encircles a spring-loaded but free-moving iron armature. Normally, the armature rests within the coil former, but when the coil is energised the resulting electromagnetic field drives the armature outwards until it reaches an end-stop. The energised solenoid thus imparts a linear thrust at one end of the armature and a linear pull at the other end. In some special types of solenoid the armature movement is rotary, rather than linear.

A solenoid's mechanical movements can be put to a variety of practical uses. In the solenoid latch the armature drives a door latch. In the solenoid clutch it drives a simple clutch mechanism. In the solenoid valve it activates a gas, steam or liquid valve; such valves are widely used in washing machines, central-heating systems, and in industrial control processes. The solenoid switch or solenoid starter is used in automobiles and activates a heavy-duty switch that connects the vehicle's battery to the starter motor.

Moving coil meters

A moving coil meter is a delicate instrument that draws current from the signal under test and uses it to drive a pointer a proportionate amount across a calibrated scale, thus giving an analog display of the current's magnitude. *Figure 2.17* shows a basic moving-coil meter movement. The test current is fed through a coil of fine copper wire that is mounted on an aluminium bobbin that also carries the meter's pointer and is supported on low-friction bearings. This assembly is mounted within the field of a powerful magnet, and the test current is fed to the coil ends via a pair of contra-wound springs fitted in such a way that their tensions balance out when the meter's pointer is in the 'zero' position. The test current generates a magnetic field around the coil, and this interacts with that of the magnet and generates a torque that rotates the coil assembly against the spring pressure and moves the pointer a proportionate amount across the scale. In most meters the coil unit is supported on jewelled bearings, but in some high-quality units it is supported on a taut-band or rod, which gives friction-free suspension.

The most important parameters of a moving-coil meter are its basic full-scale deflection (f.s.d.) current value (usually called its 'sensitivity') and its coil resistance value, *r*, which causes a proportionate voltage to be generated or 'lost' across the meter's terminals (typically

50mV to 400mV at f.s.d.). *Figure 2.18* lists typical performance details of some popular and readily-available meters. Most of these have the 'zero' current mark on the left of their scale, but 'centre zero' types have it in the middle, with negative numbers to its left and positive ones to its right. The pointer's 'zero' position is usually trimmable via a screw-headed control on the front of the meter.

A moving-coil meter's precision is expressed in terms of 'percentage of f.s.d. value' error over the 'effective range' of the instrument. The 'effective range' is defined as 'from 10% to 100% of f.s.d. value' in d.c.-indicating meters. Typical meter precision is ±2% of f.s.d. in medium-quality units, i.e., ±4% at half-scale, or ±20% at 1/10th-scale, etc.

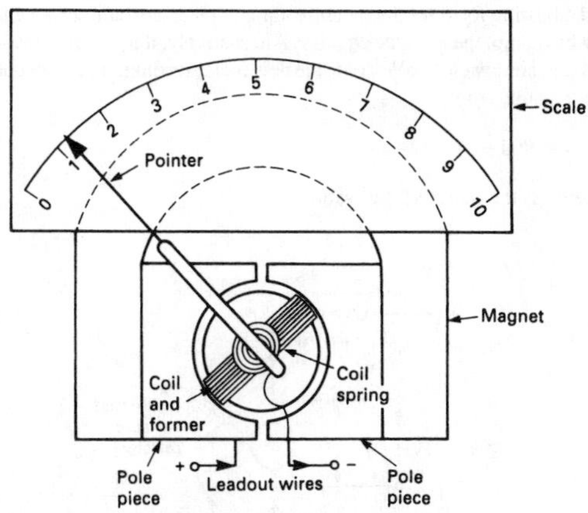

Figure 2.17. Basic moving-coil meter movement.

Meter F.S.D., (indicated)	Coil resistance (typical), OHMS	Volt drop at F.S.D.	Sensitivity, OHMS/volt
50μA	2700 – 4300	135 – 215mV	20k/V
50 – 0 – 50μA	1300 – 3000	65 – 150mV	20k/V
100μA	1300 – 3750	130 – 375mV	10k/V
100 – 0 – 100μA	1100	110mV	10k/V
200μA	750	150mV	5k/V
1mA	75 – 200	75 – 200mV	1k0/V
100mA	0.5 – 0.8Ω	50 – 80mV	N.A.
1A	0.05 – 0.1Ω	50 – 100mV	N.A.
5A	0.01Ω	50mV	N.A.
10A	0.005Ω	50mV	N.A.
5V, 10V, 15V, 100V, 300V at 1k0/V	75 – 200Ω	75 – 200mV	1k0/V

Figure 2.18. Typical ranges and performance details of some popular fixed-value moving-coil meters.

Any sensitive moving-coil meter can be made to read higher ranges of d.c. current by wiring a suitable 'shunt' resistor across its terminals, or can be made to read d.c. voltages by wiring a suitable 'multiplier' resistor in series with its terminals. To do this, however, it is first necessary to know the value of the meter's coil resistance, *r*.

Measuring the coil resistance

A meter's '*r*' value can be measured via the *Figure 2.19* circuit, in which the series value of R_1 and RV_1 equals V_{IN} multiplied by 'y', the meter's 'ohms per volt' or 1/I sensitivity value, and R_x is a calibrated variable resistor. In use, RV_1 is first adjusted, with SW_1 open, to set the meter reading at precisely f.s.d.; '*r*' can then be found by closing SW_1 and adjusting R_x to set the meter reading at precisely half-scale value, at which point the R_x value equals *r*. Alternatively, if R_x is a fixed value resistor, simply close SW_1, note the new meter reading, 'i', and deduce the '*r*' value from:

$$r = R_x(I - i)/i,$$

where I is the meter's f.s.d. value.

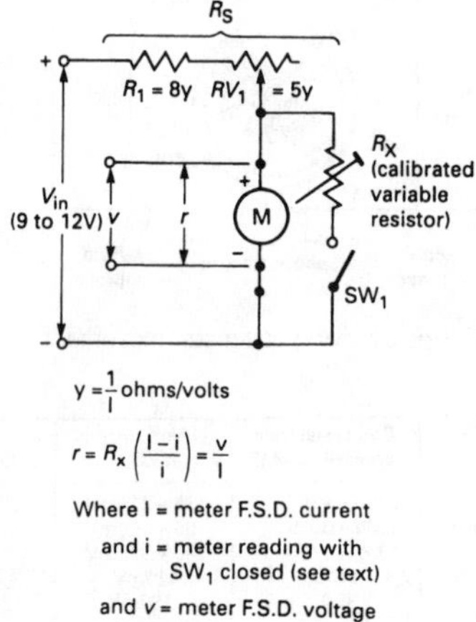

Figure 2.19. Circuit for measuring the meter's internal coil resistance, *r*.

Designing d.c. voltmeters

A moving-coil meter can be made to read d.c. voltages by feeding them to it via a series 'multiplier' resistor, as in *Figure 2.20*. The appropriate multiplier value equals V/I, where V is the desired f.s.d. voltage value and I is the meter's f.s.d. current value.

Note that this multiplier value includes r, the meter's coil resistance, and the actual value of external multiplier resistance, R_m, needed to give a desired f.s.d. voltage value is thus given by:

$$R_m = (V/I) - r.$$

In practice, r can usually be ignored in cases where R_m is at least 100 times greater than r.

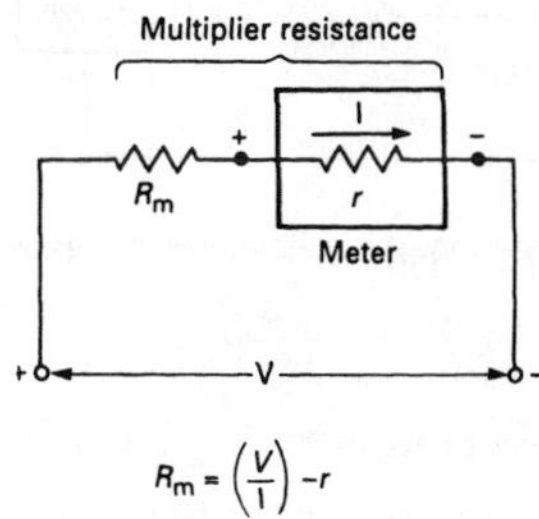

Figure 2.20. Basic d.c. voltmeter circuit.

Figures 2.21 and *2.22* show alternative ways of using a 100μA meter to give f.s.d. ranges of (a) 3V, (b) 10V, (c) 30V, (d) 100V, and (e) 300V. In each case the total multiplier resistance needed on each range is (a) 30k, (b) 100k, (c) 300k, (d) 1M0, and (e) 3M0. Note in *Figure 2.21* that the effect of 'r' has to be allowed for on both of the lower ranges, but in *Figure 2.22* it needs to be allowed for on the lowest range only. The *Figure 2.22* type of circuit is widely used in good-quality multimeters.

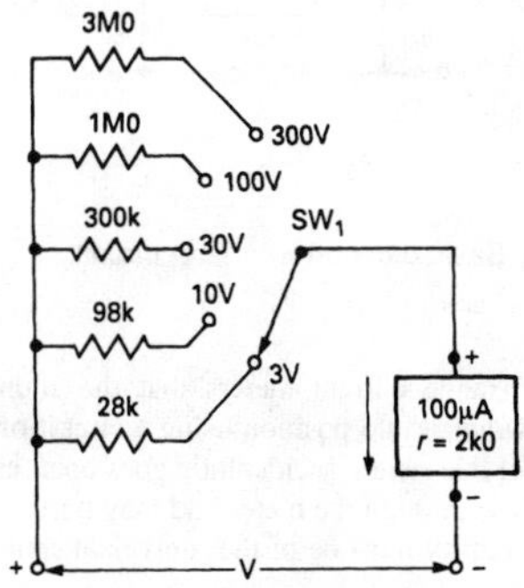

Figure 2.21. 5-range d.c. voltmeter using individual multiplier resistors.

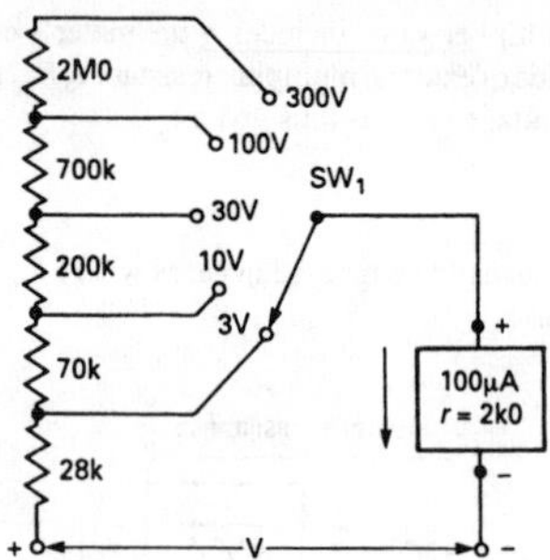

Figure 2.22. Five-range d.c. voltmeter using series-wired multiplier resistors.

Extending current ranges

A meter's effective current range can be extended by connecting a 'shunt' resistor across the basic meter, as in *Figure 2.23*, so that a known fraction of the total current passes through the shunt and the remainder passes through the meter, which can be calibrated in terms of 'total' current. The relative values of shunt and meter currents are set by the relative values of the coil and shunt resistances, and the value of shunt resistor, R_s, needed to give a particular f.s.d. current reading (I_t) is given by:

$$R_s = r/(n - 1) \text{ or } (I.r)/(I_t - I),$$

where $n = I_t/I$, the number of times by which the desired meter range is greater than the basic range. Thus, to convert the 100μA, 2k0 meter to read 100mA f.s.d., R_s needs a value of 2000/(1000 – 1) = 2.0Ω.

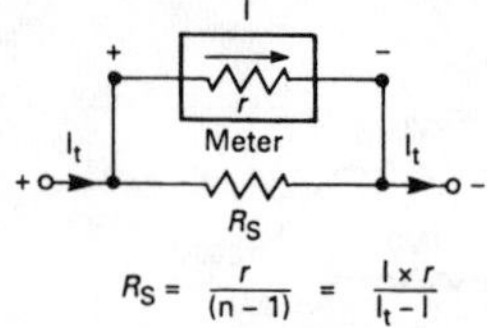

Figure 2.23. Basic d.c. current meter circuit.

Note in multi-range current meters that the shunt resistors must NEVER be switched into position using a circuit of the *Figure 2.24* type, because if this switch accidentally goes open-circuit the full test current will flow through the meter and may burn it out. Instead, the multi-range circuitry must be of the 'universal shunt' type shown in *Figure 2.25*.

Figure 2.25 shows a 100μA, 2k0 meter fitted with a universal shunt that gives d.c. current ranges of 1mA, 10mA, and 100mA. The three series-

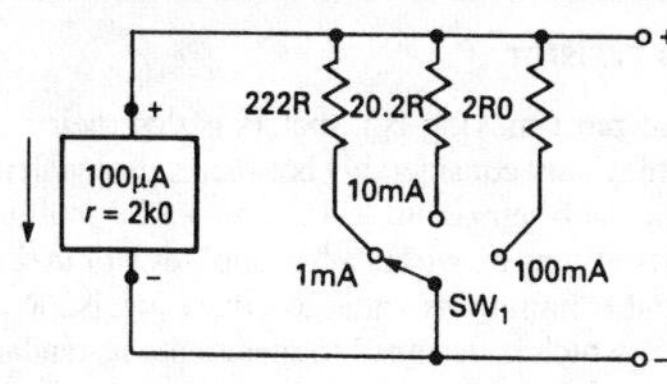

Figure 2.24. Classic example of how NOT to use shunt switching in a multi-range current meter.

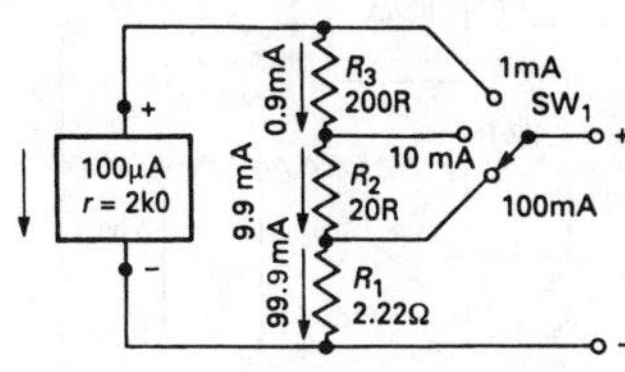

Figure 2.25. Worked example of a three-range universal shunt circuit.

connected range resistors are permanently wired across the meter, and range changing is achieved by switching the test current into the appropriate part of the series chain; the meter's accuracy is thus not influenced by variations in SW_1's contact resistances.

The procedure for designing a universal shunt follows a logical sequence. The first step is to determine the *TOTAL* resistance (R_t) of the R_1–R_2–R_3 shunt chain, which sets the f.s.d. value of the lowest (1mA) current range, using the formula:

$$R_t = r/(n - 1),$$

where r is the meter's coil resistance and n is the current multiplication factor. In the example shown

$$R_t = 2000/9 = 222.2\Omega.$$

The next step is to find the value of the *highest* current shunt (R_1), using the formula:

$$R_s = (r + R_t)/n,$$

which in this case gives a value of 2222.2/1000 = 2.22Ω for R_1. The same formula is used to find the values of all other shunts, and on the 10mA range gives 22.22Ω, but since this shunt comprises R_1 and R_2 in series, the R_2 value = 22.2 – 2.2Ω = 20Ω. Similarly, the shunt value for the 1mA range is 222.2Ω, but is made up of R_1 and R_2 and R_3 in series, so R_3 needs a value of 200Ω.

The swamp resistor

A weakness of most moving-coil meters is that their values of coil resistance (*r*) may vary considerably between individual models of the same type, and with temperature. In commercial multimeters these problems are overcome by wiring a 'swamp' resistor in series with the basic meter and trimming its value to give a precise R_{TOTAL} (= r + R_{SWAMP}) value, which is designed to match into a standard universal shunt network and at the same time give a near-zero overall temperature coefficient. *Figure 2.26* shows an example of a swamp resistor used in a 6-range current meter circuit.

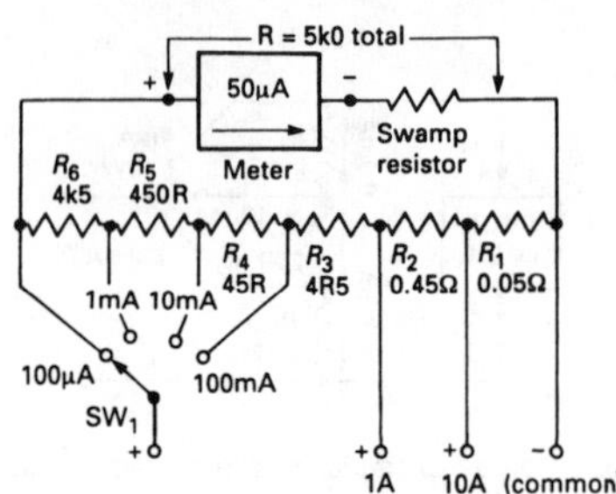

Figure 2.26. Example of a swamp resistor used in a six-range d.c. current meter.

A swamp resistor converts an ordinary moving-coil meter into a truly useful and semi-precision measuring instrument, but at the expense of an increase in its f.s.d. voltage value, i.e., a meter that has a normal f.s.d. sensitivity of 100mV will have one of 200mV if $R_{SWAMP} = r$, etc. The *Figure 2.26* circuit has an f.s.d. sensitivity of 250mV on its most sensitive (100μA) range, and of 500mV on the 10A range; note that (like many commercial multimeters) its 1A and 10A ranges are selected via terminals, thus eliminating the need to use switches with very high current ratings.

A.c. voltmeters

A moving-coil meter can be used as an a.c. voltmeter by connecting its input via a suitable rectifier and a multiplier resistor. If the rectifier is a bridge type, the voltmeter is calibrated to read r.m.s. values of a sinewave input on the assumption that the resulting meter current is 1.11 times greater than the simple d.c. equivalent current; such a voltmeter uses the basic circuit and formula of *Figure 2.27*. Note that the R_m value approximates (V/I) x 0.9, but the design formula is complicated by the fact that the forward voltage drop of the bridge rectifier (= 2 x V_f) must be deducted from the effective 'V' value, and that the meter's coil resistance (*r*) and the bridge's forward impedance (2 x Z_f) must be deducted from the simplified R_m value.

In practice, the V_f and Z_f characteristics of diodes are highly non-linear, and an a.c. voltmeter consequently gives a reasonable linear scale

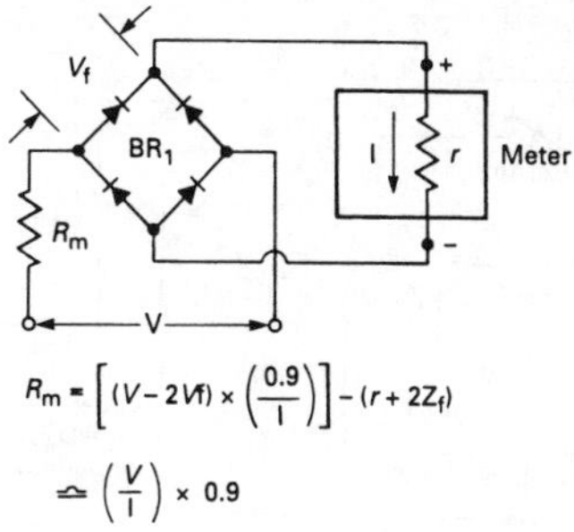

Figure 2.27. Basic a.c. voltmeter using a bridge rectifier.

reading only if the input voltage is large relative to V_f, and R_m is large relative to Z_f. For this reason, bridge-rectifier types of a.c. voltmeter usually have a maximum useful f.s.d. sensitivity of about 10V.

The bridge rectifier should ideally give a low forward voltage drop. Old (pre-1970) instruments often used copper oxide rectifiers to meet this ideal, but these were very leaky, and to overcome this snag the meters were usually operated at an f.s.d. current of 900μA (to give a high forward/reverse current ratio), and this resulted in the typical circuit of *Figure 2.28*, which has a basic a.c. sensitivity of 1k0/volt. Most modern meters use a bridge rectifier made of either Schottky or germanium diodes that are pre-tested for low reverse-leakage currents, and are able to give a sensitivity of up to 10k/volt, as in the case of the circuit of *Figure 2. 29*.

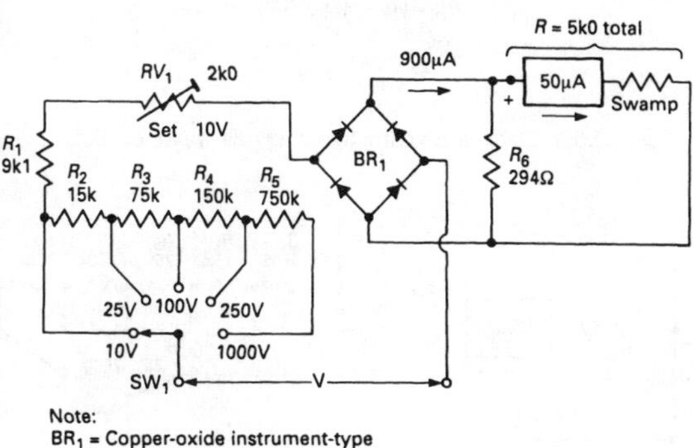

Figure 2.28. Old-style a.c. voltmeter with a sensitivity of 1k0/volt.

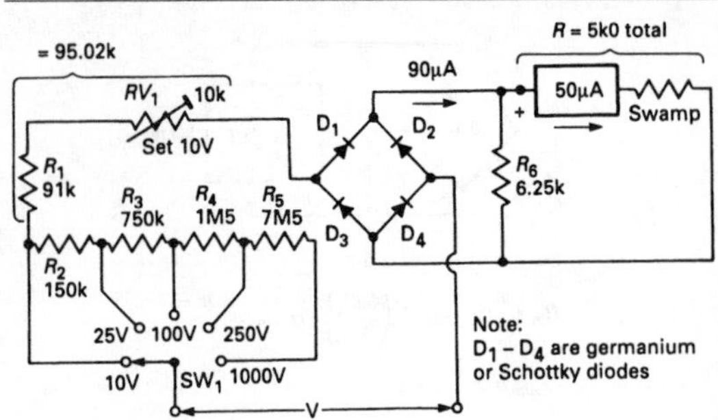

Figure 2.29. Modern-style a.c. voltmeter with a sensitivity of 10k/volt.

Some a.c. voltmeters use half-wave a.c. rectifier circuits of the type shown in *Figure 2.30*, in which the 'V_f' voltage losses are only half as great as in the bridge type, but in which a.c. sensitivity is also halved. *Figure 2.31* shows a multi-range a.c. voltmeter using this technique; here, the meter is shunted (by R_6) to give an effective f.s.d. sensitivity of 450μA, enabling the multiplier resistor (R_1 to R_5) values to be chosen on the basis of 1k0/volt.

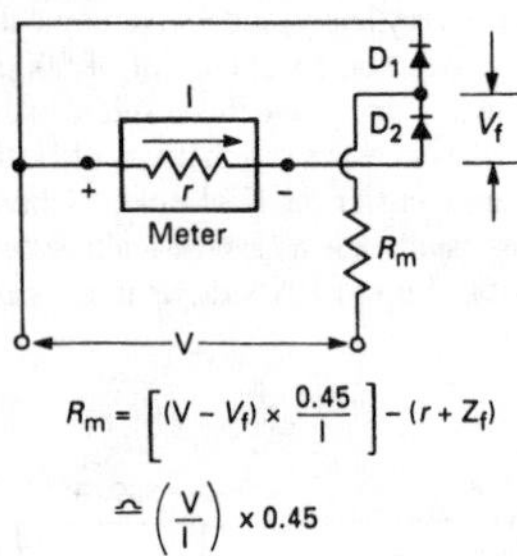

$$R_m = \left[(V - V_f) \times \frac{0.45}{I}\right] - (r + Z_f)$$

$$\simeq \left(\frac{V}{I}\right) \times 0.45$$

Figure 2.30. Basic a.c. voltmeter using half-wave rectification.

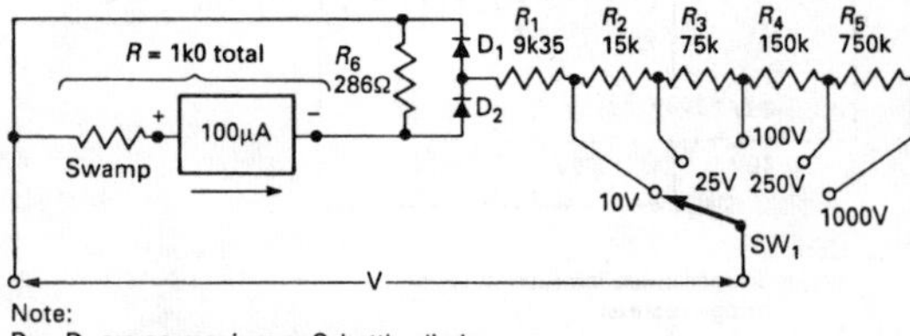

Figure 2.31. A.c. voltmeter using half-wave rectification; sensitivity = 1k0/volt.

Simple a.c. voltmeters become non-linear when measuring low voltages, so commercial units rarely have ranges lower than 10V f.s.d., or have any facility for measuring a.c. currents of any value. Alternating currents and low-value voltages are best measured by using the moving-coil meter as an analogue readout unit in an 'electronic' type of meter or multimeter.

Meter overload protection

Moving-coil meters are easily damaged by large overload currents. If the meter is used in conjunction with a swamp resistor, excellent overload protection can be gained by connecting a pair of silicon diodes as shown in *Figure 2.32*. Here, the swamp resistor is split into two parts (R_1 and R_2), with R_2's value chosen so that 200mV is generated across the diodes at f.s.d.; at overloads in excess of twice the f.s.d. value the diodes start to conduct, and thus limit the meter's overload current; R_1 limits the diode overload currents to no more than a few mA.

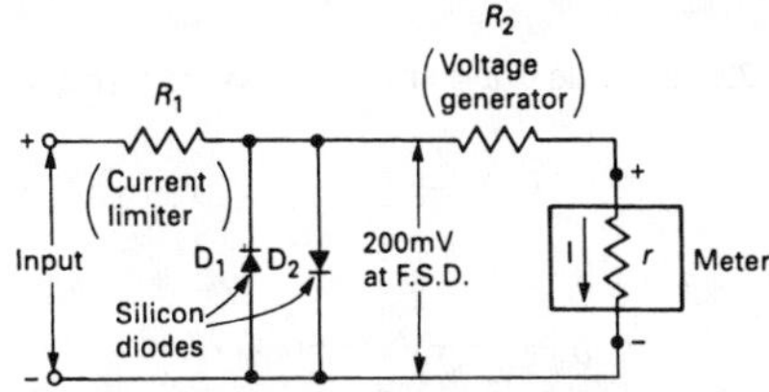

Figure 2.32. Meter overload protection given via silicon diodes.

Multi-function meters

Often, a single meter is built into an item of test gear and used as a multi-function 'V and I' meter. Here, all multiplier and shunt resistors are permanently wired into circuit, and the meter is simply switched in series or parallel with the appropriate element. If a common measuring point can be found the switching can be made via a single-pole switch, as in *Figure 2.33*, which uses one meter to monitor the output voltage and current of a regulated P.S.U. If a common measuring point can not be found, the switching must be done via a 2-pole multi-way switch, as in the circuit of *Figure 2.34*, which can monitor several independent d.c. voltage and current values.

D.C. motors

The three most widely used types of D.C. electric motor are the ordinary permanent magnet type, the servomotor type, and the multiphase 'stepper' type. The rest of this chapter is devoted to these three types of motor.

The most widely used D.C. motor is the permanent magnet 'commutator' type, which is usually known simply as a 'D.C. motor' and has a commutator that rotates when the motor is powered from an appropriate D.C. voltage. *Figure 2.35* shows the motor's circuit symbol and its simplified equivalent circuit.

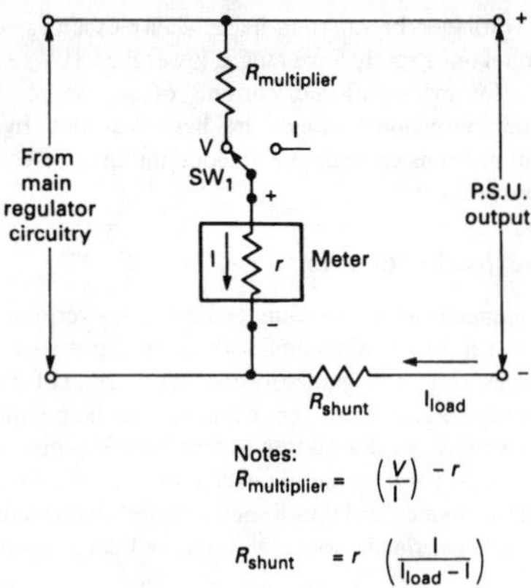

Figure 2.33. Example of a single meter used to read both V and I in a P.S.U.

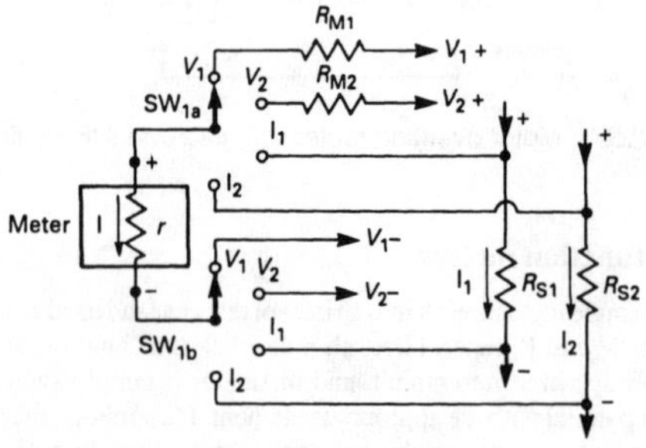

Figure 2.34. A single meter used to monitor two V and two I ranges in an instrument.

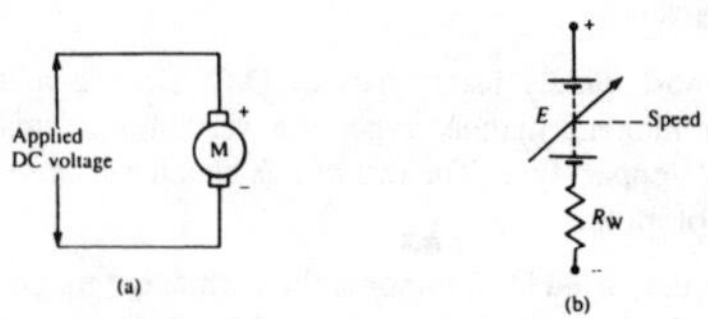

Figure 2.35. Symbol *(a)* and equivalent circuit *(b)* of a permanent-magnet type of D.C. motor.

The D.C. motor's basic action is such that an applied D.C. voltage makes a current flow through sets of armature windings (via commutator segments and pick-up brushes). This current generates electromagnetic fields that react with the fields of fixed stator magnets in such a way that the armature is forced to rotate. As it rotates, the armature's interacting fields make it generate a back-emf proportional to the armature speed and opposing the applied D.C. voltage, thus giving the equivalent circuit of *Figure 2.35(b)* in which R_w represents the total resistance of the armature windings, and E represents the speed-dependent back-emf. Important points to note about this kind of motor are as follows:

(1) When the motor is driving a fixed load, its speed is proportional to supply voltage; when it is powered from a fixed D.C. supply, its running current is proportional to its armature loading.

(2) The motor's *effective* applied voltage equals the applied D.C. voltage minus the speed-dependent back-emf. Consequently, when it is powered from a fixed voltage, motor speed tends to self-regulate, since any increase in loading tends to slow the armature, thus reducing the back-emf and increasing the *effective* applied voltage, and so on.

(3) The motor current is greatest when the armature is stalled and the back-emf is zero, and then equals V/R_w (where V is the supply voltage); this state naturally occurs under 'start' conditions.

(4) The direction of armature rotation can be reversed by reversing the motor's supply connections.

D.C. motors can be subjected to various forms of electronic power control, the main ones being in on/off switching control, in direction control, in improved speed regulation, and in variable speed control.

On/off switching

A D.C. motor can be turned on and off by wiring a control switch between the motor and its power supply. This switch can be an ordinary type as in *Figure 2.36(a)* or it can be a switching transistor as in *Figure 2.36(b)*, in which the motor is off when the input is low and is on when the input is high. Note that D_1 and D_2 are used to damp the motor's back-emf, C_1 limits unwanted RFI, and R_1 limits Q_1's base current to about 52mA with a 6V input, and under this condition Q_1 can provide a maximum motor current of about 1A.

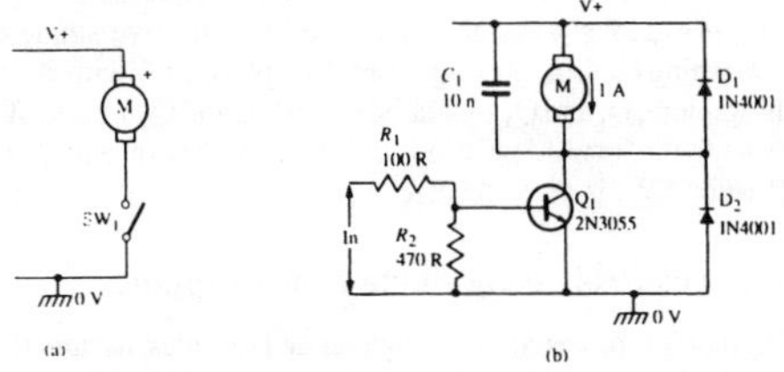

Figure 2.36. On/off motor control using *(a)* electromechanical and *(b)* transistor switching.

Direction control, using dual supplies

A D.C. motor's direction of rotation can be reversed by simply changing the polarity of its supply connections. If the motor is powered via dual (split) supplies, this can be achieved via a single-pole switch connected as in *Figure 2.37*, or via transistor-aided switching by using the circuit of *Figure 2.38*.

In *Figure 2.38*, Q_1 and Q_3 are biased on and Q_2 and Q_4 are cut off when SW_1 is set to the forward position, and Q_2 and Q_4 are biased on and Q_1 and Q_3 are cut off when SW_1 is set to the reverse position. Note that if this circuit is used with supply values greater than 12 volts, diodes must be wired in series with the Q_1 and Q_2 base-emitter junctions, to protect them against breakdown when reverse biased.

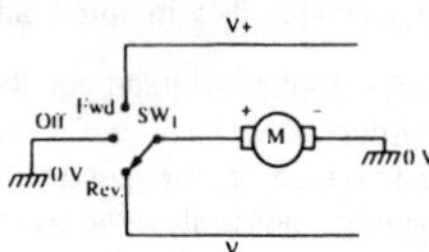

Figure 2.37. Switched motor-direction control, using dual (split) power supplies.

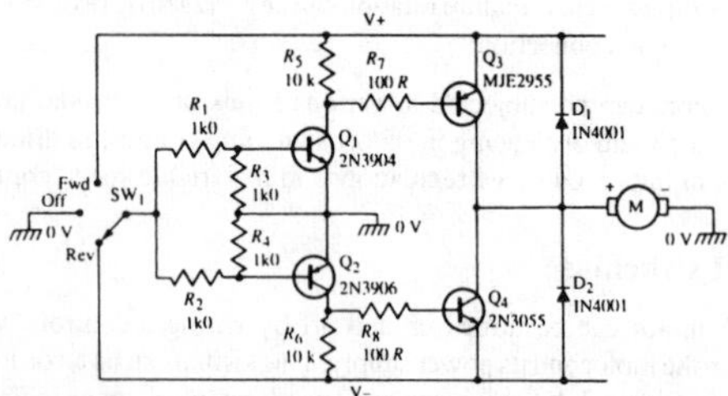

Figure 2.38. Transistor-switched direction control, using dual supplies.

Figure 2.38 uses double-ended input switching, making it difficult to replace SW_1 with electronic control circuitry in 'interfacing' applications. *Figure 2.39* shows the design modified to give single-ended input switching control, making it easy to replace SW_1 with electronic switching. Here, Q_1 and Q_3 are biased on and Q_2 and Q_4 are cut off when SW_1 is set to the forward position, and Q_2 and Q_4 are on and Q_1 and Q_3 are off when SW_1 is set to reverse.

Direction control, using single-ended supplies

If a D.C. motor is powered from single-ended supplies, its direction can be controlled via a double-pole switch connected as in *Figure 2.40*, or via a bridge-wired set of switching transistors connected in the basic form shown in *Figure 2.41*. In the latter case, Q_1 and Q_4 are turned on and Q_2 and Q_3 are off when SW_1 is set to the forward position, and Q_2 and Q_3 are on and Q_1 and Q_4 are off when SW_1 is set to reverse. Diodes D_1 to D_4 protect the circuit against possible damage from motor back-emfs, etc.

Figure 2.42 shows how the above circuit can be modified to give alternative switching control via independent forward/reverse (SW_1) and on/off (SW_2) switches. An important point to note about this configuration is that it causes Q_1 or Q_2 to be turned on at all times, with the on/off action being applied via Q_3 or Q_4, thus enabling the motor currents to collapse very rapidly (via the Q_1–D_2 or Q_2–D_1 loop) when the circuit is switched off. This so-called 'flywheel' action is vital if SW_2 is replaced by a pulse-width modulated (PWM) electronic switch, enabling the motor speed to be electronically controlled.

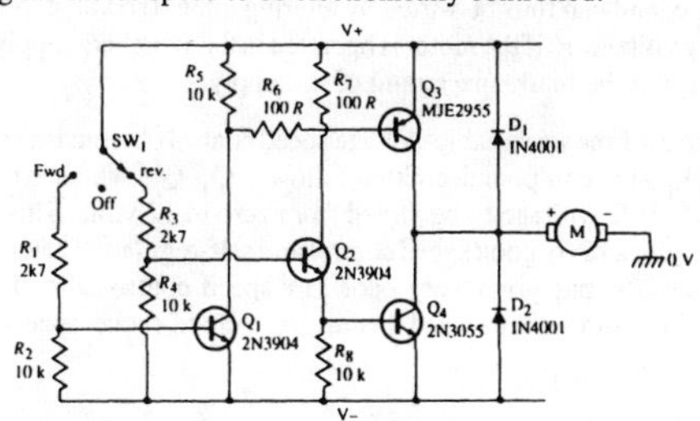

Figure 2.39. Transistor-switched motor-direction control, using dual power supplies but a single-ended input.

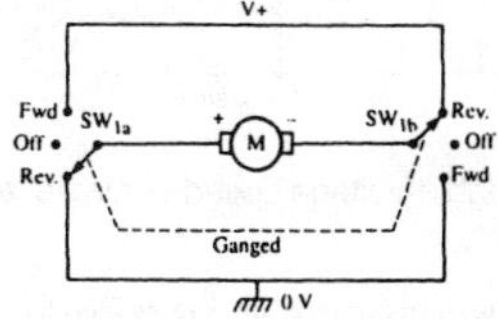

Figure 2.40. Switched motor-direction control, using a single-ended power supply.

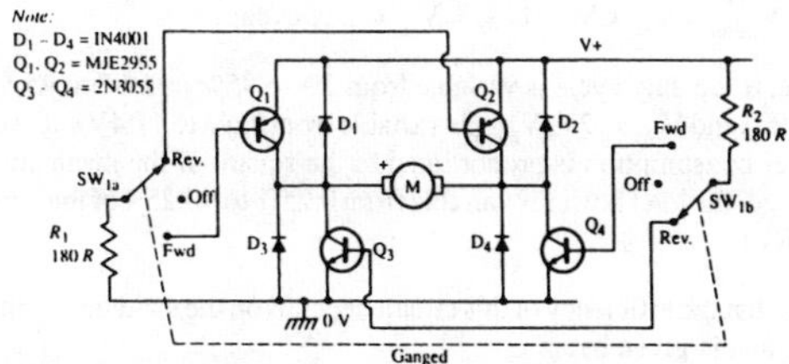

Figure 2.41. Transistor-switched motor-direction control circuit, using single-ended supplies.

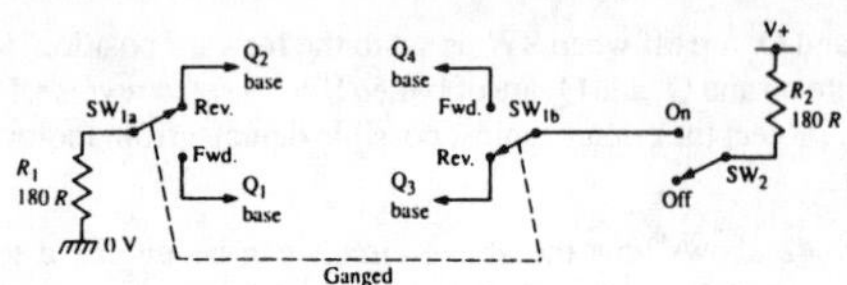

Figure 2.42. Alternative switching for the *Figure 2.41* circuit, using separate forward/reverse (SW_1) and on/off (SW_2) switches.

Motor speed control

A D.C. motor's speed is proportional to the *mean* value of its supply voltage, and can thus be varied by altering either the value of its D.C. supply voltage or, if the motor is operated in the switched-supply mode, by varying the mark-space ratio of its supply.

Figure 2.43 shows variable-voltage speed control obtained via variable pot RV_1 and compound emitter follower Q_1–Q_2, which enable the motor's D.C. voltage to be varied from zero to 12 volts. This type of circuit gives fairly good speed control and self-regulation at medium to high speeds, but gives very poor low-speed control and slow-start operation, and is thus used mainly in limited-range speed-control applications.

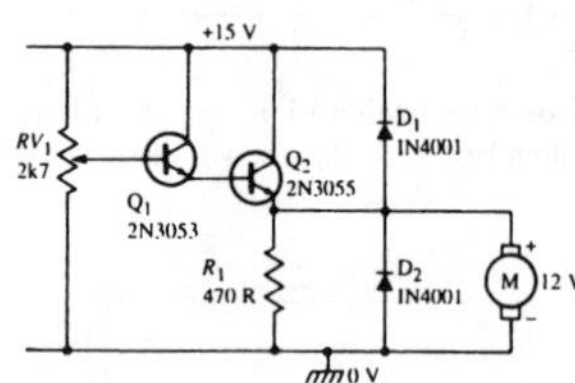

Figure 2.43. Variable-voltage speed-control of a 12V D.C. motor.

The most efficient way to control the power feed to any D.C. load is via the basic 'switched mode' circuit of *Figure 2.44*, in which power reaches the load via a solid-state switch that is activated via a square-wave generator with a variable M/S-ratio or duty cycle. The mean voltage (V_{mean}) reaching the load is given by:

$$V_{mean} = V_{pk} \times M/(M+S) = V_{pk} \times \text{duty cycle.}$$

Thus, if the duty cycle is variable from 5% to 95% (= 5:95 to 95:5 M/S-ratio) and V_{pk} = 12V, V_{mean} is variable from 0.6V to 11.4V and, since power consumption is proportional to the square of the mean supply voltage, the load power is variable from 0.25% to 90.25%of maximum via *RV*1.

Note that the efficiency of this circuit depends on the switching voltage loss, and is given by:

$$\%\ \text{efficiency} = (V_{pk} \times 100)/V_{supply}.$$

Efficiency levels of over 95% can easily be obtained.

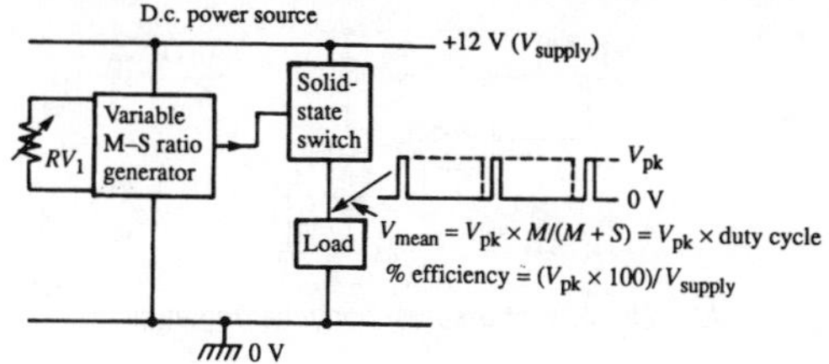

Figure 2.44. Switched-mode D.C. power-level controller.

Figure 2.45 shows a practical example of a switched-mode speed-control circuit. IC_1 acts as a 50Hz astable that generates a rectangular output with a mark-space ratio fully variable from 20:1 to 1:20 via RV_1, and this waveform is fed to the motor via Q_1 and Q_2. The motor's mean supply voltage (integrated over a 50Hz period) is thus fully variable via RV_1, but is applied in the form of high-energy pulses with peak values of 12 volts; this type of circuit thus gives excellent full-range speed control and generates high torque even at very low speeds; its degree of speed self-regulation is proportional to the mean value of applied voltage.

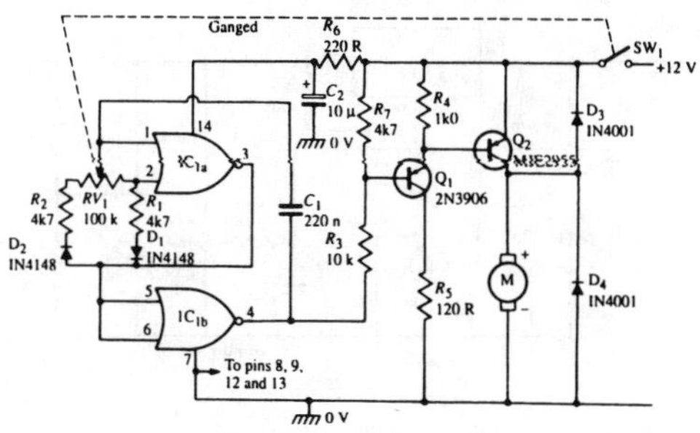

Figure 2.45. Switched-mode speed-control of a 12V D.C. motor.

Motor speed regulation

Motor speed regulators are meant to keep motor speed fairly constant in spite of wide variations in supply voltage and motor loading conditions, etc. *Figure 2.46* shows one that is designed to simply keep the motor's applied voltage constant in spite of wide variations in supply voltage and temperature. It is designed around a 317K 3-terminal variable voltage regulator IC which (when fitted to a suitable heat sink) can supply output currents up to 1.5 amps and has an output that is fully protected against short circuit and overload conditions. With the component values shown, the output is variable from 1.25V to 13.75V via RV_1 if the supply voltage is at least 3V greater than the desired output value.

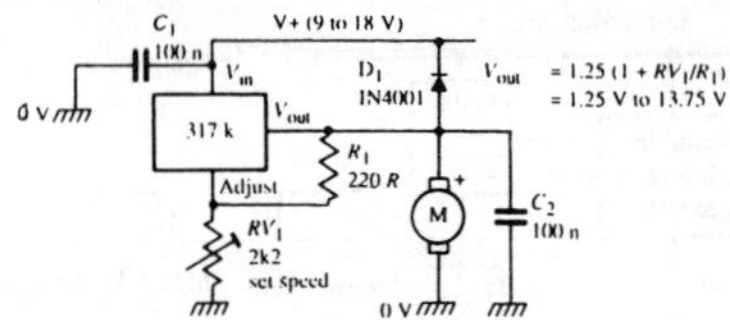

Figure 2.46. Simple motor-speed controller/regulator.

Figure 2.47 shows a high-performance variable-speed regulator that can be used to control 12V D.C. mini-drills, etc. The motor is again powered via the output of a 317K variable voltage regulator IC, but in this case the motor current is monitored via R_5–RV_2, which feed a proportional voltage to the input of the IC_2–Q_1 non-inverting D.C. amplifier, to generate a Q_1 emitter voltage that is directly proportional to the motor's load current. This circuit's output voltage equals the normal output value of the 317K IC (variable from 1.25V to 13.75V via RV_1) plus the voltage on Q_1's emitter; consequently, any increase in motor loading makes the circuit's output voltage rise, to automatically increase the motor drive and hold its speed reasonably constant. To initially set up the circuit, simply set the motor speed to about one-third of maximum via RV_1, then lightly load the motor and set RV_2 so that the speed remains similar in both loaded and unloaded states.

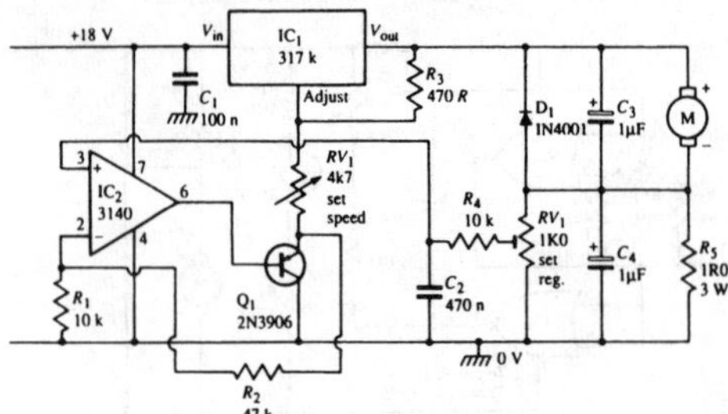

Figure 2.47. High-performance variable-speed regulator circuit.

Servomotor systems

A servomotor is a conventional electric motor with its output coupled (via a speed-reduction gearbox) to a movement-to-data translator such as a potentiometer or a tachogenerator. *Figure 2.48* shows a controller that can be used to give proportional movement (set via RV_2) of a servomotor with a pot (RV_1) output; the motor can be any 12V to 24V type that draws less than 700mA. Here, RV_1 and RV_2 are wired as a Wheatstone bridge, and the IC (a dual 4 watt power amplifier) is wired as a bridge-configured motor-driving difference amplifier. The circuit action is such that any movement of RV_2 upsets the bridge balance and generates a RV_1–RV_2 difference voltage that is amplified and fed to the motor, making its shaft rotate and move RV_1 to restore the bridge balance; RV_1 thus 'tracks' the movement of RV_2, which can thus be used to remote-control the shaft position.

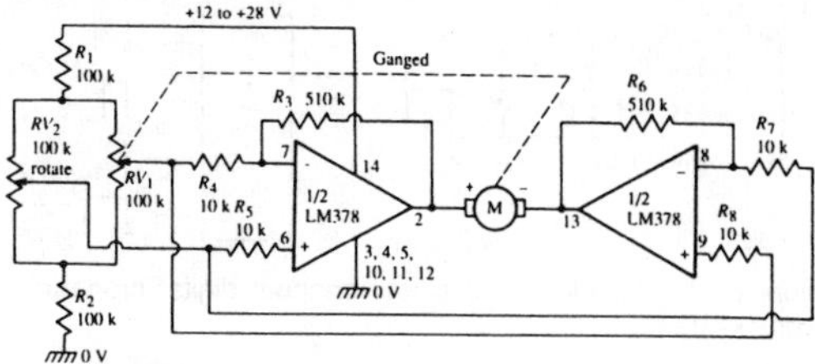

Figure 2.48. Proportional-movement (servomotor) controller.

One of the best known types of servomotor is that used in digital proportional remote control systems; these consist of a special IC plus a motor and a reduction gearbox that drives a pot and gives a mechanical output; *Figure 2.49* shows the block diagram of one of these systems, which is driven via a variable-width (1mS to 2mS) input pulse that is repeated once every 15mS or so (the frame time). The input pulse width controls the position of the servo's mechanical output; at 1mS the servo output may (for example) be full left, at 1.5mS neutral, and at 2mS full right.

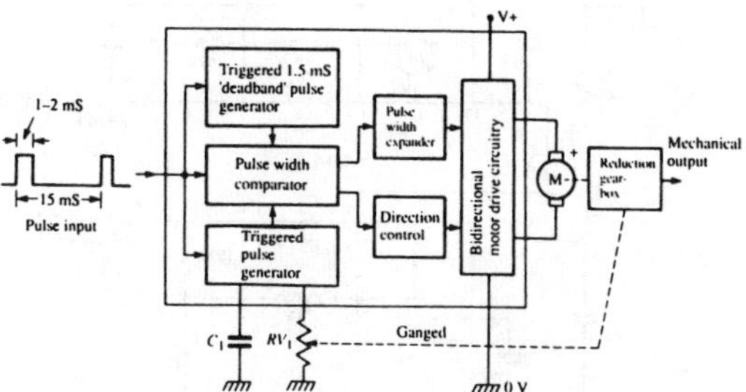

Figure 2.49. Block diagram showing basic digital proportional-control servomotor system.

Each input pulse triggers a 1.5mS 'deadband' pulse generator and a variable-width pulse generator controlled (via RV_1) by the gearbox output; these three pulses are fed to a width comparator that gives one output that gives direction control of the motor drive circuitry, and another that (when fed through a pulse-width expander) controls the motor speed, thus making the servomotor's RV_1-driving mechanical output rapidly follow any variations in the width of the input pulse.

Servomotors of the above type are usually used in multi-channel remote control systems, as in the basic 4-channel system of *Figure 2.50*. Here, a serial data input is fed (via some form of data link) to the input of a suitable decoder; each input frame comprises a 4ms synchronisation pulse followed by four variable-width (1ms to 2ms) sequential 'channel' pulses. The decoder converts the four channel pulses to parallel form, enabling each pulse to be used to control a servomotor.

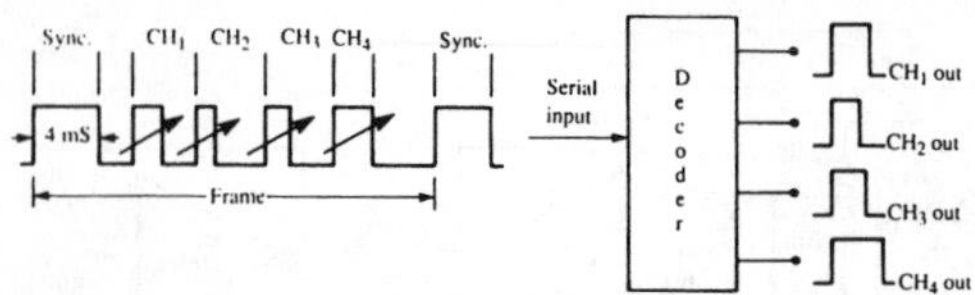

Figure 2.50. Waveforms of a four-channel digital proportional-control system.

Digital servomotor circuits

Digital proportional servomotor units are widely available in both kit and ready-built forms, and are usually designed around either the Ferranti ZN409CE or the Signetics NE544N servo amplifier ICs. *Figures 2.51* and *2.52* show practical application circuits for both of these IC types, with component values suitable for input pulse lengths in the 1ms to 2ms range and a frame length of about 18ms nominal.

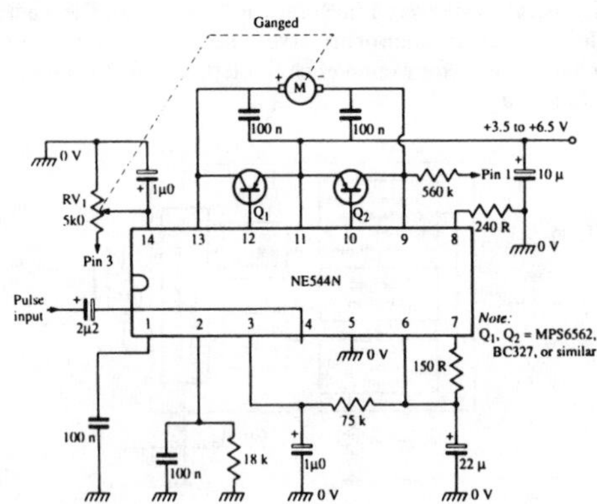

Figure 2.51. Digital proportional servo driver using the ZN409CE IC.

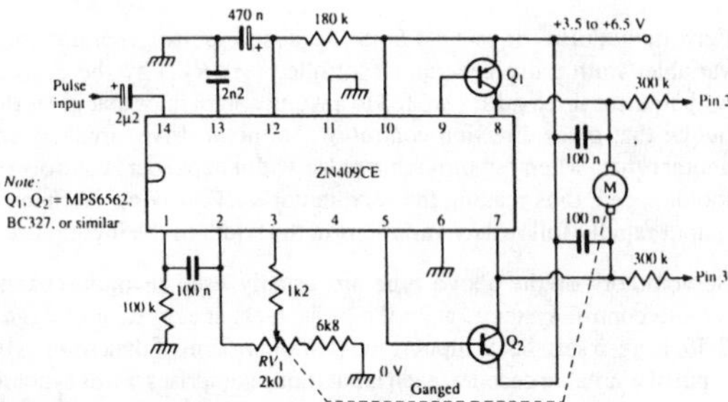

Figure 2.52. Digital proportional servo driver using the NE544N IC.

Stepper motors

Stepper motors have several phase windings, and each time these are electrically pulsed in the appropriate sequence the output spindle rotates by a precise step angle (usually 7.5 degrees). By applying a suitable sequence of pulses, the spindle can be turned a precise number of steps backwards or forwards, or can be made to rotate continuously at any desired speed in either direction. These motors can easily be controlled via a microprocessor or a dedicated stepper motor driver IC, and are widely used in all applications where precise amounts of angular movement are required, such as in the movement of robot arms, in daisy wheel character selection, or the movement control of the printhead and paper feed in an electronic typewriter.

Most stepper motors have four phases or coil windings, which may be available via eight independent terminals, as shown in *Figure 2.53(a)* or via two sets of triple terminals, as shown in *Figure 2.53(b)*. The phases are usually designed for unipolar drive, and must be connected in the correct polarity.

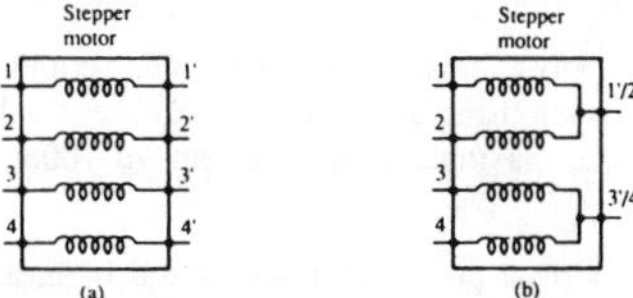

Figure 2.53. Four-phase stepper motors usually have either *(a)* eight or *(b)* six external connection terminals.

Figure 2.54 shows the basic way of transistor driving a normal 4-phase stepper motor, and *Figure 2.55* shows the normal full-step switching sequence. Note that the motor can be repeatedly stepped or rotated clockwise by repeating the 1–2–3–4 sequence or anticlockwise by repeating the 4–3–2–1 sequence.

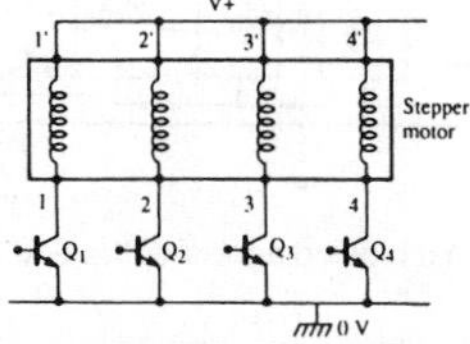

Figure 2.54. Basic transistor-driven stepper motor circuit.

Step No.	Q_1	Q_2	Q_3	Q_4
0	On	Off	On	Off
1	Off	On	On	Off
2	Off	On	Off	On
3	On	Off	Off	On
4	On	Off	On	Off
5	Off	On	On	Off

Above sequence repeating → (from step 4)

Clockwise ↓ Anticlockwise ↑

Figure 2.55. Full-step mode sequencing of the *Figure 2.54* circuit.

The SAA1027 driver IC

A number of dedicated 4-phase stepper motor driver ICs are available, and the best known of these is the SAA1027, which is designed to operate from supplies in the 9.5V to 18V range and to give full-stepping 4-phase motor operation at total output drive currents up to about 500mA. *Figure 2.56* shows the internal block diagram of the SAA1027, which is housed in a 14-pin DIL package, and *Figure 2.57* shows its basic application circuit. Note (in *Figure 2.56*) that each of the IC's four phase-driving output transistors operates in the open-collector mode and is protected against damage from motor back-emfs via a diode.

The IC has two sets of supply rail pins, with one set (pins 13 and 12) feeding the high-current output circuitry, and the other (pins 14 and 5) feeding its low current sections. In use, pins 5 and 12 are grounded, and the positive (usually 12V) rail is fed directly to pin 13 and via decoupling component R_1–C_1 to pin 14. The positive rail must also be fed to pin 4 via R_x, which sets the maximum drive current capacity of the four output transistors; the R_x value is given by:

$$R_x = (4E/I) - 60\Omega,$$

where E is the supply voltage and I is the desired maximum motor phase current. Thus, when using a 12V supply, R_x needs values of 420Ω, 180Ω, or 78Ω for maximum output currents of 100mA, 200mA, or 350mA.

The SAA1027's reset pin is normally biased high, and under this condition the IC's outputs change state each time the count terminal transitions from the low to the high state, as shown in the output sequencing table of *Figure 2.58*. The sequence repeats at 4-step intervals, but can be reset to the zero state at any time by pulling the reset

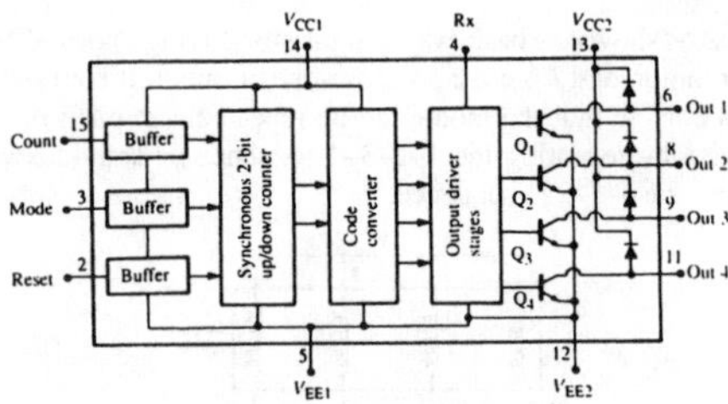

Figure 2.56. Internal block diagram of the SAA1027.

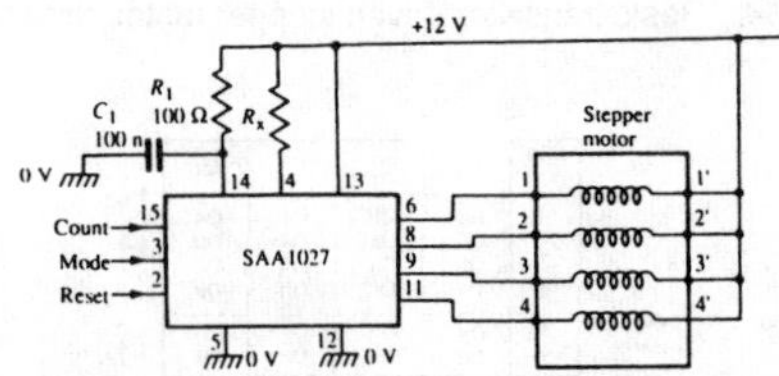

Figure 2.57. Basic SAA1027 application circuit.

pin low. The sequence repeats in one direction (normally giving clockwise motor rotation) when the mode input pin is tied low, and in the other (normally giving anticlockwise motor rotation) when the mode input pin is tied high.

	Counting sequence	Mode = low Q_1	Q_2	Q_3	Q_4	Mode = High Q_1	Q_2	Q_3	Q_4
	0	On	Off	On	Off	On	Off	On	Off
	1	Off	On	On	Off	On	Off	Off	On
	2	Off	On	Off	On	Off	On	Off	On
	3	On	Off	Off	On	Off	On	On	Off
Above sequence repeating	0	On	Off	On	Off	On	Off	On	Off
	Reset low	On	Off	On	Off	On	Off	On	Off

Figure 2.58. SAA1027 output sequencing table.

Figure 2.59 shows a practical drive/test circuit that can be used to activate 4-phase stepper motors with current rating up to about 300mA. The motor can be manually sequenced one step at a time via SW_3 (which is effectively 'debounced' via R_4–C_5), or automatically via the 555/7555 astable oscillator, by moving SW_2 to either the step or free-run position; the motor direction is controlled via SW_4, and the stepping sequence can be reset via SW_5.

The astable's operating speed is fully variable via RV_1, and is variable in three switch-selected decade ranges via SW_1. In the slow (1) range, the astable frequency is variable from below 5Hz to about 68Hz via RV_1; on a 48-step (7.5 degree step angle) motor this corresponds to a speed range of 6 to 85 RPM. SW_1 ranges 2 and 3 give frequency ranges that are ten and one-hundred times greater than this respectively, and the circuit thus gives a total speed control range of 6 to 8,500 RPM on a 48-step motor.

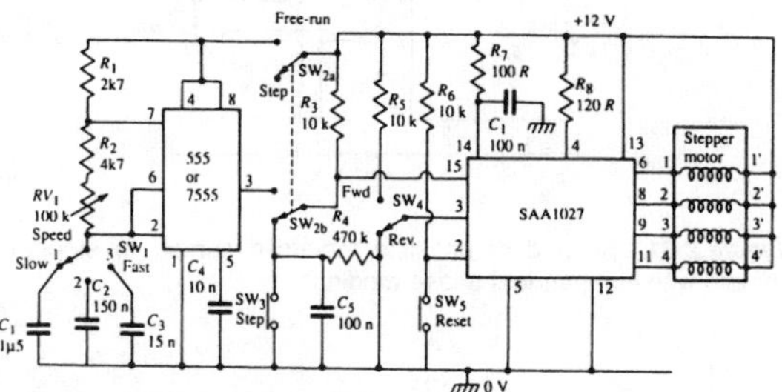

Figure 2.59. Complete stepper motor drive/test circuit.

Circuit variations

The basic *Figure 2.59* circuit can be varied in a several ways. *Figure 2.60* shows how it can be driven via a microprocessor or computer output port with voltages that are below 1V in the logic 0 state and above 3.5V in the logic-1 state. Note that this circuit reverses the normal polarity of the input control signals; thus, the step input is pulsed by a high-to-low transition, the stepping sequence is reset by a high input, and a low mode input gives forward motor rotation and a high input gives reverse rotation.

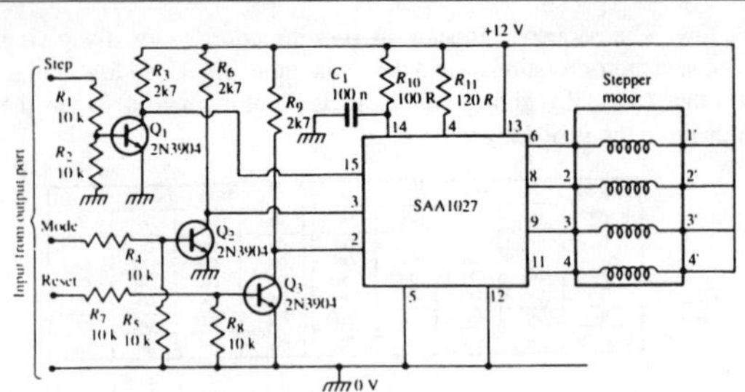

Figure 2.60. Stepper-motor-to-microprocessor interface.

The circuits of *Figures 2.59* and *2.60* are designed to give maximum output drive currents up to 300mA. Their outputs can be boosted to a maximum of 5A by using the circuits of *Figures 2.61* or *2.62*, which each show the additional circuitry needed to drive one of the four output phases of the stepper motor; four such driver stages are needed per motor. The circuit of *Figure 2.61* can be used to drive motors with fully independent phase windings, and *Figure 2.62* can be used in cases where two windings share a common supply terminal. Diodes D_1 and D_2 are used to damp the motor back-emfs.

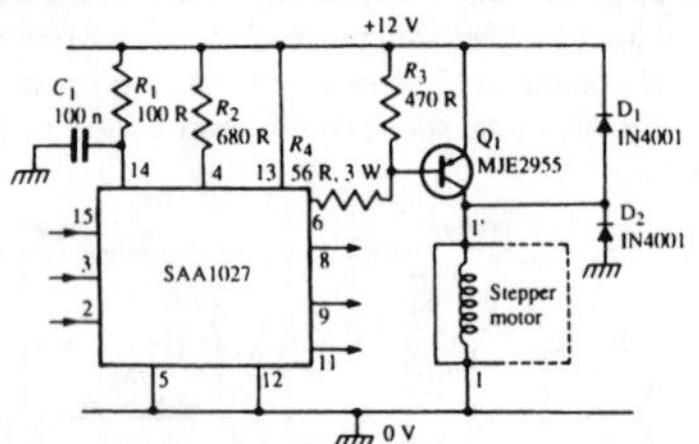

Figure 2.61. Method of boosting the drive current to stepper motors with independent phase windings.

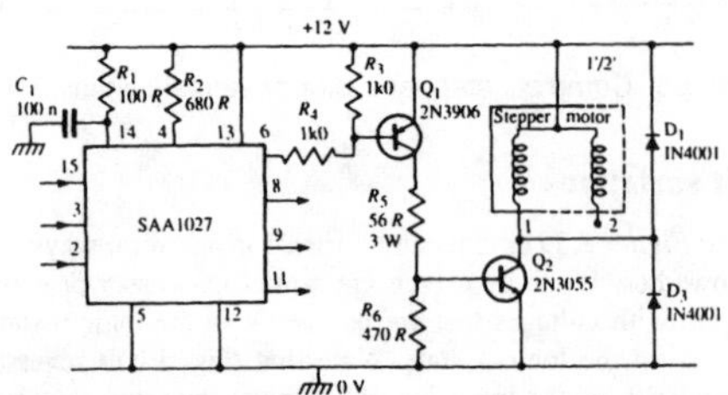

Figure 2.62. Method of boosting the drive current to stepper motors with coupled phase windings.

3 Modern sensors & transducers

The modern electronics design engineer has ready access to a considerable armoury of useful sensors and transducers that enable him to interface his circuits with the outside world. This chapter gives a concise run-down on some of the best known and most useful of these devices. Before looking at these, however, it is necessary to first define both a 'sensor' and a 'transducer'.

A **sensor** is, in the context of this volume, defined as any device that directly converts some physical quantity (such as heat, light, pressure) into a proportional *electrical* quantity. A **transducer**, on the other hand, is defined as any device that directly converts one physical quantity into another physical quantity; thus, any device that (for example) converts electrical current into mechanical movement (such as a relay or electric motor) is a transducer.

From the above definition it is obvious that although all sensors are transducers, it is not true that all transducers are sensors. Consequently, to tie the two groups of devices together in a logical way, this chapter is split into sections that first categorise the devices by function (electroacoustic, optoelectronic, thermoelectric, or piezoelectric), and then describe the individual sensors and transducers that come under that heading. One major category of transducer, the electromechanical type, has already been described in Chapter 2, and at this point it is worth noting that there is often considerable flexibility in the use of generic titles when describing transducers; an electric bell, for example, is undoubtedly an electromechanical device, but is most noted for its noise-making qualities and is thus best described as an electroacoustic device.

Electroacoustic devices

An electroacoustic device is one that converts electrical or electromechanical power into acoustic energy, or *vice versa.* Among the best known of such devices are electric bells and buzzers, piezo 'sounders' and sirens, loudspeakers, earphones, and microphones.

Bells and buzzers

Electric bells and buzzers are crude but effective noise-makers. *Figure 3.1* shows their typical basic construction and electrical equivalent circuit. They consist of a free-moving iron armature that is surrounded by a solenoid that can be energised via a pair of normally-closed switch contacts and a leaf spring. Normally, the armature is forced out of the solenoid by a light coil spring. When a D.C. supply is connected to the circuit the solenoid is energised and its electromagnetic field drags the armature downwards until its hammer striker hits a sounding board (in a buzzer) or dome (in a bell); at this point a claw at the other end of the armature pulls open the switch contacts via the leaf spring, and the armature shoots outwards again under the pressure of the coil spring until the switch contacts close again, and the process then repeats *ad infinitum.*

From the engineers point of view, the important things to note about electric bells and buzzers is that they are highly inductive self-interrupting devices that are not 'polarity conscious'. They can be activated via a single switch wired in series with the supply, or by any number of such switches connected in parallel, as shown in *Figure 3.2*. Bells have their acoustic energy concentrated into a narrow 'tone' band and are thus reasonably efficient and can generate very high sound levels. Electric buzzers generate a broad 'splurge' of sound and are very inefficient, and have now been superseded by electronic units designed around piezo 'sounders'.

Piezo sounders and buzzers

Piezoelectric 'sounders' are super-efficient low-level sound-making devices. They consist of a thin slice of electroconstrictive (piezo) ceramic material plus two electrical contacts, and act as a semi-tuned

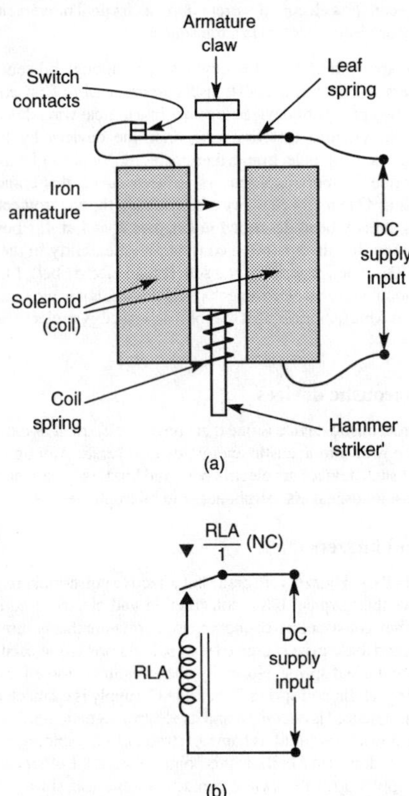

Figure 3.1. Typical basic construction *(a)* and equivalent circuit *(b)* of an electric bell or buzzer.

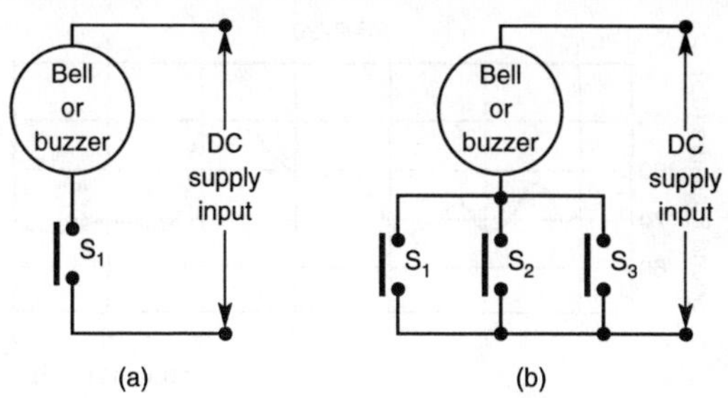

Figure 3.2. A bell or buzzer can be activated by a single switch *(a)* or by any of several switches wired in parallel *(b)*.

(typically 1kHz to 5kHz) electric-to-acoustic power converter. They give typical power conversion efficiencies of 50 percent, compared to about 0.5 percent for conventional loudspeakers. The Toko PB2720 is a typical piezo 'sounder' and houses its actual transducer in an easy-to-use plastic casing, as shown in *Figure 3.3*. Its input terminals appear to the outside world as a simple capacitor of about 20nF. The device has a highly non-linear frequency response that peaks at about 4.5kHz, as shown in *Figure 3.4*.

The most effective and cheapest way to use a PB2720 or similar piezo unit as a sound-making 'buzzer' is to feed it with square waves from a driver that can source and sink currents with equal ease and has a current-limited (short-circuit proof) output. CMOS oscillator/drivers fit this bill nicely. *Figures 3.5* and *3.6* show two inexpensive ways of driving the PB2720 (or any similar device) from a 4011B CMOS

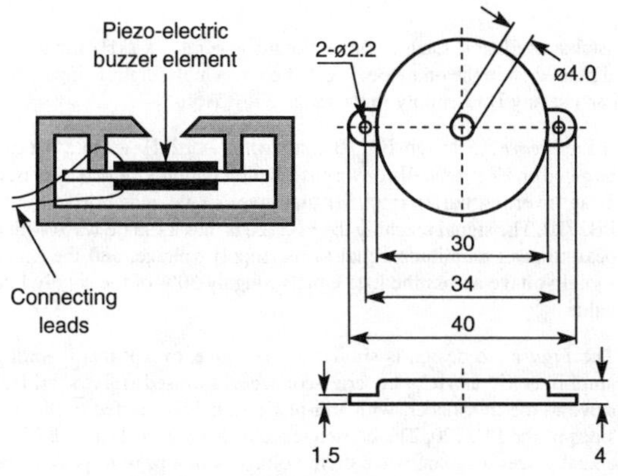

Figure 3.3. Basic construction and dimensions (in mm) of the PB2720 piezo sounder.

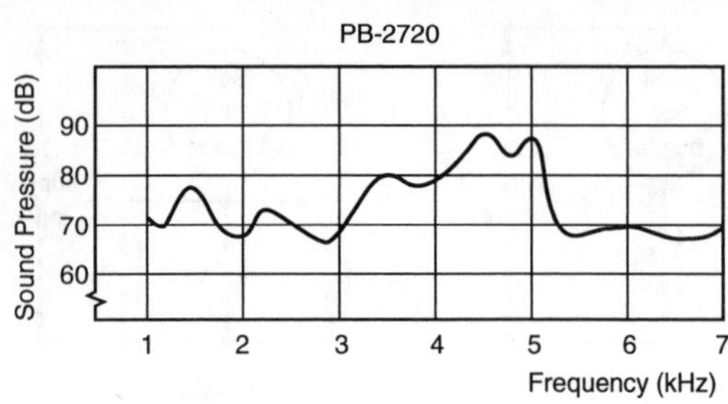

Figure 3.4. Frequency response of the PB2720 with a 1.5V r.m.s. input: sound pressure is measured at 10 cm.

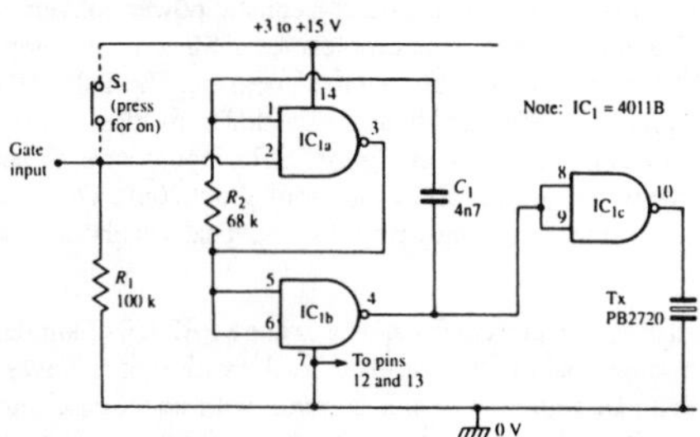

Figure 3.5. Gated 2kHz piezo buzzer with single-ended output.

astable oscillator. Each of these circuits generates a 2kHz monotone signal when in the on mode, is gated on by a high (logic 1) input, and can use any D.C. supply in the range 3 to 15volts.

In the *Figure 3.5* design, IC_{1a}–IC_{1b} are wired as a 2kHz astable that can be gated on electronically or via push-button switch S_1, and IC_{1c} is used as an inverting buffer/amplifier that gives single-ended drive to the PB2720. The signal reaching the PB2720 is thus a square wave with a peak-to-peak amplitude equal to the supply voltage, and the r.m.s. signal voltage across the load equals roughly 50% of the supply line value.

The *Figure 3.6* design is similar to the above, except that inverting amplifiers IC_{1c} and IC_{1d} are series-connected and used to give a 'bridge' drive to the transducer, with anti-phase signals being fed to the two sides of the PB2720. The consequence is that the load (the PB2720) actually sees a square wave drive voltage with a peak-to-peak value equal to twice the supply voltage value, and an r.m.s voltage equal to the supply value, and thus gives four times more acoustic output power than the *Figure 3.5* design.

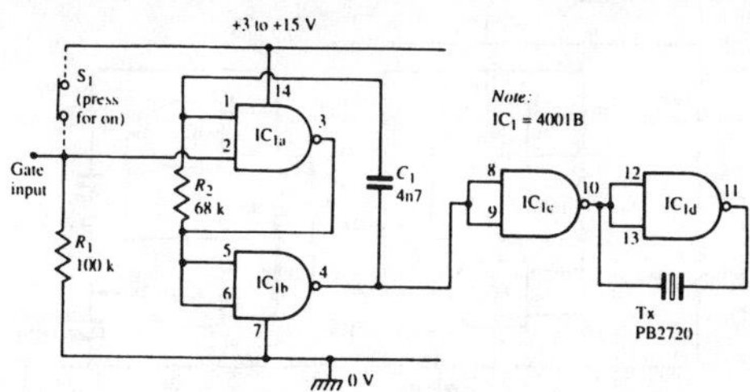

Figure 3.6. Gated 2kHz piezo buzzer with bridge-drive output.

Figure 3.7 shows the above basic bridge-driven circuit modified so that it can be gated on by a low (logic O) input by simply using a 4001B CMOS IC in place of the 4011B type.

Note that many commercial piezo buzzer units use circuits similar to those shown in *Figures 3.5* to *3.7* but with the 'gate input' terminal eliminated and S_1 replaced with a short circuit, so that the units have only two connecting leads.

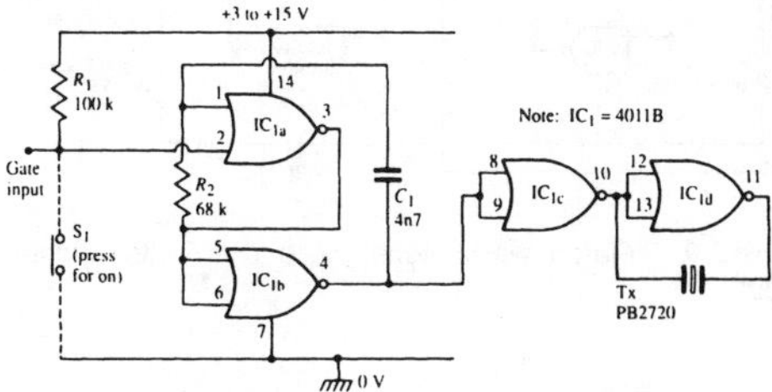

Figure 3.7. Alternative version of the gated bridge-drive circuit.

Piezo alarms and sirens

Gated CMOS oscillator/driver circuits can be used in a variety of ways to produce alarm or siren sounds from a piezo sounder. Two examples are shown in *Figures 3.8* and *3.9*. *Figure 3.8* shows a single 4011B used to make a pulsed-tone (bleep-bleep) alarm circuit with direct drive to the PB2720. Here, IC_{1a}–IC_{1b} are wired as a gated 6Hz astable which is used to gate the IC_{1c}–IC_{1d} 2kHz astable on and off. This circuit is gated on by a high input; if low-input gating is wanted, simply swap the 4011B for a 4001B and transpose the positions of S_1 and R_1.

Finally, *Figure 3.9* shows a warble-tone (dee-dah-dee-dah) gated 'siren' that generates a sound similar to a British police car siren and has a bridge-driven output. Here, 1Hz astable IC_{1a}–IC_{1b} is used to modulate the frequency of the IC_{1c}–IC_{1d} astable; the depth of frequency modulation depends on the R_3 value, which can vary from 120k to 1MΩ.

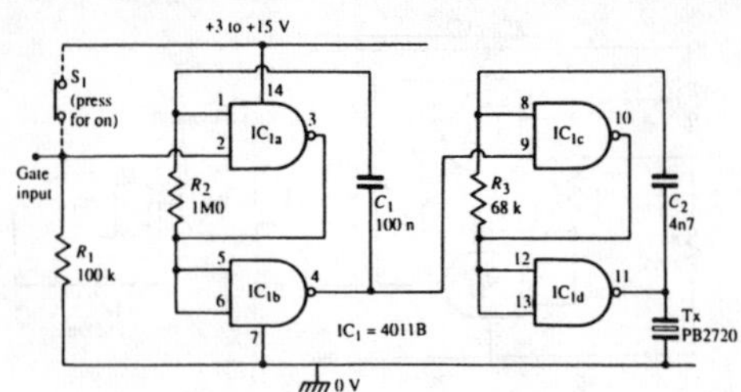

Figure 3.8. Gated pulsed-tone (6Hz and 2kHz) piezo alarm.

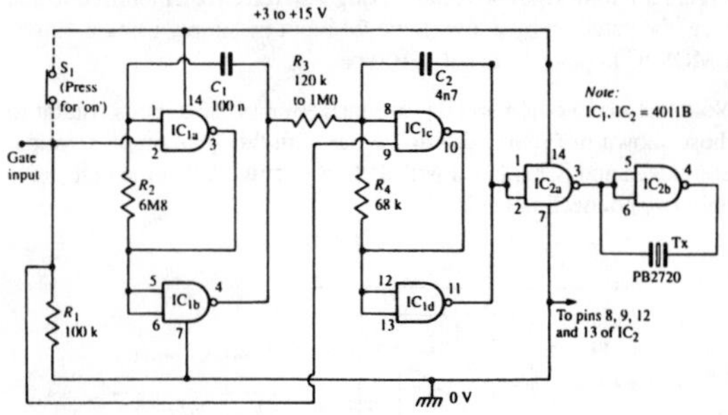

Figure 3.9. Gated warble-tone piezo siren with bridge-drive output.

Loudspeakers

Most modern loudspeakers operate on the moving coil principle and use the basic form of construction shown in *Figure 3.10*, in which the coil (usually called a 'voice' or 'speech' coil) is mounted within the field of a powerful magnet and is attached to a rigid cone that is flexibly suspended within the speaker's chassis. When the coil is driven from the output of an audio power amplifier it and the cone move in sympathy with the audio signal, and the cone generates a corresponding acoustic output. Most loudspeakers have electrical-to-acoustic conversion efficiency values of only 0.2 to 0.5 percent.

Most modern loudspeakers (other than miniature types) have nominal impedances, at 400Hz or 1kHz, of either 4, 8, (the most popular value), or 16Ω. They have the basic electrical equivalent circuit of *Figure 3.11*. The impedance of a loudspeaker inevitably varies with frequency, and *Figure 3.12* shows the typical frequency/impedance response curve of a 4Ω type; it shows distinct self-resonance at about 100Hz.

Modern audio power amplifiers gave a low-impedance voltage output; because of the loudspeaker's impedance variation, its power absorbtion

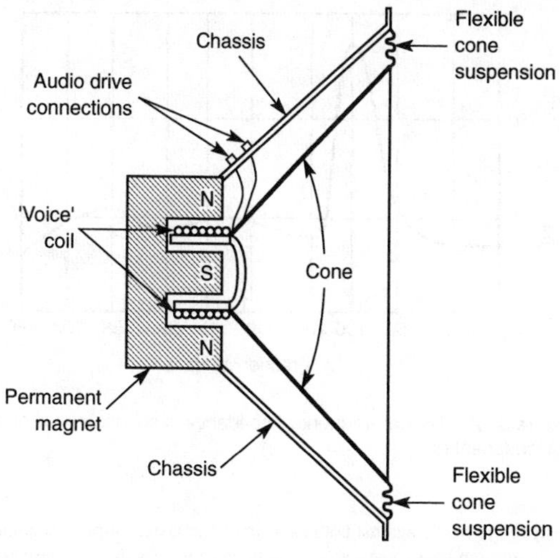

Figure 3.10. Basic construction of a moving coil loudspeaker.

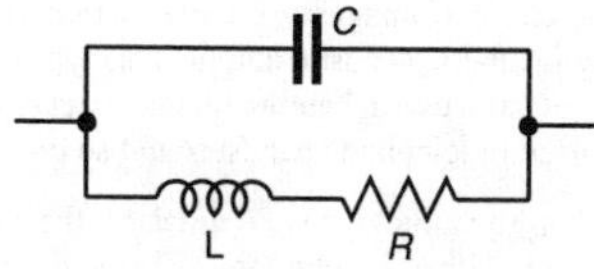

Figure 3.11. Electrical equivalent of a loudspeaker.

and acoustic output power also varies, when voltage-driven, with frequency. The power absorbtion, W, is given by:

$$W = E.I.\cos\theta,$$

where W is in watts, E is the signal drive voltage, I is the signal current through the 'speaker, and cosθ is the cosine of the phase angle between E and I. Typically, the phase angle varies from about –50° at 100Hz, to 0° at 250Hz, to +55° at 4kHz, giving cosθ values of 0.643, 1.000, and 0.574 respectively. Thus, voltage-driven loudspeakers have a highly non-linear basic acoustic response. Most of this non-linearity can be ironed out by fitting the loudspeaker into a custom-designed enclosure.

Earphones and headphones

An earphone can be simply described as a small electrical-to-acoustic transducer that is designed to fit against or inside one ear; a headphone is simply a pair of such units joined together by a headband and

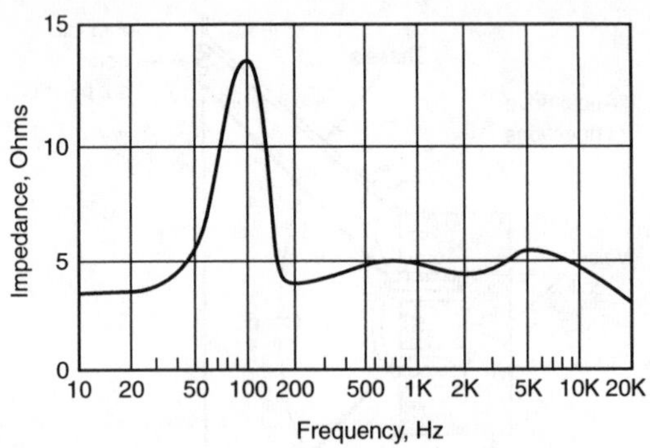

Figure 3.12. Typical frequency/impedance response curve of a 4Ω loudspeaker.

designed to nestle against both ears simultaneously. Earphones come in a variety of basic types, and the three most widely used of these are illustrated in *Figures 3.13* to *3.15*.

In the moving-iron type *(Figure 3.13)* the electrical input signal is fed to a pair of coils mounted on a lightly magnetised iron core, and produces a modulated electromagnetic effect on a thin iron diaphragm that generates a sympathetic acoustic output. This type of earphone is cheap and robust but has a useful bandwidth that extends to only a few kHz; it is widely used in telephone handsets and so on.

In the moving coil or 'dynamic' type *(Figure 3.14)* the electrical input is fed to a coil that is either attached to, or printed on (using PCB techniques), a light diaphragm and placed within the field of a strong permanent magnet, so that the diaphragm moves and generates an acoustic output in sympathy with the input signal. Such units often have excellent fidelity and a wide bandwidth; they usually have an impedance in the 4 to 32Ω range.

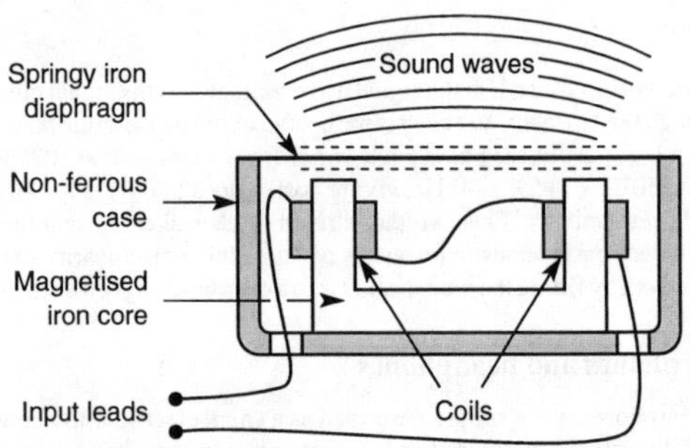

Figure 3.13. Basic moving-iron earphone.

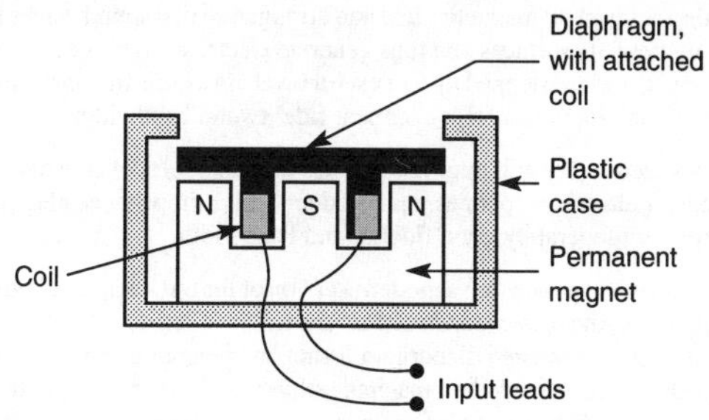

Figure 3.14. Basic moving coil earphone.

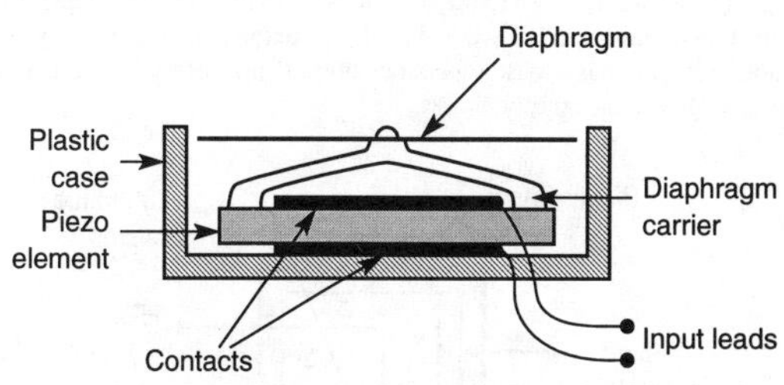

Figure 3.15. Basic piezoelectric (crystal) earphone.

In the piezoelectric or 'crystal' type the input signal is applied to printed contact 'plates' on the two faces of the piezo element, which expands and contracts in sympathy with the signal; these physical movements are coupled, via a lightweight carrier, to a diaphragm that produces an acoustic output. This type of earphone has moderately good fidelity and bandwidth, and has a very high impedance (it acts like a capacitance with a value of a few nF).

Microphones

Microphones are acoustic-to-electrical transducers. The four best known variations of these are the moving coil ('dynamic'), ribbon, piezo-electric ('crystal'), and electret types.

The moving coil type uses the same form of construction as the moving coil earphone (see *Figure 3.14*). Sound waves cause a diaphragm and coil to move within the field of a strong magnet, and the coil generates a sympathetic electrical output. Units of this type offer good fidelity and many high-quality types have a built-in transformer that gives a voltage-boosted and impedance-matched output.

Ribbon microphones are a variation of the moving coil type, with the coil replaced by a ribbon of corrugated aluminium foil that is suspended within a powerful magnetic field and arranged so that sound waves fall easily upon its surfaces and thus generate electrical signals across the ribbon; these are stepped up to a useful level via a built-in transformer. Ribbon microphones offer excellent fidelity and bandwidth.

Piezoelectric (crystal) microphones use the same form of construction as piezo-electric earphones, and produce a high-impedance electrical output of moderately good fidelity and bandwidth.

Electret microphones are a modern version of the old 'capacitor' types, which use the basic construction shown in *Figure 3.16*. Here, a lightweight metallised diaphragm forms one plate of a capacitor, and the other plate is fixed; the capacitance thus varies in sympathy with the acoustic signal. The capacitor acquires a fixed charge, via a high-value resistor, from a high voltage supply (typically 200V) and, since the voltage across a capacitor is equal to its charge divided by its capacitance, it thus generates an output voltage in sympathy with the acoustic signal. This signal is fed to the outside world via a coupling capacitor. This type of microphone is capable of producing excellent fidelity and bandwidth, but has a high impedance output and suffers from the need for a high voltage supply.

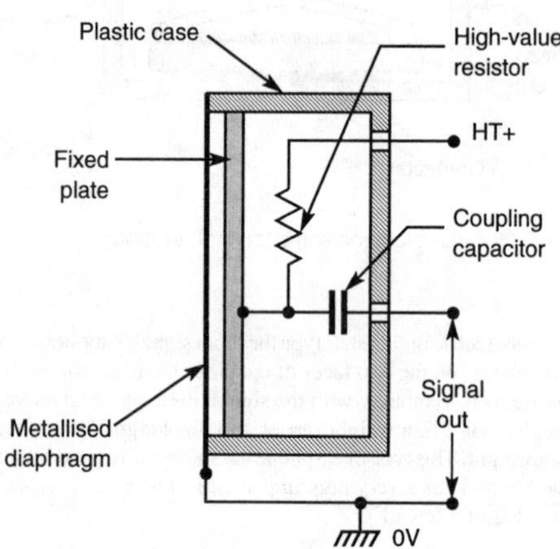

Figure 3.16. Basic elements of a capacitor microphone.

Figure 3.17 illustrates the basic construction of a modern electret microphone. Here, the 'fixed' plate of the capacitor is metallised onto the back of a slab of insulating material known as 'electret', which holds an electrostatic charge that is built in during manufacture and can be held for an estimated 100-plus years. The output of the resulting 'capacitor microphone' is coupled to the outside world via a built in IGFET transistor, which needs to be powered externally from a battery (from 1.5V to about 9V) via a 1k0 resistor, as shown. Electret microphones are robust and inexpensive and give a good performance up to about 10kHz; they are often built into cassette recorders and so on.

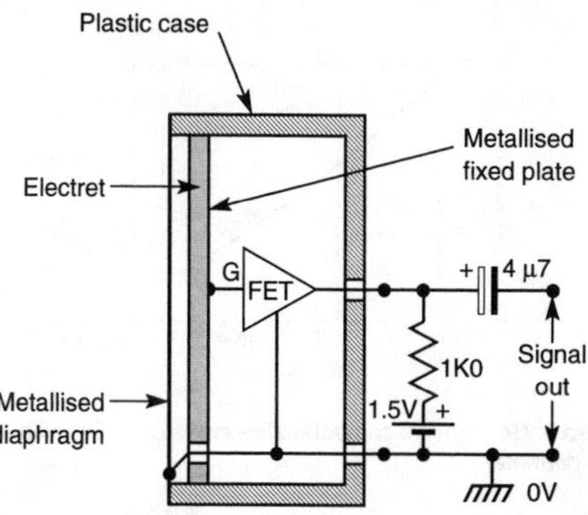

Figure 3.17. Basic elements of a modern electret microphone.

Optoelectronic devices

An optoelectronic transducer is one that converts visible or invisible light energy into electrical energy, or *vice versa.* The best known 'passive' versions of such devices are LDRs (light-dependent resistors), filament lamps, and solar cells. The best known 'active' versions are photodiodes, phototransistors, and LEDs, and these are dealt with in detail in Chapter 13.

LDR basics

An LDR is also known as a cadmium-sulphide (CdS) photocell. It is a passive device with a resistance that varies with visible light intensity. *Figure 3.18* shows the device's circuit symbol and basic construction, which consists of a pair of metal film contacts separated by a snake-like track of light-sensitive cadmium sulphide film; the structure is housed in a clear plastic or resin case. *Figure 3.19* shows the typical photoresistive graph that applies to an LDR with a face diameter of about 10mm; the resistance may be several megohms under dark conditions, falling to about 900R at a light intensity of 100 Lux (typical of a well lit room) or about 30R at 8000 Lux (typical of bright sunlight).

Figures 3.20 to *3.22* show three light-sensitive relay-output switching circuits that will each work with virtually any LDR with a face diameter in the range 3mm to 12mm.

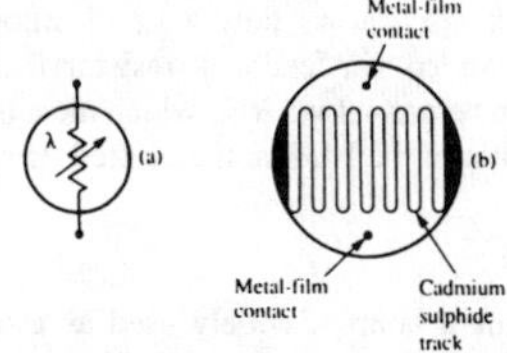

Figure 3.18 *(a)* LDR symbol and *(b)* basic structure.

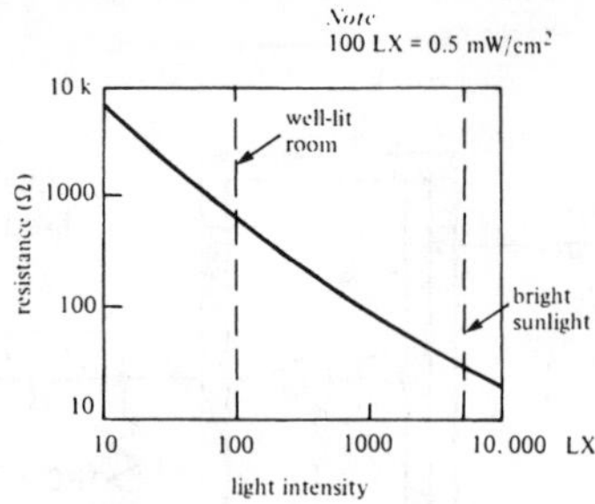

Figure 3.19. Typical characteristics curve of a LDR with a 10mm face diameter.

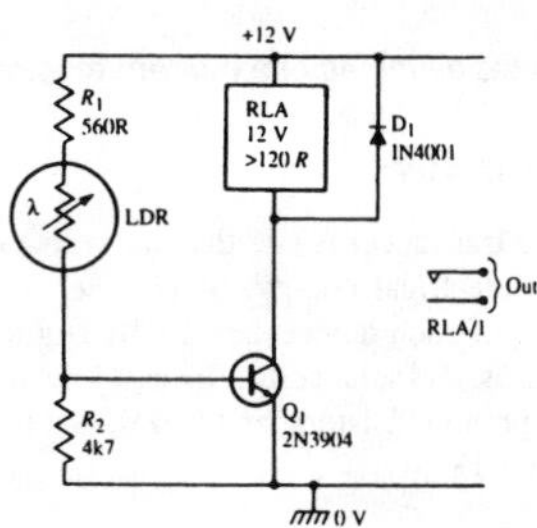

Figure 3.20. Simple non-latching light-activated relay switch.

The transistor circuit of *Figure 3.20* activates the relay when light enters a normally-dark area such as the inside of a safe or cabinet. The LDR (plus R1) and R_2 act as a simple voltage divider that controls Q_1's bias, and drives Q_1 and the relay on only when the light intensity exceeds a certain limit.

Figures 3.21 and *3.22* show circuits that are suitable for use in 'precision' light-sensing applications. Here, the LDR and RV_1–R_1–R_2 are connected in the form of a Wheatstone bridge, and the op-amp and Q_1 are wired as a precision bridge-balance detector. In *Figure 3.21* the relay turns on when the light level exceeds a pre-set value; the action can be reversed, so that the relay turns on when the light level falls below a pre-set value, by transposing RV_1 and LDR as in the *Figure 3.22* circuit, which also shows how a small amount of switching hysteresis can by added via feedback resistor R_5. In all cases the switching point can be pre-set via RV_1, which must have an 'adjusted' value that equals that of the LDR at the desired 'trip' level.

Filament lamps

The ordinary filament lamp is widely used as a light generator. It consists of a coil of tungsten wire (the filament) suspended in a vacuum- or gas-filled glass envelope and externally connected via a

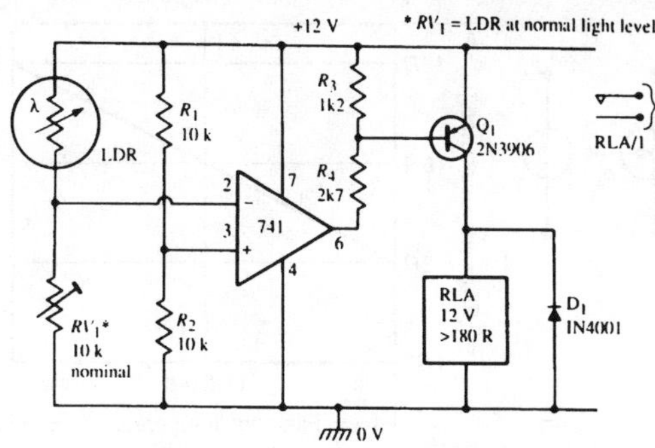

Figure 3.21. Precision light-activated relay switch.

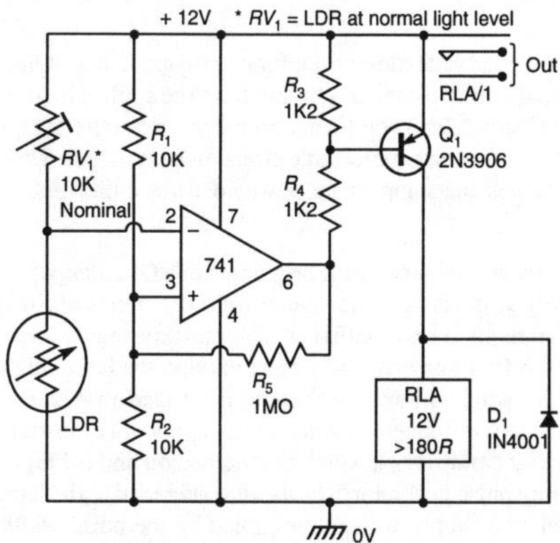

Figure 3.22. Precision dark-activated relay switch with hysteresis.

pair of metal terminals; the filament runs white hot when energised via a suitable A.C. or D.C. voltage, thus generating a bright white light. *Figure 3.23(a)* shows the standard symbols used to represent such a lamp.

The filament lamp has, from the electronic engineer's point of view, two notable characteristics. Firstly, its resistance varies with filament temperature: *Figure 3.23(b)* shows the typical variation that occurs in a 12V 12 watt lamp; the resistance is 12Ω when the filament is operating at its normal 'white' heat, but is only 3Ω when the filament is cold. This 4:1 resistance variation is typical of all filament lamps, and causes them to have switch-on 'inrush' current values about four times greater than the normal 'running' values.

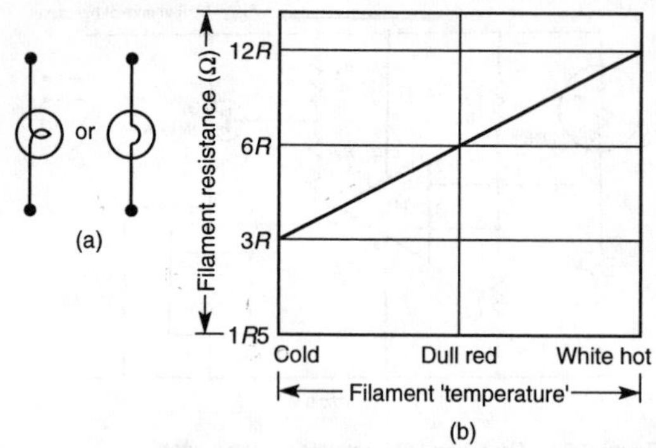

Figure 3.23. Filament lamp symbols *(a)* and graph *(b)* showing typical variation in filament resistance with filament temperature for a 12V, 12W lamp.

The second notable feature of the filament lamp is that it has a fairly long thermal time constant; power thus has to be applied to (or removed from) the filament for a significant time (tens or hundreds of milliseconds) before it has any appreciable effect on light output. This characteristic enables the lamp to be powered from either A.C. or D.C. supplies.

The simplest way of varying the brightness of a D.C. lamp is to vary its D.C. supply voltage, as in the emitter follower circuit of *Figure 3.24*, but this technique is very inefficient, since all unwanted power is 'lost' across Q_1. A far more efficient way of varying the lamp brightness is to use the 'switched-mode' technique illustrated in *Figure 3.25*, in which the lamp is connected to the D.C. supply via electronic switch SW_1 (usually a transistor), which is switched on and off rapidly via a free-running pulse generator. The *mean* voltage fed to the lamp in this case equals the supply voltage multiplied by the pulse-width/frame-width ratio, and can easily be varied (via the pulse-width control) between 5 and 95 percent of the D.C. supply-voltage value; if the frame frequency is greater than 50Hz the lamp shows no sign of flicker as its brightness is varied.

Figure 3.25 shows a practical switched-mode D.C. lamp dimmer that can control 12 volt lamps with rating up to 24 watts. CMOS gates IC_{1a}–IC_{1b} form a 100Hz astable with a variable mark/space-ratio output that

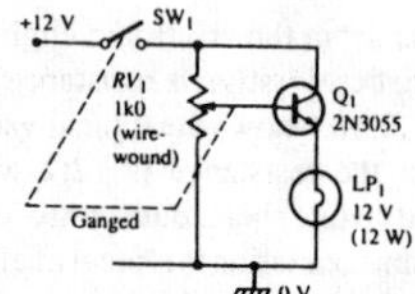

Figure 3.24. Variable voltage D.C. lamp brightness control circuit.

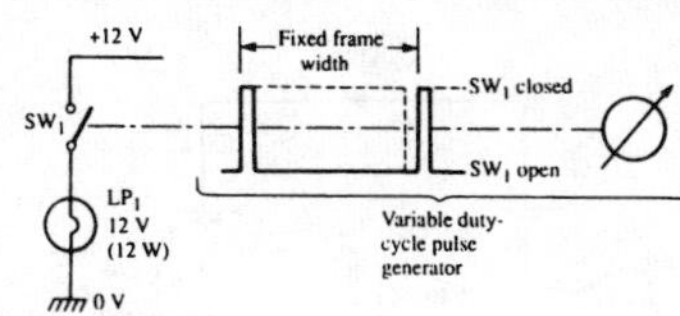

Figure 3.25. Basic switched-mode D.C. lamp brightness control circuit.

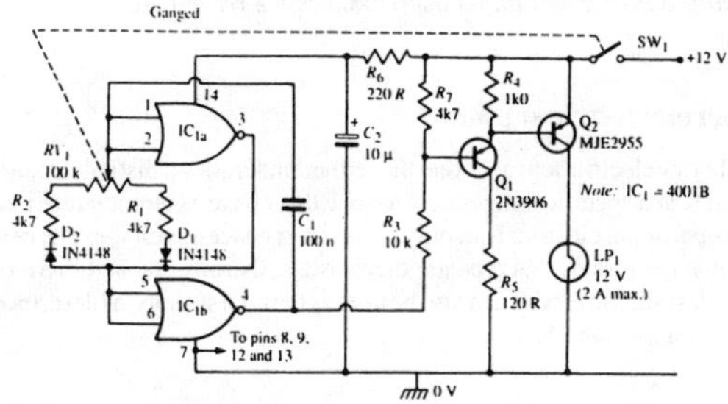

Figure 3.26. Switched-mode D.C. lamp dimmer.

is fed to the lamp via Q_1–Q_2, enabling its brightness to be varied over a wide range via RV_1; SW_1 enables the lamp and circuit to be turned fully off, and R_6–C_2 protect the CMOS circuitry from supply-line transients.

Solar cells

Solar cells are photovoltaic units that convert light directly into electrical energy. *Figure 3.27* shows the symbol used to represent a single solar cell, which typically generates an open circuit voltage of about 500mV (depending on light intensity) when active. Individual cells can be connected in series to increase the available terminal voltage, or in parallel to increase the available output current; banks of cells manufactured ready-wired in either of these ways are called solar panels. *Figure 3.28* shows how a bank of 16 to 18 cells can be used to autocharge a 6V ni-cad battery via a germanium diode.

The available output current of a solar cell depends on the light intensity, on cell efficiency (typically only a few percent), and on the size of the active area of the cell face. The naturally-available light energy at sea level in temperate geographic zones is typically in the range 0.5 to $2kW/m^2$ on bright sunny days.

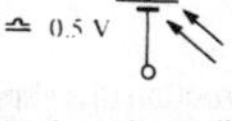

≏ 0.5 V

Figure 3.27. Symbol of single solar cell.

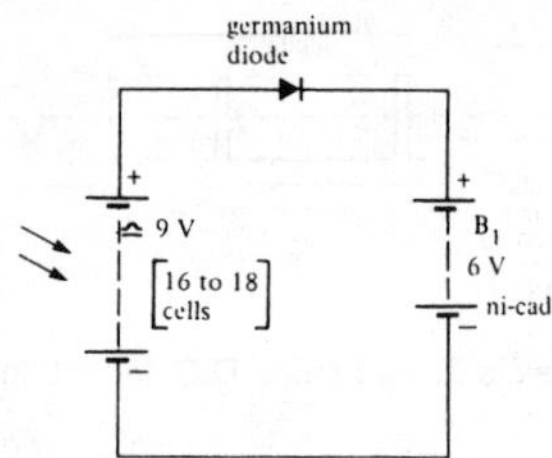

Figure 3.28. Solar panel used to charge a 6V ni-cad.

Thermoelectric sensors

A thermoelectric sensor is one that either undergoes a distinct change of state at a specific temperature, or one that converts temperature into a proportional electrical quantity such as resistance or voltage. The best known devices of this type are thermostats, thermistors, and thermocouples; silicon diodes can also be used as thermal sensors, as described in Chapters 7 and 8.

Thermostats

Thermostats are simple temperature-activated on/off switches that usually work on the 'bimetal' principle illustrated in *Figure 3.29(a),* in which the bimetal strip consists of two bonded layers of highly conductive metal with dissimilar coefficients of thermal expansion, thus causing the strip to bend in proportion to temperature and to make (or break) physical and electrical contact with a fixed switch contact at a specific temperature. In practice, the bimetal element may be in strip, coiled, or snap-action conical disc form, depending on the application, and the thermal 'trip' point may or may not be adjustable. *Figures 3.29(c)* and *(d)* show the symbols used to represent fixed and variable thermostats.

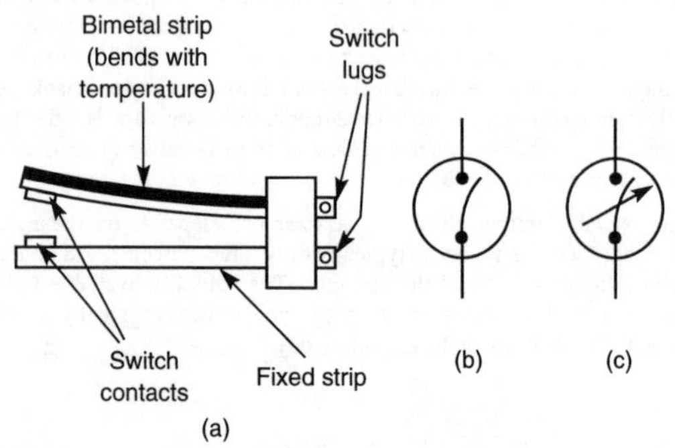

Figure 3.29. Basic construction of a simple bimetal thermostat *(a),* and symbols for fixed *(b)* and variable *(c)* thermostats.

A variety of thermostats are readily available, and can easily be used to give automatic control of electric room and water heaters, etc. Their only significant disadvantage is that they suffer from hysteresis; typically, a good quality adjusted thermostat may close when the temperature rises to (say) 21°C but not re-open again until it falls to 19.5°C.

Figure 3.30 shows an adjustable thermostat used in conjunction with a master on/off switch and an electric heater to give simple automatic control of room temperature; this circuit is adequate for use in most domestic applications. *Figure 3.31* shows a more sophisticated circuit, for use in commercial offices and so on; in this case TH_2 controls the normal office temperature, but is enabled via the programmable time switch only during the specified hours of occupancy; outside these hours the temperature is prevented from falling below (say) 5°C via TH1.

Thermistors

A thermistor is a form of resistor in which the resistance value is highly sensitive and proportional to temperature. Practical thermistors are available in rod, disc, and bead forms, and with either positive or

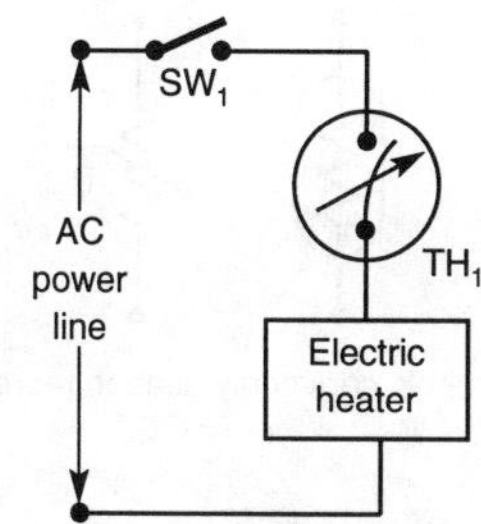

Figure 3.30. Simple example of a thermostat used to control room temperature.

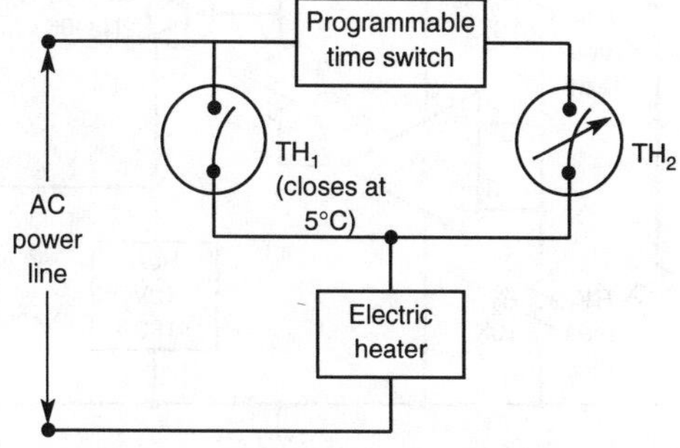

Figure 3.31. Example of a smart room-temperature control system using two thermostats.

negative temperature coefficients (known as PTC and NTC types respectively). Unlike thermostats, they do not suffer from hysteresis problems, and are thus suitable for use in a variety of precision control and measurement applications. *Figure 3.32* shows two alternative symbols that can be used to represent them.

Figure 3.33 shows how the precision light-activated relay switch of *Figure 3.21* can be converted into a precision over-temperature switch. The thermistor (TH_1) can be any NTC type with a resistance in the range 1k0 to 20k at the required trigger temperature. RV_1 enables the trip temperature to be varied over a wide range, and must have the same value as TH_1 at the trip point. The circuit can be made to act as a precision 'ice' or under-temperature switch by simply transposing TH_1 and RV_1.

Thermocouples

When a junction is formed between two dissimilar metals a thermo-electric or temperature-dependent voltage is generated across the junction. Thermocouples are devices in which the two types of metal are specially chosen to exploit this effect for temperature-measurement

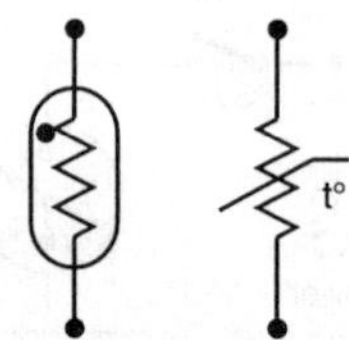

Figure 3.32. Symbols commonly used to represent a thermistor.

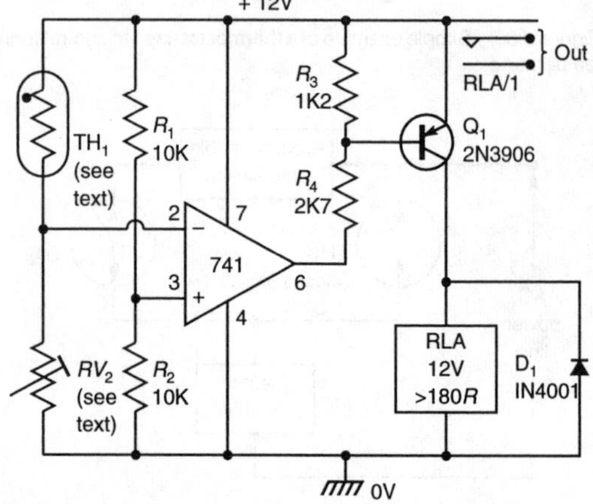

Figure 3.33. Precision over-temperature relay switch.

purposes. A thermocouple using a copper and copper-nickel junction, for example, has a useful 'measurement' range from –100°C to +250°C, and has a typical sensitivity of 42μV per °C over the positive part of that range. Some thermocouples using other types of metal have useful measurement ranges well above +1100°C. *Figure 3.34(a)* shows the symbol used to denote a normal thermocouple.

In some special types of thermocouple device the junction can be heated via a d.c. or r.f. current passed through a pair of input terminals; the thermocouple output can then be used to indicate the magnitude of the input current or power. Devices of this type use the symbol shown in *Figure 3.34(b)*.

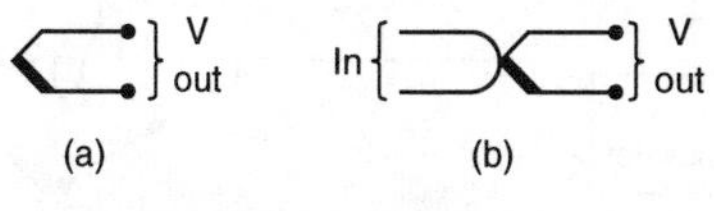

V out (a)

In V out (b)

Figure 3.34. Symbols of *(a)* a conventional and *(b)* an electrically-heated thermocouple device.

Piezoelectric transducers

A piezoelectric transducer is an electroconstrictive device that converts a varying electrical signal into a sympathetic set of fine mechanical variations, or *vice versa*. Piezo sounders, earphones, and microphones are 'audio frequency' examples of such devices and have already been described in this chapter. Some other types are specifically designed to operate at high frequencies and act as IF (455kHz or 10.7MHz, say) filters, etc. Other types of particular interest are the so-called 'ultrasonic' transducers, and ordinary quartz crystals.

Ultrasonic piezo transducers

These are sharply tuned low-power devices that are designed to resonate at an 'ultrasonic' frequency of about 40kHz. They are supplied in matched pairs, with one optimised for use as a 'transmitter' and the other for use as a receiver. They are useful in many remote control, distance measurement, and intrusion alarm applications. *Figure 3.35* shows, in block diagram form, one way of using them in a 'doppler effect' intruder alarm system.

The *Figure 3.35* alarm system consists of three main elements. The first is a transmitter (T_x) that floods the room with 40kHz ultrasonic signals, which bounce back and forth around the room. The second is a receiver (R_x) that picks up and amplifies the reflected signals and passes them on to a phase comparator, where they are compared with the original 40kHz signals. If nothing is moving in the room the T_x and R_x signal frequencies will be the same, but if an object (an intruder) is moving in the room the R_x signal frequency is doppler-shifted by an amount proportional to the rate of object movement (= about 66Hz at 10 inch/sec in this case). The a.f. output of the comparator is passed in to the third system element, the alarm activator; this consists of a signal conditioner that rejects spurious and 'out of limits' signals, etc., and activates the alarm-call generator only if an intruder is genuinely present.

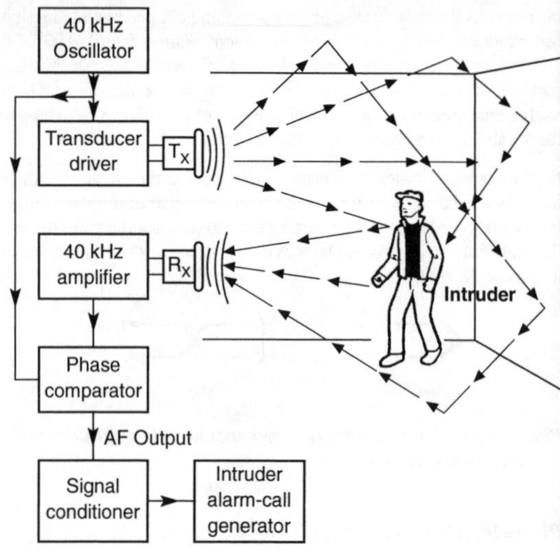

Figure 3.35. Block diagram of an ultrasonic (40kHz) doppler effect intruder alarm system.

Quartz crystals

Quartz crystals act as electromechanical resonators or tuned circuits with typical effective Qs of about 100,000 and with roughly 1000 times greater frequency stability than a conventional L–C tank circuit; they are widely used in precision filters and oscillators. The crystal's resonant frequency (which may vary from a few kHz to 100MHz) is determined by its mechanical dimensions. All quartz crystals have both series and parallel natural-resonance modes, and are cut to provide ca*librated* resonance in only one of these modes; series-mode devices present a low impedance at resonance, while parallel-mode devices present a high impedance at resonance.

Figure 3.36 shows the symbol and typical equivalent electrical circuit of a quartz crystal; it also shows the typical frequency/impedance response curve of a crystal that is cut to give series resonance at 465kHz; note that this particular crystal also has an uncalibrated parallel resonance mode at 505kHz.

Figure 3.37 shows one of the many ways of using a series-mode crystal in a precision oscillator circuit. This is a Colpitts oscillator in which Q_1 is wired as a voltage follower and the crystal provides the voltage gain needed for oscillation. Note that a small 'trimmer' capacitor (C_3) is wired in series with the crystal and enables its frequency to be varied over a narrow range. A variety of other types of crystal oscillator circuit are shown in Chapter 11.

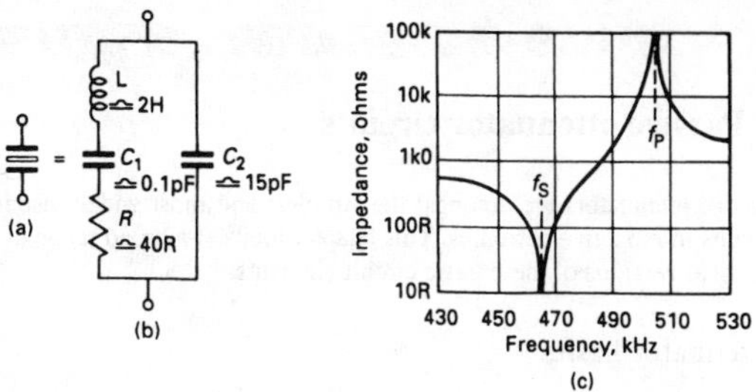

Figure 3.36. Quartz crystal symbol, typical equivalent circuit, and typical response curve of a 465kHz series-resonant crystal.

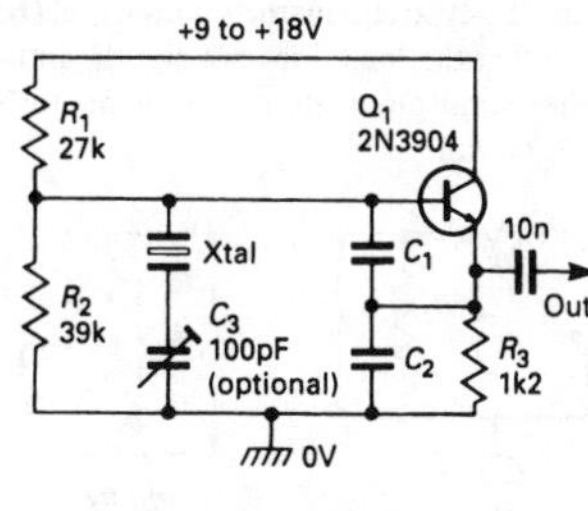

f_{range}	C_1	C_2
100kHz – 2MHz	2n7	1n0
2MHz – 10MHz	220p	220p
10MHz – 20MHz	110p	100p

Typical C_1 – C_2 values

Figure 3.37. Untuned Colpitts oscillator using series-resonant crystal.

Miscellaneous sensors/transducers

A variety of special-purpose sensors and transducers of value to the electronics engineer but not so-far mentioned in this chapter are also fairly readily available. Amongst the most useful of these are radioactive 'smoke detector' elements, humidity sensors, strain gauges, hall-effect devices that respond to magnetic field strength (flux density), and 'gas sensors' that react to gases such as propane, butane, methane, isobutane, petroleum gas, natural gas, and 'town' gas. Details of these devices are available in several specialist publications.

4 Passive attenuator circuits

Passive attenuators are amongst the simplest and most widely used circuits in modern electronics. This chapter looks at a broad range of practical versions of these basic circuit elements.

Attenuator basics

Attenuators are used to reduce awkward values of input signal voltage to lower and more convenient output levels. The simplest attenuator is the 'L'-type (so named because its diagram resembles an inverted 'L') which, as shown in *Figure 4.1*, consists of two resistors (R_1 and R_2) wired in series. Its attenuation *(a)* is set by the ratio of $R_2/(R_1 + R_2)$, as shown. Note that the 'L'-type attenuator's output must be fed to a fairly high impedance, so that the load does not significantly shunt R_2 and thereby increase the overall attenuation. The attenuator's input impedance equals $R_1 + R_2$ (= R_T).

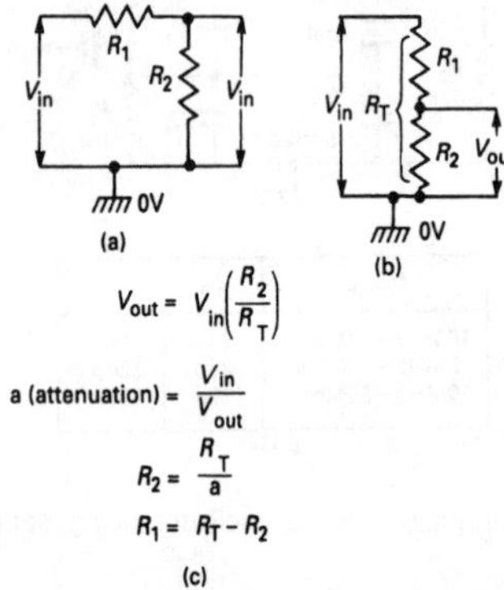

Figure 4.1. The basic L-type attenuator *(a)* is really a simple potential divider *(b)*; its design is controlled by the formulae in *(c)*.

To design an 'L'-type attenuator with a desired value of attenuation 'a' and total resistance R_T, first work out the R_2 value, and then the R_1 value, thus:

(i) $R_2 = R_T/a$, and
(ii) $R_1 = R_T - R_2$.

For example, in a unit with an R_T value of 10kΩ and an 'a' value of 10 (= 20dB), R_2 needs a value of 10kΩ/10 = 1k0, and R_1 needs a value of 10kΩ–1k0 = 9k0.

The simplest type of variable attenuator is the 'pot' type of *Figure 4.2*, which is often used as a 'volume' control, and is merely a variation of the basic 'L'-type attenuator.

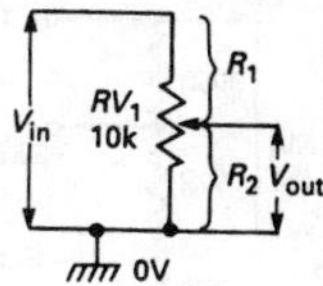

Figure 4.2. This pot attenuator is a fully-variable version of the L-type attenuator.

Another 'L' attenuator variation is the switched type of *Figure 4.3*, which gives a range of attenuation values. The procedure for designing this is similar to that described above (using obvious variations of the (i) and (ii) formulae), except that a separate calculation is made for each attenuation position, starting with the greatest. Thus, the *Figure 4.3* attenuator has an R_T value of 10k, so the first design step is to work out the R_3 value needed to give '÷100' attenuation, which is 10k/100 = 100Ω. Similarly, the 'lower arm' (i.e., R_2+R_3) value needed in the '÷10' position equals 1k0, but 100Ω of this is already provided by R_3, so R_2 needs a value of 1k0–100Ω = 900Ω. R_1 needs a value of 10k–1k0 = 9k0, as shown. This basic design procedure can be expanded up to give as many attenuator steps as are needed in any particular application.

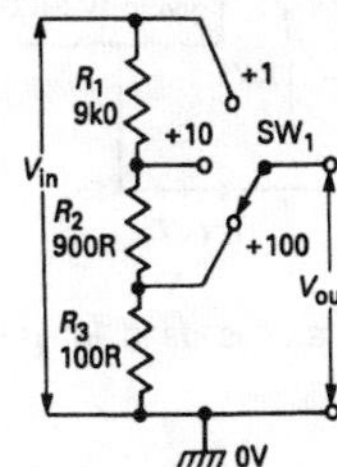

Figure 4.3. The design of this switched attenuator is fully described in the text.

Figure 4.4 shows how modified versions of the *Figures 4.2* and *4.3* circuits (with greatly reduced resistance values) can be combined to make a fully-variable wide-range attenuator that can serve as the output of an audio sinewave generator, say; RV_1's scale should be hand-calibrated.

Voltage ranging

A common 'attenuator' application is as a 'voltage ranger' at the input of an electronic voltmeter, as in *Figure 4.5*. Here, the 1 volt f.s.d. meter is 'ranged' to indicate other f.s.d. values by feeding inputs to it via a switched 'L' attenuator. The V_{IN}/V_{OUT} attenuation ratios are chosen on the basis of:

'a' = desired f.s.d./actual f.s.d.

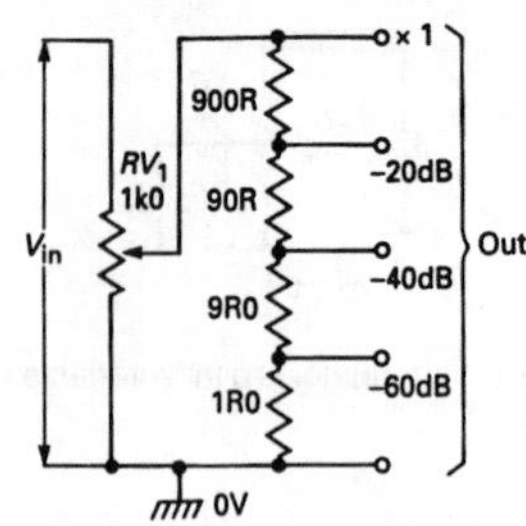

Figure 4.4. This fully-variable wide-range attenuator can be used in the output of a simple sinewave generator, etc.

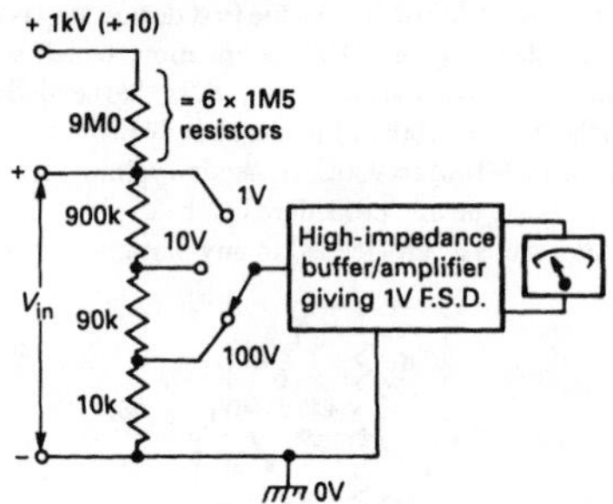

Figure 4.5. This attenuator is used for ranging an electronic voltmeter.

Thus, the *Figure 4.5* attenuator gives output ranging of 1–10–100 volts, which in this case correspond to 'a' values of 1, 10 and 100. Note that the meter's range is extended to 1000 volts f.s.d. via a separate input terminal (marked '1kV' and '÷10') and a 9M0 resistance made of six series-wired 1M5 resistors, thus ensuring that (at f.s.d.) a maximum of only 150 volts appears across any resistor or pair of switch contacts.

Figures 4.6 and *4.7* show a selection of useful 'L'-type voltage-ranging attenuators that can be used to drive any high-impedance 1V f.s.d. meter, and *Figure 4.8* shows a 1M0 attenuator that gives an output that is variable from 0dB to –20dB in 2dB steps.

Note that all attenuators shown in *Figure 4.3* to *Figure 4.8* can be made with alternative total resistance (R_T) values by simply multiplying or dividing all resistor values by a proportionate amount. Thus, any of the '1M0' designs can be adapted to give an R_T value of 10k by dividing all '*R*' values by 100, and so on.

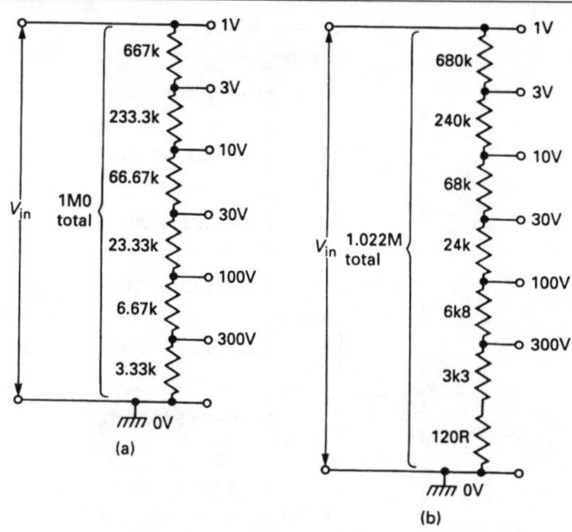

Figure 4.6. These attenuators give 1–3–10, etc., voltage ranging; the one in *(a)* uses odd-ball resistor values, which total 1M0; the one in *(b)* uses standard values, which total 1.022MΩ.

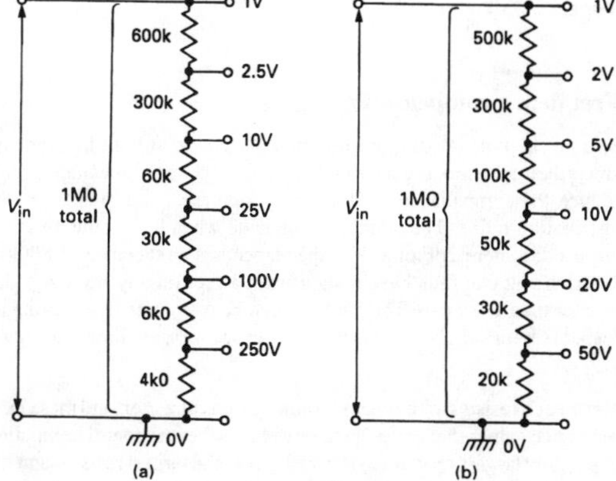

Figure 4.7. These 1M0 attenuators give (a) 1–2.5–10, etc., and (b) 1–2–5–10, etc., voltage ranging. For alternative total resistance values, simply multiply or divide all resistors by a proportionate amount (e.g. divide by 100 for 10k total).

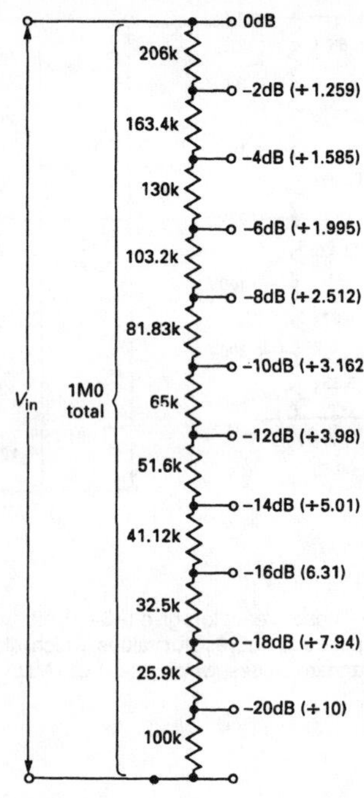

Figure 4.8. The output of this 1M0 attenuator is variable in 2dB steps.

Frequency compensation

Simple 'L'-type attenuators are often inaccurate at high frequencies, since their resistors are inevitably shunted by stray capacitances that reduce their impedances as frequency increases, and thus affect the attenuation ratios. The effect is most acute when high-value resistors are used; a mere 2pF of stray capacitance has a reactance of 800k at 100kHz and can thus have a significant affect on any resistor value greater than 10k or so. This problem can be overcome by shunting all resistors with suitable 'compensation' capacitors, as shown in *Figure 4.9.*

Here, each resistor of the chain is shunted by a capacitor, and these have reactance values that are in the same ratios as the resistors. The smallest capacitor (largest reactance) is wired across the largest resistor and has a typical value of 15 to 50pF, which is large enough to swamp strays but small enough to present an acceptably high impedance to input signals. The attenuator's frequency compensation is set up by feeding a good square wave to its input, taking its ÷100 or ÷1000 output to the input of a oscilloscope, and then trimming C_1 to obtain a good square wave picture, as shown in *(b)* in the diagram.

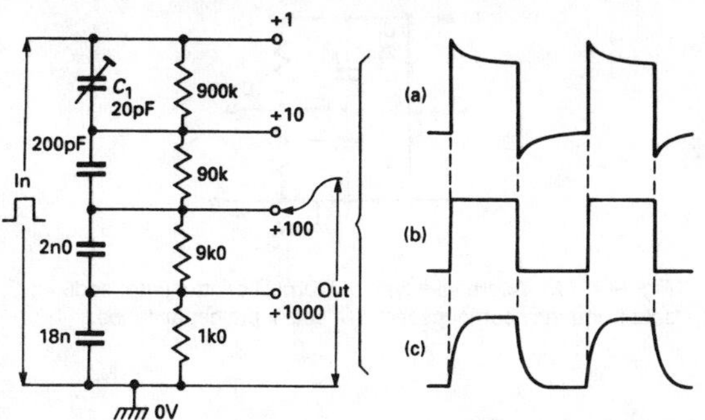

Figure 4.9. A basic compensated wide-range L-type attenuator, showing square wave output waveforms when the C_1 trimmer setting is *(a)* over compensated, *(b)* correctly compensated, and *(c)* under compensated.

Oscilloscopes invariably use compensated 'L'-type attenuators at the input of their 'Y' amplifiers. *Figure 4.10* shows part of a typical example, in which an individually trimmed 1M0 attenuator section is used on each range. *Figure 4.11* shows a variation of one of these sections; C_1 is used to set the section's frequency compensation, and C_2 adjusts the section's input capacitance so that the 'Y'-channel attenuator presents a constant input impedance on all ranges.

Figure 4.12 shows how a 2-range compensated 'primary' attenuator and a low-impedance uncompensated 6-range 'secondary' attenuator can be used together to help make an a.c. millivoltmeter that spans 1mV to 300V f.s.d. in 12 ranges. The primary attenuator gives zero attenuation in the 'mV' position and ÷1000 attenuation in the 'V' position.

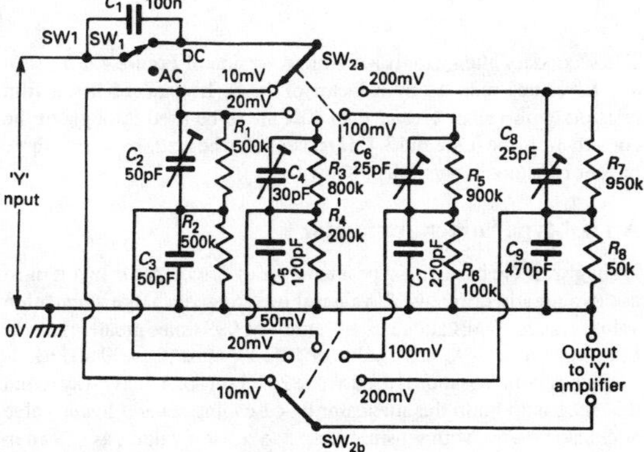

Figure 4.10. Part of a typical oscilloscope Y amplifier attenuator.

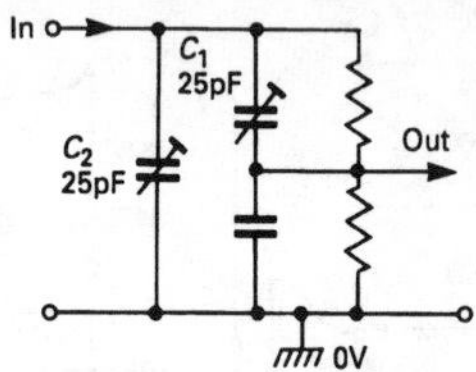

Figure 4.11. Alternative type of Y amplifier attenuator section; C_1 sets frequency compensation; C_2 sets input capacitance.

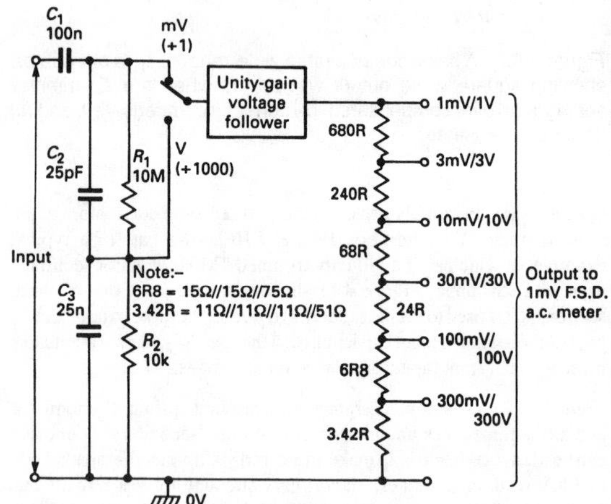

Figure 4.12. Use of primary and secondary attenuators in an a.c. millivoltmeter.

The secondary attenuator is a modified version of *Figure 4.6(b)*, with all *'R'* values reduced by a factor of 1000. If standard metal film resistors with values greater than 10Ω are to be used throughout the construction, the 6.8Ω and 3.42Ω resistors can be made by wiring three or four resistors in parallel, as shown.

An 'L'-type ladder attenuator

A snag with the basic 'L'-type attenuator of *Figure 4.1* is that it must use two greatly different *'R'* values if used to give a large attenuation value, e.g., for 60dB attenuation R_1 must be 999 times greater than R_2. In this example, if R_2 has a value of 10Ω, R_1 must be 9k99 and needs frequency compensation if used above 20kHz or so. An easy way round this snag is to build the attenuator by cascading several lower-value attenuator stages, with sensibly restricted resistor values, as shown in the practical circuit of *Figure 4.13*. Such a circuit is known as a *ladder* attenuator.

The *Figure 4.13* ladder attenuator consists of three cascaded 20dB attenuator stages, each with a maximum resistance value of 820R and with a useful uncompensated bandwidth extending to hundreds of kHz. Note that the right-hand (1mV) stage has 'R_1–R_2' (see *Figure 4.1*) values of 820R and 91R, and that these shunt the lower leg of the middle (10mV) attenuator and reduce its effective value to 91Ω. Similarly, the middle attenuator shunts the lower leg of the first (100mV) attenuator and reduces its effective value to 91R. Thus, each stage effectively consists of an 820R/91R 20dB attenuator that is accurate within +0.2%. The odd-ball 101Ω resistors are made by series-wiring 33R and 68R resistors.

The *Figure 4.13* attenuator is an excellent design that can be used as the output section of a variety of audio and pulse generator circuits. Its output is fully variable via RV_1. The attenuator's output impedance, on all but the '1V' range, is less than 90Ω, so its output voltage is virtually uninfluenced by load impedances greater than a few kilohms.

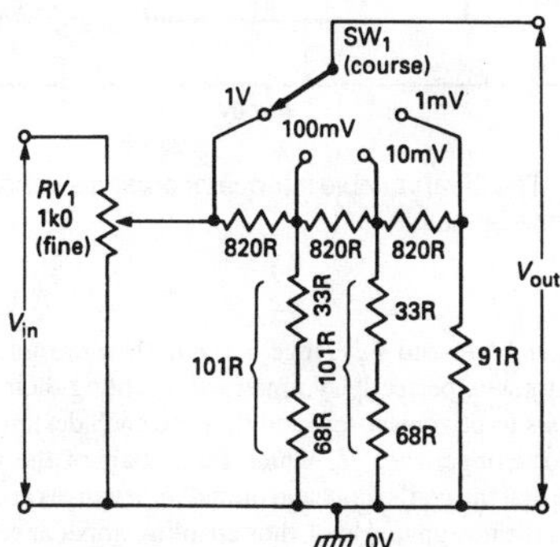

Figure 4.13. This fully-variable attenuator uses an L-type ladder network and makes an excellent wide-band output section for audio and pulse generators, etc.

Matched-resistance attenuators

Weaknesses of the 'L'-type attenuator are that its output impedance varies with the attenuator setting, and its input impedance varies with external output loading. The significance of this latter effect is illustrated in *Figure 4.14*, where the attenuator is represented by the load on the output of the waveform generator, which has an output impedance of 100Ω. If the generator is set to give 1 volt output into a 1k0 load, the output varies between 1.048 volts and 0.917 volts if the load is then varied between 2k0 and 500Ω, thus invalidating the generator's calibration.

It follows from the above that, as illustrated in *Figure 4.15*, the 'ideal' variable attenuator should have input and output impedances that remain constant at all attenuation settings. Such attenuators do exist and are designed around switch-selected fixed-value attenuator pads; these pads come in various types, and the five most popular of these are

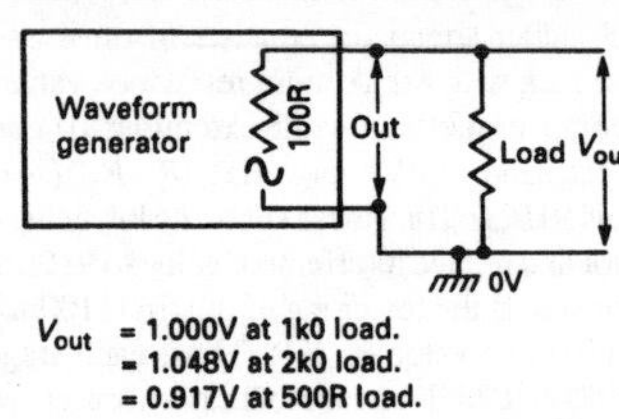

Figure 4.14. The output voltage of a generator varies with changes in its load impedance.

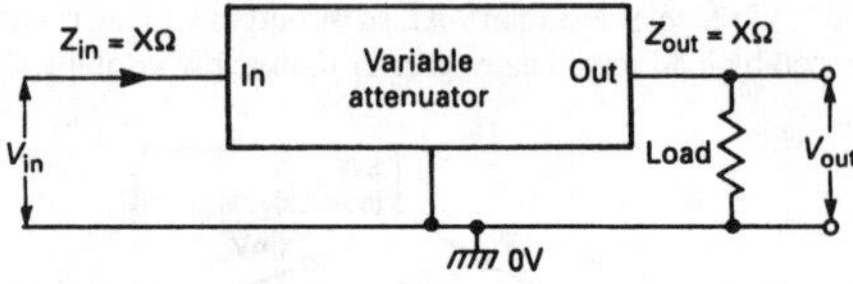

Figure 4.15. The ideal variable attenuator presents constant input and output impedances.

shown in *Figures 4.16* and *4.17*, together with their design formulae. These attenuators are perfectly symmetrical, enabling their input and output terminals to be transposed, and they are each designed to feed into a fixed load impedance, Z, which forms part of the attenuator network. Note that the pad's input and output impedances are designed to equal that of the designated load, thus enabling impedance-matched pads of any desired attenuation values to be cascaded in any desired combination, as shown in *Figure 4.18*.

The two most popular types of pad attenuator are the 'T' and 'π' types; the 'H' and 'O' types are simply 'balanced input' versions of these, and the 'bridged-T' type is a derivative of the basic 'T' type.

Figure 4.19 shows a π-type attenuator example, designed to give a matched impedance of 1k0 and 20dB (=÷10) of attenuation. Working through this design from the back, note that the 1k0 load shunts R_2 and brings its effective impedance down to 550Ω, which then acts with R_1 as an 'L'-type attenuator that give the 20dB of attenuation and has an input impedance (into R_1) of 5501Ω, which is shunted by R_3 to give an actual input impedance of 1000Ω. Note that the output load forms a vital part of the attenuator, and that if it is removed the pad's attenuation falls to only ÷5.052, or –14.07dB.

Figure 4.20 shows a T-type attenuator example that gives a matched impedance of 1k0 and 20dB of attenuation. Working through this design from the back, note that R_3 and the 1k0 load form an 'L'-type '÷1.8182' attenuator with an input impedance (into R_3) of 1818.2Ω. R_1 and R_2 also form an 'L'-type attenuator, but R_2 is shunted by the above 1818.2Ω impedance and has its effective value reduced to 181.8Ω, so

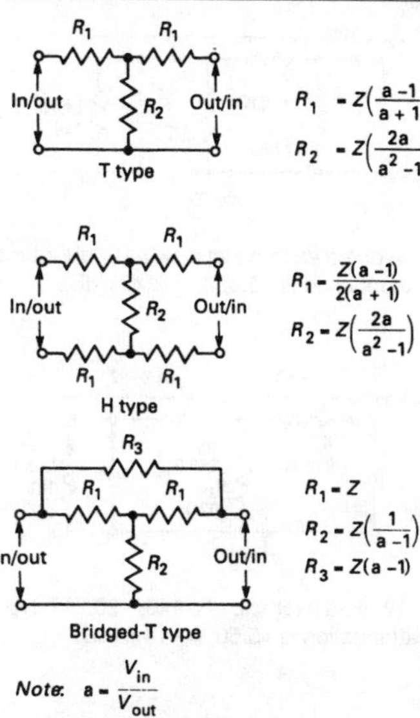

Figure 4.16. Circuits and design formulae of the basic T type attenuator and its H and Bridged-T derivatives.

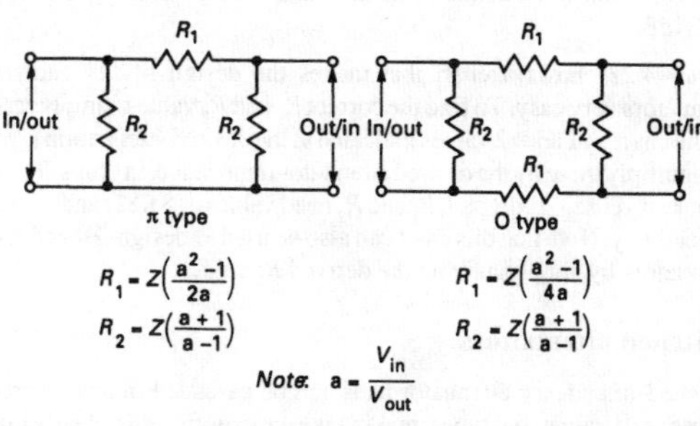

Figure 4.17. Circuits and design formulae for the basic π type attenuator and its 0 type derivative.

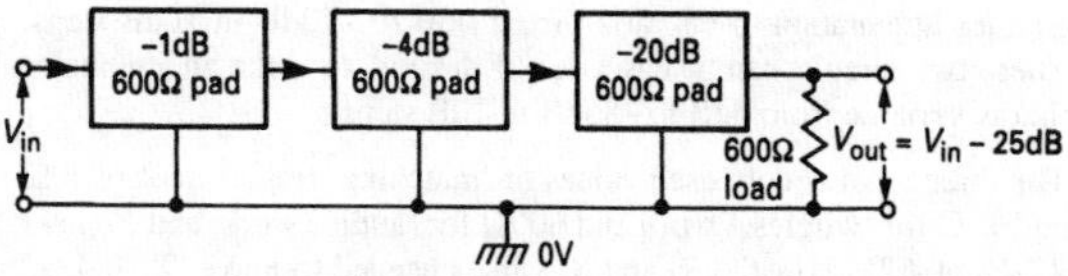

Figure 4.18. Matched attenuator pads can be cascaded in any combination.

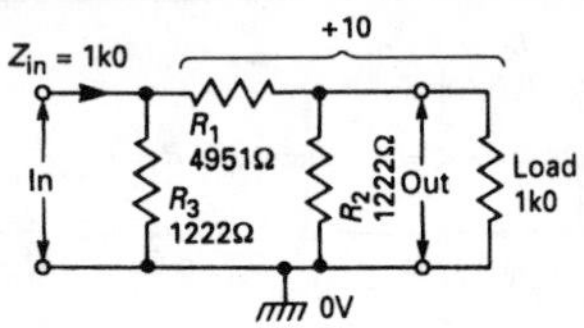

Figure 4.19. Worked example of a 1k0, –20dB π-type attenuator; its unloaded attenuation is ÷5.052, = –14.07dB.

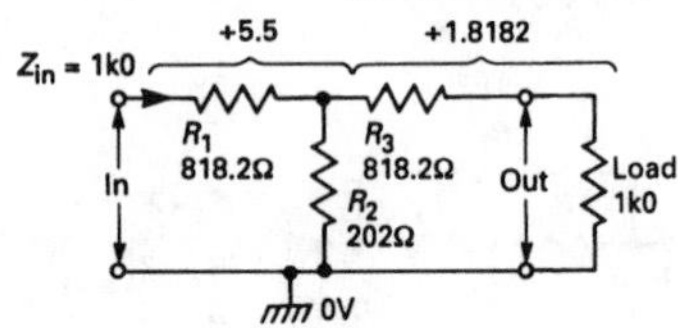

Figure 4.20. Worked example of a 1k0, –20dB T-type attenuator; its unloaded attenuation is ÷5.50, = –14.81dB.

this stage gives an attenuation of ÷5.5 and has an input impedance of 1000Ω. Thus, the T-type attenuator actually consists of a pair of cascaded L-types, which in this example give individual 'a' ratios of 1.8182 and 5.5, or ÷10.00 overall. Note that if the output load is removed from this attenuator its attenuation falls to only ÷5.50, or -14.81dB.

Figure 4.21 shows a chart that makes the design of 'T' and 'π' attenuators very easy. To find the correct R_1 and R_2 values, simply read off the chart's *r*1 and *r*2 values indicated at the desired attenuation level and multiply these by the desired attenuator impedance, in ohms. Thus, to make a 100Ω, –20dB pad, R_1 and R_2 need values of 81.8Ω and 20.2Ω respectively. Note that this chart can also be used to design 'H' and 'O' attenuators by simply halving the derived R_1 value.

Switched attenuators

Matched-impedance attenuator pads can be cascaded in any desired sequence of values and types, making it easy to design switched-value attenuator networks and 'boxes', as shown in *Figures 4.22* and *4.23*. *Figure 4.22* shows four binary-sequenced (1–2–4–8) attenuator pads cascaded to make an attenuator that can be varied from 0dB to –15dB in 1dB steps, and *Figure 4.23* shows an alternative arrangement that enables attenuation to be varied from 0dB to –70dB in 10dB steps. These two circuits can be cascaded, if desired, to make an attenuator that is variable from 0dB to –85dB in 1dB steps.

The three most widely used values of 'matching' impedance are 50Ω and 75Ω for 'wireless' work and 600Ω for 'audio' work, and *Figures 4.24* and *4.25* show the R_1 and R_2 values needed to make 'T' and 'π' pads of these impedances at various attenuation values. When designing attenuator pads note that the R_1 or R_2 values may be adversely

dB Loss	a (V_{in}/V_{out})	'T'-type r1	'T'-type r2	'π'-type r1	'π'-type r2
0	1.000	0	∞	0	∞
0.1	1.012	0.00576	86.9	0.0115	174
0.2	1.023	0.0115	43.4	0.0230	86.9
0.3	1.035	0.0173	28.9	0.0345	57.9
0.4	1.047	0.0230	21.7	0.0461	43.4
0.5	1.059	0.0288	17.4	0.0576	34.8
0.6	1.072	0.0345	14.5	0.0691	29.0
0.8	1.096	0.0460	10.8	0.0922	21.7
1.0	1.122	0.0575	8.67	0.115	17.4
1.5	1.188	0.0861	5.76	0.174	11.6
2	1.259	0.115	4.30	0.232	8.72
3	1.413	0.171	2.84	0.352	5.85
4	1.585	0.226	2.10	0.477	4.42
5	1.778	0.280	1.64	0.608	3.57
6	1.995	0.332	1.34	0.747	3.01
7	2.239	0.382	1.12	0.896	2.61
8	2.512	0.431	0.946	1.057	2.32
9	2.818	0.476	0.812	1.23	2.10
10	3.162	0.520	0.703	1.43	1.92
12	3.981	0.598	0.536	1.86	1.67
14	5.01	0.667	0.416	2.41	1.50
15	5.62	0.698	0.367	2.72	1.43
16	6.31	0.726	0.325	3.08	1.38
18	7.94	0.776	0.256	3.91	1.29
20	10.00	0.818	0.202	4.95	1.22
25	17.78	0.894	0.113	8.86	1.12
30	31.62	0.939	0.0633	15.8	1.07
32	39.81	0.951	0.0503	19.89	1.052
35	56.23	0.965	0.0356	28.1	1.04
40	100.0	0.980	0.0200	50.1	1.02
45	177.8	0.989	0.0112	88.9	1.011
50	316.2	0.994	0.00632	158	1.006
55	562.3	0.996	0.00356	281	1.0036
60	1000	0.998	0.00200	500	1.0020
64	1585	0.9987	0.001262	800	1.00126

Figure 4.21. T and π attenuator design chart. To find the correct R_1–R_2 values, read the r_1 and r_2 values indicated at the desired attenuation value and multiply by the desired attenuator impedance.

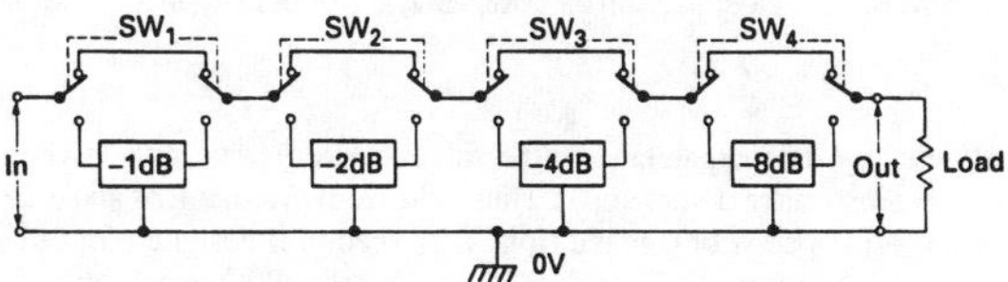

Figure 4.22. This switched attenuator is variable from 0 to –15dB in 1dB steps.

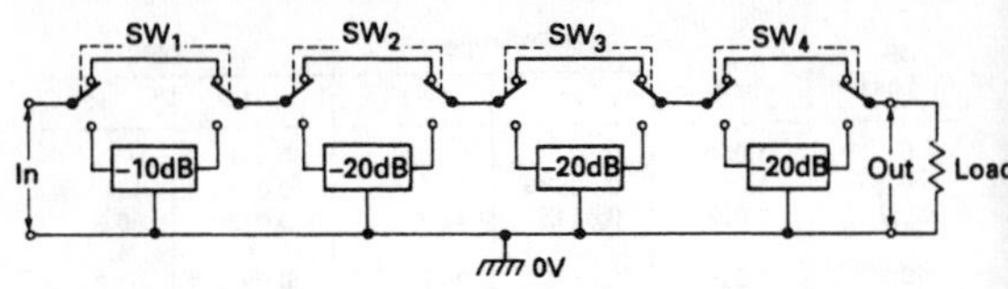

Figure 4.23. This switched attenuator is variable from 0 to –70dB in 10dB steps.

'T'-type attenuator pads

dB Loss	50Ω impedance		75Ω impedance		600Ω impedance	
	R_1 (Ω)	R_2 (Ω)	R_1 (Ω)	R_2 (Ω)	R_1 (Ω)	R_2 (Ω)
1	2.875	433.5	4.312	650.2	34.50	5202
2	5.750	215.0	8.625	322.5	69.00	2580
4	11.30	105.0	16.95	150.0	135.6	1260
8	21.55	47.30	32.33	70.95	258.6	567.6
10	26.00	35.15	39.00	52.73	312.0	421.8
16	36.30	16.25	54.45	24.37	435.6	195.0
20	40.90	10.10	61.35	15.15	490.8	121.2
32	47.55	2.515	71.32	3.772	570.6	30.18

Figure 4.24. Design chart for 50Ω, 75Ω, and 600Ω T-type attenuator pads.

'π'-type attenuator pads

dB Loss	50Ω impedance		75Ω impedance		600Ω impedance	
	R_1 (Ω)	R_2 (Ω)	R_1 (Ω)	R_2 (Ω)	R_1 (Ω)	R_2 (Ω)
1	5.750	870.0	8.625	1305	69.00	10,440
2	11.60	436.0	17.40	654.0	139.2	5,232
4	23.85	221.0	35.78	331.5	286.2	2,652
8	52.85	116.0	79.27	174.0	634.2	1,392
10	71.50	96.0	107.2	144.0	858.0	1,152
16	154.0	69.0	231.0	103.5	1848	828
20	247.5	61.0	371.2	91.5	2970	732
32	994.5	52.6	1492	78.9	11,934	631.2

Figure 4.25. Design chart for 50Ω, 75Ω, and 600Ω π-type attenuator pads.

affected by stray capacitance if the values are very large, or by switch contact resistance if very small. Thus, note from *Figures 4.24* and *4.25* that a –1dB pad is best made from a 'π' section if designed for 50Ω matching, but from a 'T' section if intended for 600Ω matching.

If large (greater than –32dB) values of pad attenuation are needed, it is best to make the pad from two or more cascaded attenuator networks.

If the multi-stage pad is made from *identical* 'π'-type stages, as shown in *Figure 4.26(a)*, an economy can be made by replacing adjoining pairs of R_2 resistors with a single resistor with a value of $R_2/2$, as in *Figure 4.26(b)*.

The *Figure 4.26(b)* ladder attenuator is actually a set of cascaded 'L'-type attenuators with a shunt across its main input terminals. An ingenious development of this circuit is the switched ladder attenuator, a 5-step version of which is shown in *Figure 4.27*, together with its design formulae and with worked values for giving ÷10 (= 20dB) steps and a 1k0 matching impedance. Note that the input signal's *effective* source impedance forms a vital part of this attenuator network, and needs a value of 2Z.

To understand the operation of the *Figure 4.27* attenuator, imagine it first without the external load and then work through the design from right to left. Thus, the 5th (output) section (R_2–R_4) acts as a ÷10 'L' attenuator with an 11k input impedance; this impedance shunts R_3 of the preceding section and reduces its effective value to 1100Ω, so that section (the 4th) also gives ÷10 attenuation and an 11k input impedance; sections 2 and 3 act in the same way. The '1' input 'L'-type section consists of the generator's source impedance (2k0) and R_1, which (since it is shunted by the 11k input impedance of the '2' section) has an effective impedance of 2k0; this section thus has an *effective* attenuation of ÷2.

Now imagine the effect of connecting the external 1k0 load to any one of the attenuator's output terminals. If it is connected to the output of the 5th section it shunts R_4 and increases that section's attenuation to ÷19.9 and reduces its input impedance to 10424Ω, thereby also increasing the attenuation of the preceding section by 0.5%, the net result being that the attenuation at that output terminal increases by a factor of 1.995, or exactly 6dB. Similarly, if the load is connected to the output of any of the '2' to '4' sections, the attenuation at that point increases by 6dB; if it is connected to the output of section '1', that section's attenuation increases by a factor of 2.000 (to ÷4), or 6.021dB.

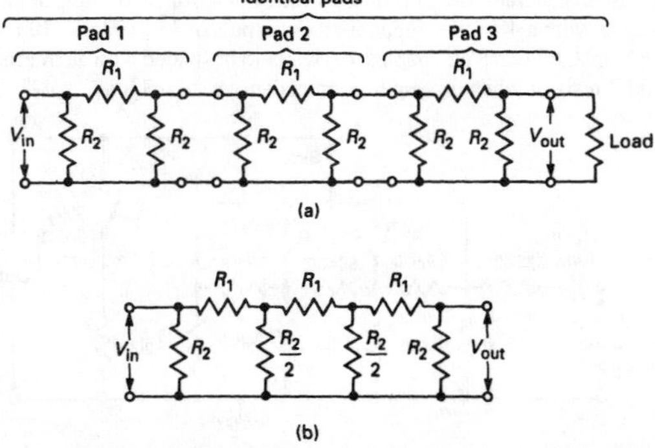

Figure 4.26. The three-stage π-type ladder attenuator of *(b)* is a simple development of the three-pad circuit shown in *(a)*.

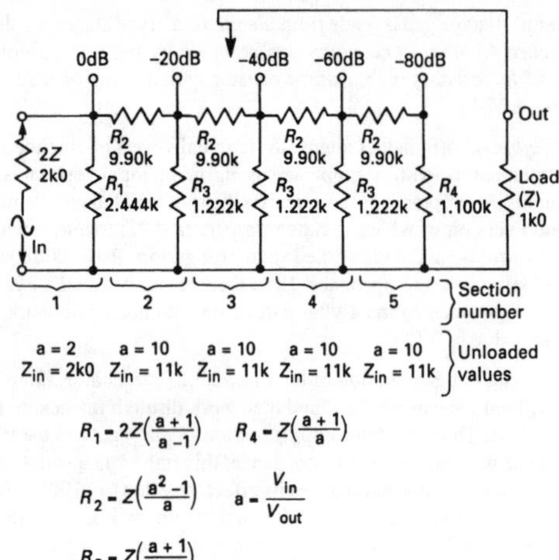

Figure 4.27. A worked example of a five-step switched ladder attenuator, giving –20dB steps into 1k0, with design formulae.

The big thing to note here is that, since the load is connected to some part of the circuit at all times, it does not affect the *step* attenuation of the overall network. Thus, if the load is shifted a 20dB step down the line, from the output of section '2' to that of section '3', the output of section '2' (and the input of section '3') rises by 6dB but that of section '3' increases by 6dB (to –26dB), to give an overall step change of precisely –20dB. This step change accuracy is maintained on all except the 1st step position, where a trivial error of +0.25% (0.021dB) occurs. Consequently, this type of switched attenuator is widely used in the output of audio and RF generators.

Figure 4.28 shows a practical 4-step 600Ω ladder attenuator suitable for audio generator use. It is meant to be driven from a low-impedance source; with a 4V r.m.s. input, it gives outputs of 1V, 100mV, 10mV, and 1mV. Switch SW_2 enables the output to be loaded with an internal 600Ω resistor when driving high-impedance external loads.

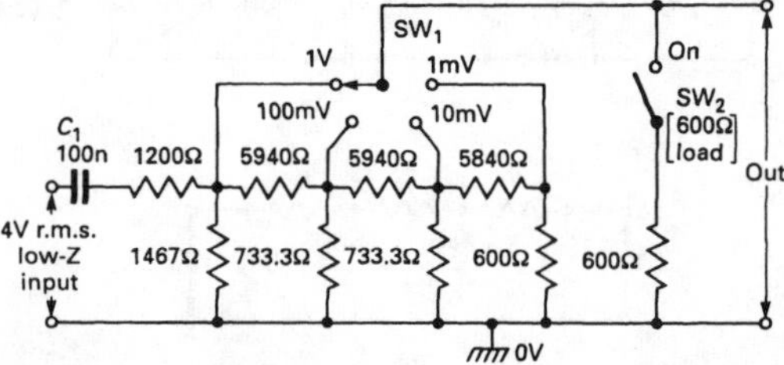

Figure 4.28. Practical 600Ω four-step switched ladder attenuator for use in an audio generator.

5 Passive and active filter circuits

Simple passive *C–R* and L–C filters are widely used to either 'condition' existing a.c. signals or to form the basis of signal-generating circuits. The performances of simple filters can be greatly enhanced with the aid of 'feedback' amplifiers, the resulting combination being known as an 'active' filter circuit. This chapter looks at a variety of passive and active filter designs.

Passive *C–R* filters

Filter circuits are used to reject unwanted frequencies and pass only those wanted by the designer. In low-frequency applications (up to 100kHz) the filters are usually made of *C–R* networks, and in high-frequency (above 100kHz) ones they usually use L–C components.

A simple *C–R* low-pass filter *(Figure 5.1)* passes low-frequency signals but rejects high-frequency ones. Its output falls by 3dB (to 70.7% of the input value) at a 'break', 'crossover', or 'cutoff' frequency (f_c) of 1/(2πRC), and then falls at a rate of 6dB/octave (= 20dB/decade) as the frequency is increased. Thus, a 1kHz filter gives about 12dB of rejection to a 4kHz signal, and 20dB to a 10kHz one. The phase angle (ø) of the output signal lags behind that of the input and equals -arctan (2π*fCR*), or –45° at f_c.

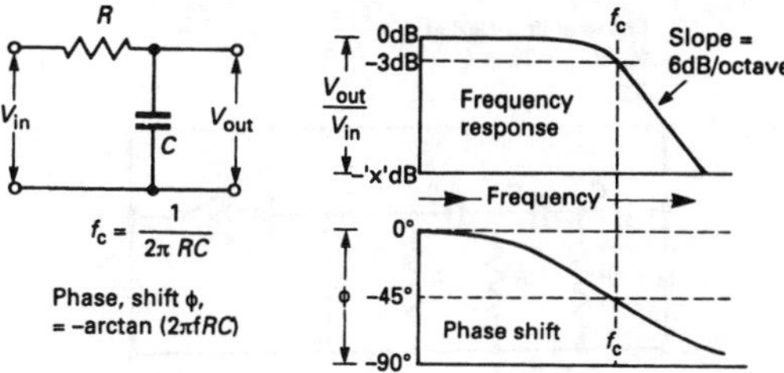

Figure 5.1. Passive *C–R* low-pass filter.

A simple *C–R* high-pass filter *(Figure 5.2)* passes high-frequency signals but rejects low-frequency ones. Its output is 3dB down at a break frequency of 1/(2πRC), and falls at a 6dB/octave rate as the frequency is decreased below this value. Thus, a 1kHz filter gives 12dB of rejection to a 250Hz signal, and 20dB to 100Hz one. The phase angle of the output signal leads that of the input and equals arctan 1/(2π*fCR*), or +45° at f_c.

Each of the above two filter circuits uses a single *C–R* stage and is known as a '1st-order' filter. If a number (*n*) of such filters are cascaded the resulting circuit is known as an '*n*th-order' filter and has a slope, beyond f_c, of (*n* x 6dB)/octave. Thus, a 4th-order 1kHz low-pass filter has a 24dB/octave slope and gives 48dB of rejection to a 4kHz signal and 80dB to a 10kHz one.

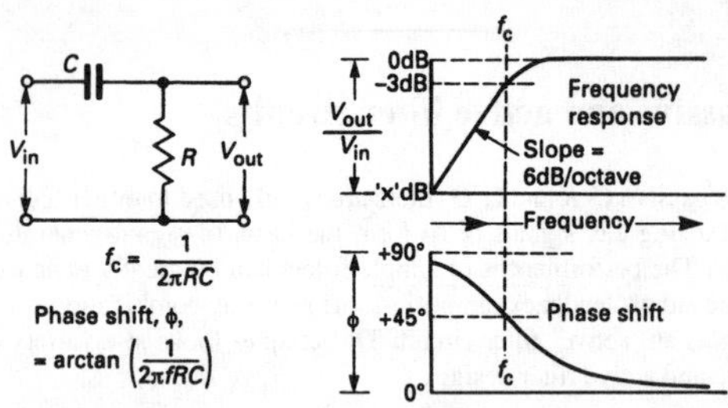

Figure 5.2. Passive *C–R* high-pass filter.

In practice, 'simple' cascaded passive filters are hard to design and are rarely used; instead, they are *effectively* cascaded by incorporating them in the feedback networks of 'active' filters. One instance where they are used, however, is as the basis of a so-called 'phase-shift' oscillator, as shown in basic form in *Figure 5.3*. Here, the filter is inserted between the output and input of the inverting (180° phase shift) amplifier; the filter gives a total phase shift of 180° at a frequency, f_o, of about 1/(14RC), so the complete circuit has a loop shift of 360° under this condition and oscillates at f_o if the amplifier has enough gain (about x29) to compensate for the filter's losses and thus give a loop gain greater than unity.

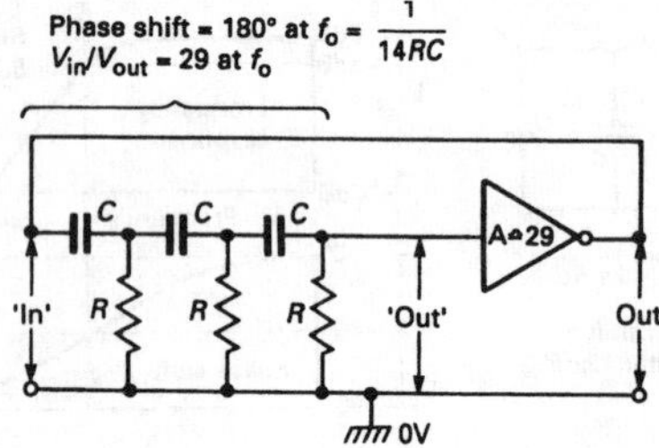

Figure 5.3. Third-order high-pass filter used as the basis of a phase-shift oscillator.

Figure 5.4 shows a practical 800Hz phase-shift oscillator. RV_1 must be adjusted to give reasonable sinewave purity; the output level is variable via RV_2.

Band-pass and notch filters

A 'band-pass' filter is one that accepts a specific band or spread of frequencies but rejects or attenuates all others. A simple version of such a circuit can be made by cascading a pair of *C–R* high-pass and low-pass filters, as shown in *Figure 5.5*. The high-pass component values determine the lower break frequency, and the low-pass values set the upper break frequency.

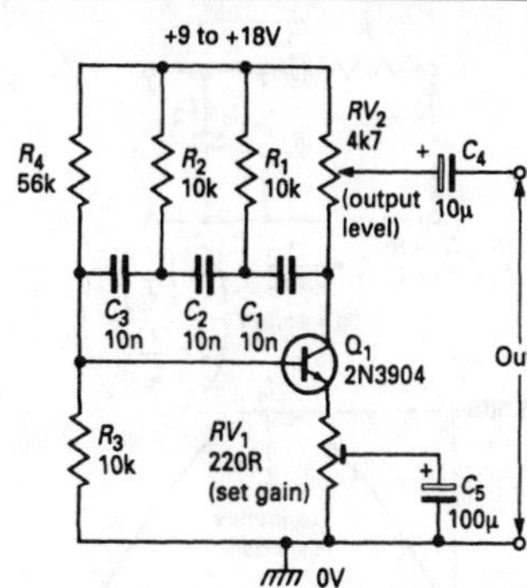

Figure 5.4. 800Hz phase-shift oscillator.

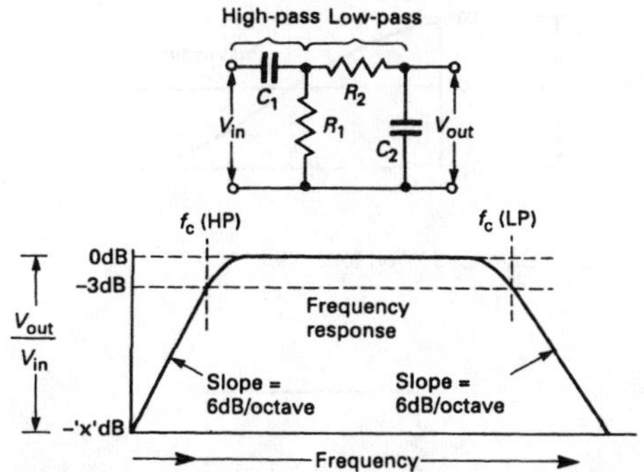

Figure 5.5. High-pass and low-pass filters cascaded to make a band-pass filter.

If both filters in the above circuit have identical break frequencies the circuit acts as a tone-select filter, and gives minimum attenuation to a single frequency. *Figure 5.6* shows a popular variation of this type of circuit, the 'balanced' Wien tone filter, in which $R_1 = R_2$, and $C_1 = C_2$. The balanced Wien filter gives an attenuation factor of 3 (=–9.5dB) at f_c, and its output phase shift varies between +90° and –90°, and is precisely 0° at f_c; consequently, it can be used as the basis of a sinewave generator by connecting its output back to its input via a non-inverting amplifier with a gain of x3 (to give a loop gain of unity), as shown in basic form in *Figure 5.7*.

A 'notch' filter is one that gives total rejection of one specific frequency, but accepts all others. Such a filter can be made by wiring the Wien network into the bridge configuration of *Figure 5.8*. Here, R_1–R_2 act as a x3 (nominal) attenuator; consequently, the attenuator and the Wien filter outputs are identical at f_c, and the output (which equals the difference between the two signals) is thus zero under this condition. In practice, the value of R_1 or R_2 must be carefully trimmed to give precise nulling at f_c.

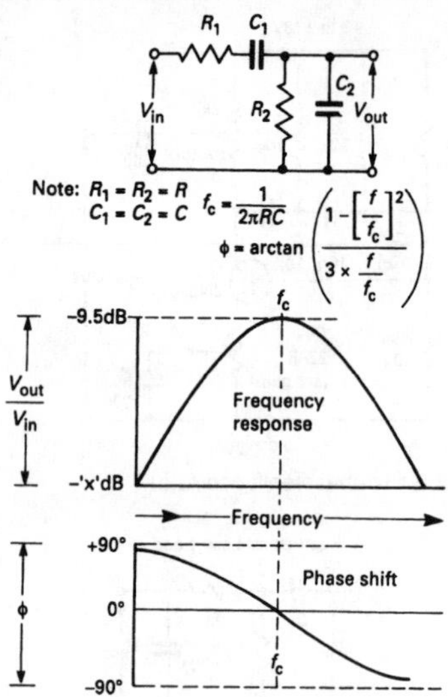

Figure 5.6. Balanced Wien tone filter.

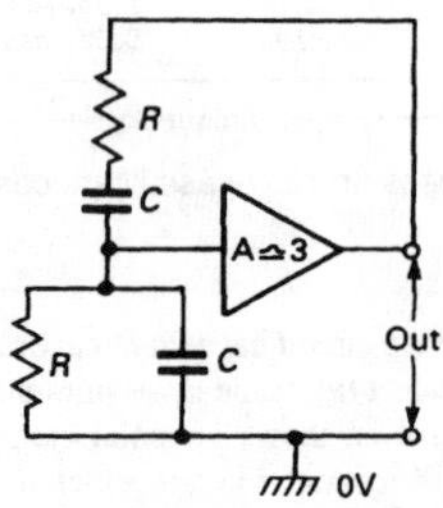

Figure 5.7. Basic Wien-based oscillator.

The Wien bridge network can be used as the basis of an oscillator by connecting it as in *Figure 5.9(a)*, which can be redrawn as in *Figure 5.9(b)* to more clearly indicate that the op-amp is actually used as a x3 non-inverting amplifier, and that this circuit is similar to that of *Figure 5.7*. In practice these circuits must be fitted with some form of automatic gain control if they are used to generate good-quality sinewaves.

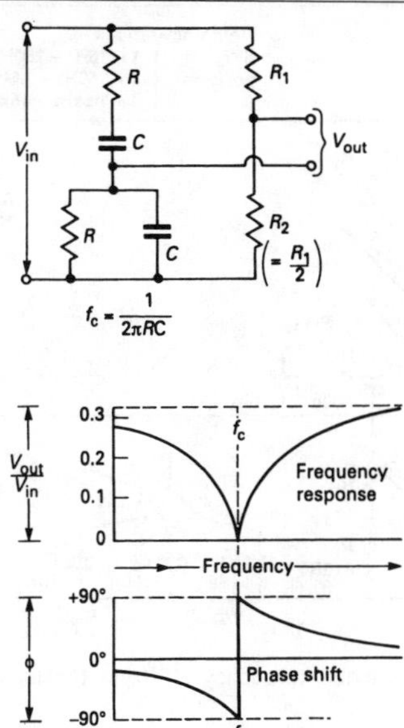

Figure 5.8. Wien Bridge notch filter.

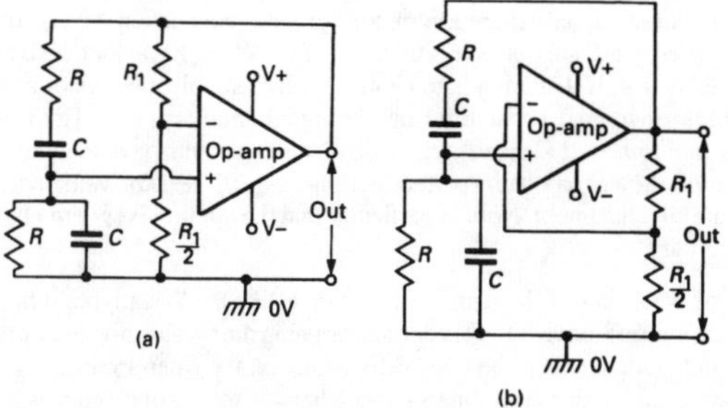

Figure 5.9. The Wien Bridge oscillator in *(a)* is the same as that of *(b)*.

The 'tuned' frequency of the Wien bridge network can easily be changed by simultaneously altering its two R or two C values. *Figure 5.10* shows this technique used to make a wide-range (15Hz to 15kHz) variable notch filter in which fine tuning and decade switching are available via RV_1 and SW_1, and null trimming is available via RV_2.

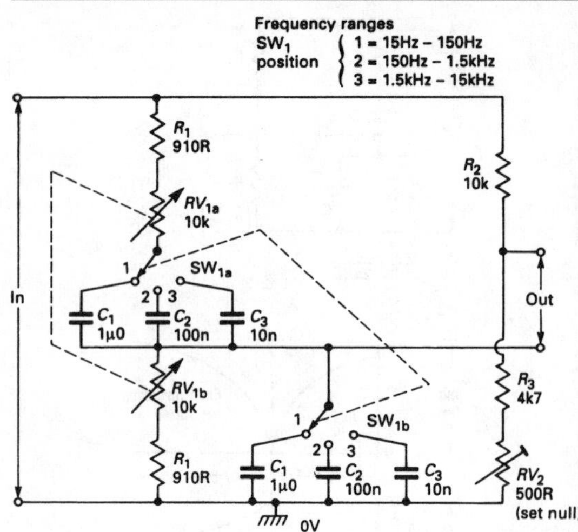

Figure 5.10. Variable frequency (15Hz to 15kHz) Wien Bridge notch filter.

The 'twin-T' filter

Figure 5.11 shows another type of notch filter, the 'twin-T' circuit. Advantages of this filter are that (unlike the Wien bridge type) its input and output signals share a common 'ground' connection, and its off-frequency attenuation is less than that of the Wien. Its major disadvantage is that, if its tuning is to be made fully variable, the values of all three resistors (or capacitors) must be varied simultaneously. This filter is said to be a 'balanced' type when its components have the precise ratios shown; to obtain perfect nulling, the R/2 resistor value needs careful adjustment. Note in particular that the circuit gives zero phase shift at f_c.

One weakness of the twin-T filter is that (like the Wien type) it has a very low effective 'Q'. Q is defined as being the f_c value divided by the bandwidth between the two –3dB points on the filter's transmission curve, and in this case equals 0.24. What this means in practice is that the filter subjects the second harmonic of f_c to 9dB of attenuation, whereas an ideal notch filter gives it zero attenuation. This weakness can easily be overcome by 'bootstrapping' the common terminal of the filter, as shown in basic form in *Figure 5.12*. This technique enables high effective Q values to be obtained, with negligible attenuation of the second harmonic of f_c.

The action of the balanced twin-T filter is fairly complex, as indicated in *Figure 5.13*. It consists of a parallel-driven low-pass $f_c/2$ and a high-pass $2f_c$ filter, with their outputs joined by an *R–C* voltage divider. This output divider loads the two filters and affects their phase shifts, the consequence being that the signals at points A and B have identical

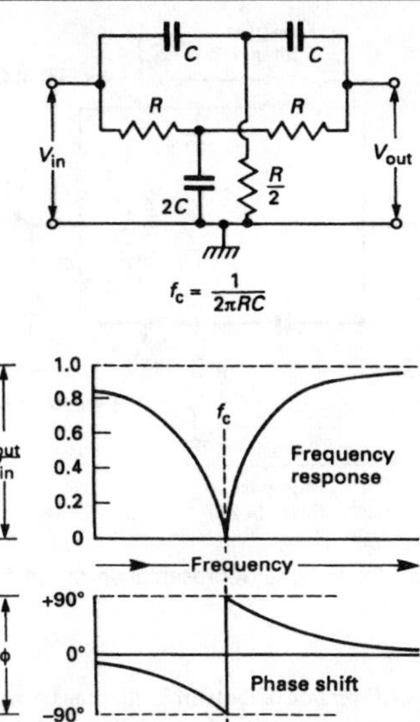

Figure 5.11. Balanced twin-T notch filter.

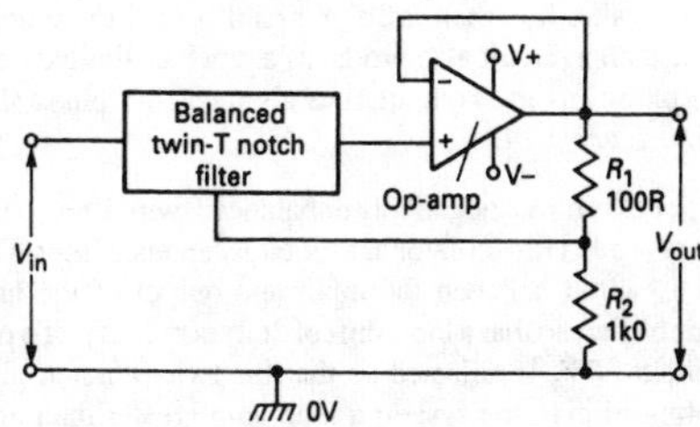

Figure 5.12. Bootstrapped high-Q notch filter.

amplitudes but have phase shifts of –45° and +45° respectively at f_c; simultaneously, the impedances of the *R* and *C* sections of the output divider are identical and give a 45° phase shift at f_c. Consequently, the divider effectively cancels the A and B phase differences under this condition and gives a 'zero' output, this being the phase-cancelled difference between the A and B signals.

Thus, a perfectly balanced twin-T filter gives zero output and zero phase shift at f_c. At frequencies fractionally below f_c the output is dominated by the actions of its low-pass filter, and is phase-shifted by –90°; at frequencies fractionally above f_c the output is dominated by the actions of its high-pass filter, and is phase shifted by +90° (see *Figure 5.11*).

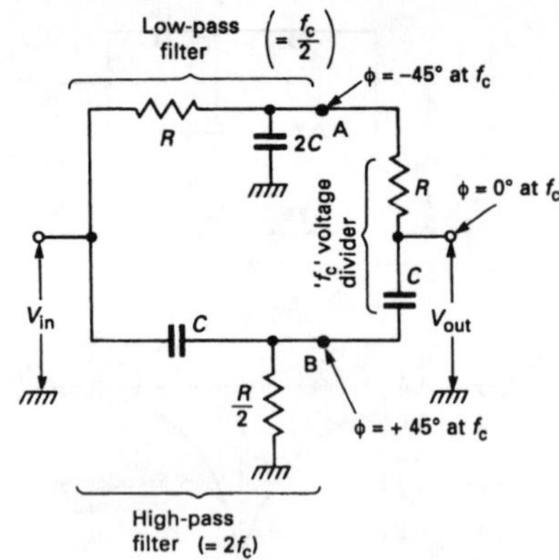

Figure 5.13. Operational representation of the balanced twin-T filter.

An 'unbalanced' version of the twin-T filter can be made by giving the '*R*/2' resistor a value other than the ideal. If this resistor value is greater than *R*/2 the circuit can be said to be positively unbalanced; such a circuit acts in a manner similar to that described, except that its notch has limited depth; it gives zero phase shift at f_c. Alternatively, if the resistor has a value less than *R*/2 the circuit is said to be negatively unbalanced; such a circuit also produces a notch of limited depth, but generates a phase-inverted output, thus giving a 180° phase shift at f_c, as shown in *Figure 5.14*.

Figure 5.15 shows how a negatively unbalanced twin-T notch filter can be used to make a 1kHz oscillator or a tuned acceptance filter. The twin-T network is wired between the input and output of the high-gain inverting amplifier, so that a loop shift of 360° occurs at f_c. To make the circuit oscillate, RV_1 is adjusted so that the twin-T notch gives just enough output to give the system a loop gain greater than unity; the circuit generates an excellent sinewave under this condition. To make the circuit act as a tone filter, RV_1 is adjusted to give a loop gain less than unity, and the audio input signal is fed in via C_1 and R_1; R_1 and the twin-T filter then interact to form a frequency-sensitive circuit that gives heavy negative feedback and low gain to all frequencies except f_c, to which it gives little negative feedback and high gain; the tuning sharpness is variable via RV_1.

C–R component selection

Single-stage *C–R* low-pass and high-pass filters and balanced Wien and twin-T networks all use the same formula to relate the f_c value to that of *R* and *C*, i.e., $f_c = 1/(2\pi RC)$. *Figure 5.16* shows this formula transformed to enable the values of *R* or *C* to be determined when f_c and one component value is known. When using these formulae it is often easiest to work in terms of kHz, kΩ, and μF, as indicated.

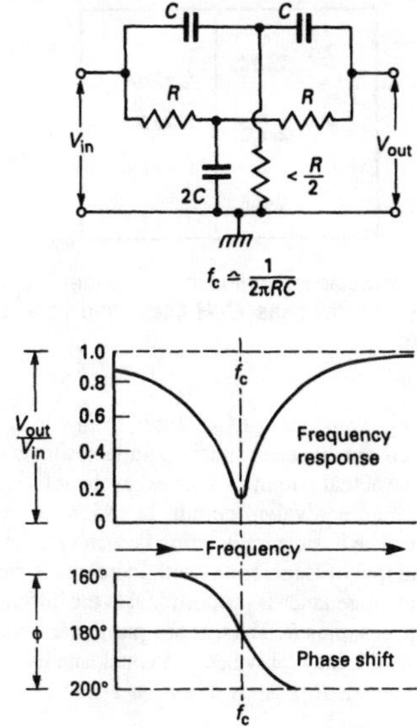

Figure 5.14. Negatively unbalanced twin-T filter gives 180° phase shift at f_c.

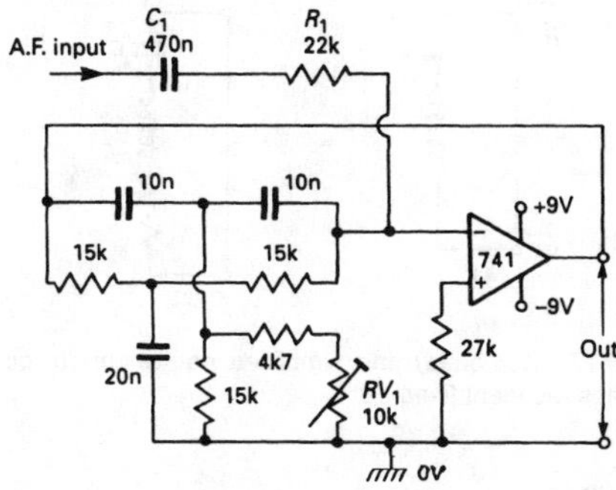

Figure 5.15. 1kHz oscillator/acceptance-filter using negatively unbalanced twin-T network.

Resonant L–C filters

L–C filters are used primarily, but not exclusively, in high-frequency applications. Like *C–R* types, they can easily be designed to give low-pass, high-pass, band-pass, or 'notch' filtering action, but have the great advantage of offering at least 12dB per octave of roll-off, compared to the 6dB/octave of *C–R* types.

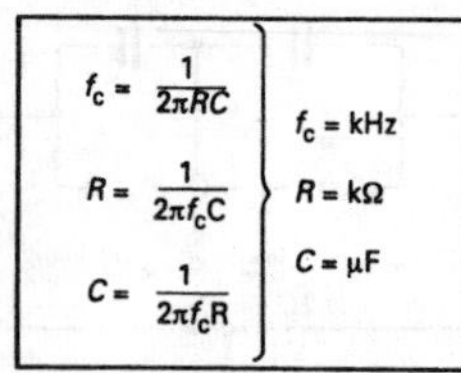

Figure 5.16. Formulae for finding the component values of single-stage high-pass or low-pass *C–R* filters and balanced Wien or twin-T networks.

The two most important types of L–C filter, from which all others are ultimately derived, are the series and the parallel resonant types. *Figure 5.17(a)* shows the actual circuit of a series resonant filter, and *5.17(b)* shows its simplified equivalent circuit, in which '*R*' represents the coil's resistance. The basic circuit action is such that the reactance of *C* decreases, and that of L increases, with increases in frequency, and the circuit's input impedance is proportional to the difference between these two reactances, plus *R*. Thus, at one particular frequency, f_c, the reactances of *C* and L inevitably become equal, and the circuit's input impedance then equals *R*, as shown in *Figure 5.18(a)*.

Suppose that f_c occurs when the reactances of *C* and L are each 1000Ω, and that *R* equals 10Ω. In this case the input impedance falls to 10Ω, and the entire signal voltage is generated across *R*. *R*'s signal currents,

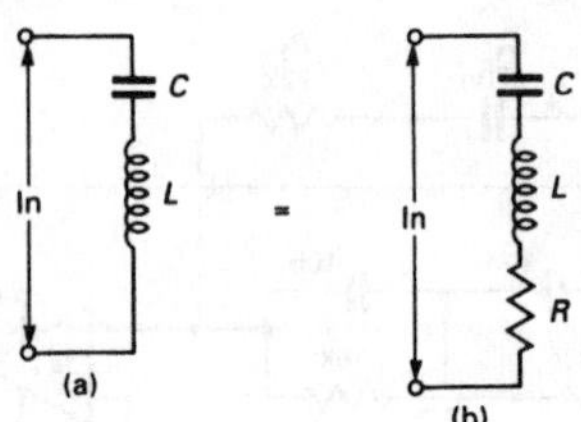

Figure 5.17. Actual *(a)* and simplified equivalent *(b)* circuit of L–C series resonant filter.

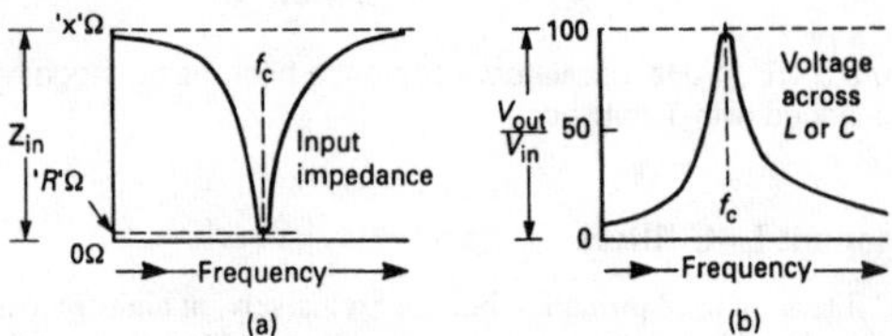

Figure 5.18. Graphs showing how *(a)* the input impedance and *(b)* the L or C signal voltage of the series resonant filter varies with frequency.

however, flow via C and L, which each have a reactance 100 times greater than R; consequently the signal voltage generated across C and across L is 100 times greater than the actual input signal voltage, as shown in *Figure 5.18(b)*; this voltage magnification is known as the circuit's 'Q'. The f_c impedance of L (or C) is known as the circuit's 'characteristic impedance, Z_o', and equals $\sqrt{(L/C)}$.

Figure 5.19 shows two basic ways of using a series resonant L-C filter. In *(a)* the 2k2 resistor (R_x) and the filter act together as a frequency-selective attenuator that gives very high attenuation at f_c, and low attenuation at all other frequencies, i.e., the circuit acts as a notch rejector. In *(b)* the input signal is applied directly to the filter, and the output is taken from across L; this circuit thus acts as a notch acceptor that gives high gain at f_c and low gain at all other frequencies.

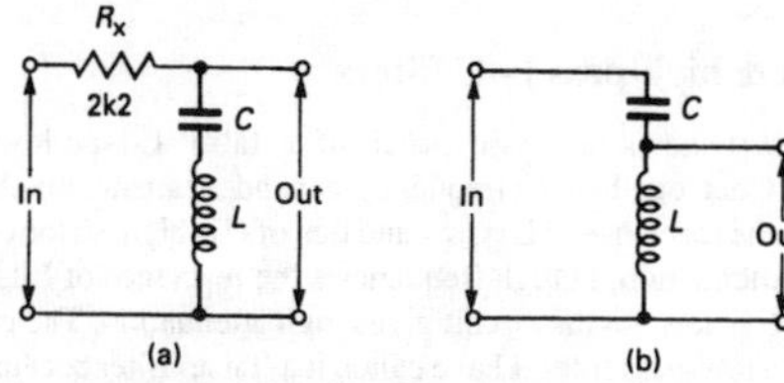

Figure 5.19. Ways of using a series resonant filter as *(a)* a notch rejector and *(b)* a notch acceptor.

Figure 5.20 lists the major formulae that apply to the *Figure 5.17* series resonant circuit, and also to all other types of L–C filter described in this chapter.

Figures 5.21(a) and *5.21(b)* show the actual and the simplified equivalent circuits of a parallel resonant filter; 'R' represents the coil's resistance. The basic circuit action is such that C's reactance decreases, and that of L increases, with increases in frequency; L and C each draw a signal current proportionate to reactance, but the two currents are in antiphase, and the total signal current thus equals the difference between the two. At f_c the L and C reactance are equal, and the total signal current falls to near-zero, and the filter thus acts as a near-infinite impedance under this condition. In reality the presence of 'R' modifies the action slightly, and reduces the fc impedance to Z_o^2/R. Thus, if Z_o equals 1000Ω and R equals 10Ω, the actual f_c impedance is 100k. *Figure 5.21(c)* shows how the input impedance varies with frequency. All the formulae of *Figure 5.20* apply to the parallel resonant filter.

$$f_c = \frac{1}{2\pi\sqrt{LC}} \qquad Z_o = \sqrt{\frac{L}{C}}$$

$$L = \frac{Z_o}{2\pi f_c} \qquad C = \frac{1}{2\pi f_c Zo}$$

$$Q = \frac{X_L}{R} = \frac{Z_o}{R}$$

Figure 5.20. Basic design formulae for all the L–C filters shown in this chapter.

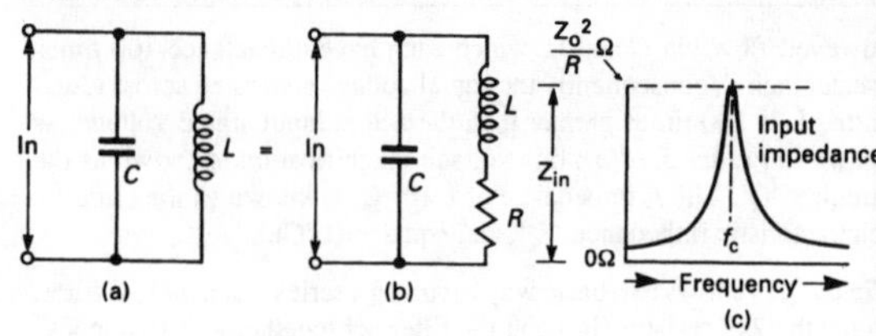

Figure 5.21. Actual *(a)* and equivalent *(b)* circuits of the parallel resonant filter, together with a graph *(c)* showing how its input impedance varies with frequency.

Low-pass & high-pass L–C filters

Figure 5.22(a) shows the basic circuit of a 'false' L-type low-pass filter. L and *C* act together as a frequency-dependent attenuator. At low frequencies the reactance of L is low and that of *C* is high, so the circuit gives little attenuation; at high frequencies the reactance of L is high and that of *C* is low, so the circuit gives high attenuation. The circuit thus acts as a low-pass filter. I have called it a 'false' filter because the circuit will only work correctly if it is driven from a source impedance equal to Z_o, but there is no indication of this fact in the diagram. The circuit is actually a series resonant filter (like *Figure 5.17*) with its output taken from across *C*. If the circuit is driven from a low-impedance source the output will consequently produce a huge signal peak at f_c, as shown in *Figure 5.22(b)*. The magnitude of this peak is proportional to the circuit's Q value.

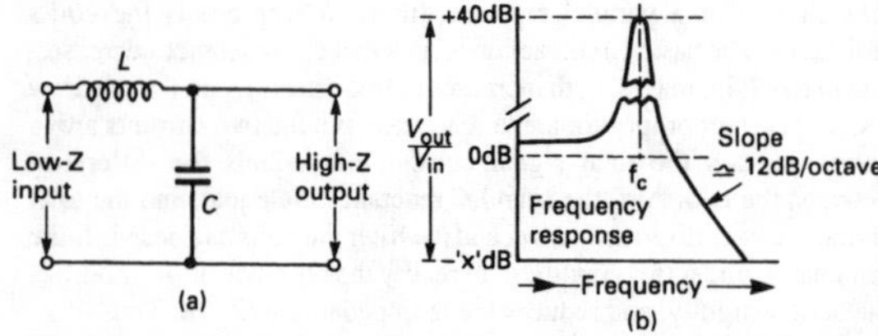

Figure 5.22. Circuit *(a)* and performance graph *(b)* of a false L-type low-pass filter.

Figure 5.23(a) shows how the above circuit can be modified to act as a genuine L-type low-pass filter, by wiring R_x in series with the circuit's input, so that the sum of R_x and R_s (the input signal's source impedance) and *R* (the resistance of L) equals the circuit's characteristic impedance, Z_o. The addition of this resistance reduces the circuit's Q to precisely unity, and the low-pass filter consequently generates the clean output shape shown in *Figure 5.23(b)*.

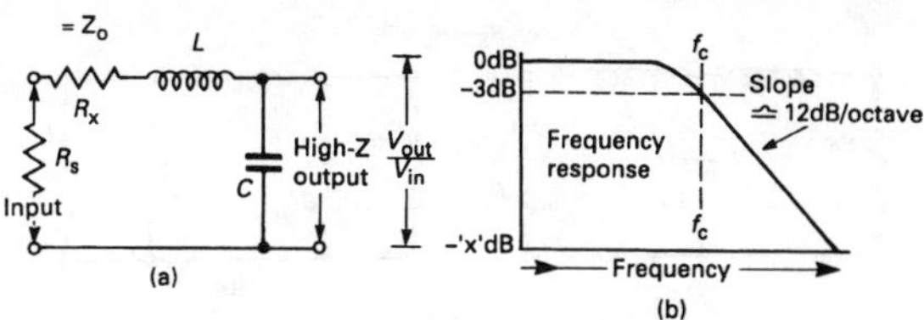

Figure 5.23. Circuit *(a)* and performance graph *(b)* of a genuine L-type low-pass filter.

Figure 5.24 shows how the above principle can be used to make a good L-type high-pass filter, with the output taken from across the inductor. Note in both circuits that R_x can be reduced to zero if the filter's Z_0 value is designed to match R_s, as shown in the design formulae of *Figure 5.20*, and that the outputs of these filters must (like those of the series and parallel resonant types) feed into high-impedance loads.

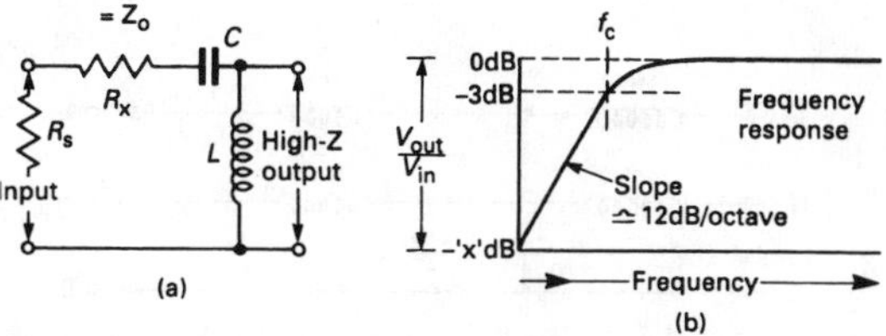

Figure 5.24. Circuit *(a)* and performance graph *(b)* of an L-type high-pass filter.

The most widely used types of low-pass and high-pass L–C filters are balanced, matched impedance types that are designed to be driven from, and have their output loaded by, a specific impedance value. Such filters can readily be cascaded, to give very high levels of signal rejection. Amongst the most popular of these filters are the T-section and π-section low-pass types shown in *Figure 5.25*, and the T-section and π-section high-pass types shown in *Figure 5.26*. Note that all of these types give an output roll-off of about 12db/octave (= 40dB/decade), and must have their outputs correctly loaded by a matching filter section or terminating load; their design formulae is complemented by *Figure 5.20*.

One useful application of the T-section low-pass filter is as a power-line filter, as shown in *Figure 5.27*. This prevents line-borne interference from reaching sensitive equipment, or equipment-generated hash from reaching the power line; it gives a useful performance up to about 25MHz.

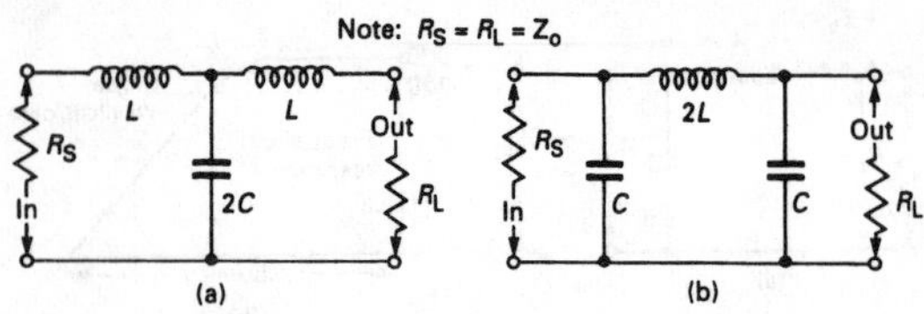

Figure 5.25. *(a)* T-section and *(b)* π-section low-pass filters (see *Figure 5.20* for design formulae).

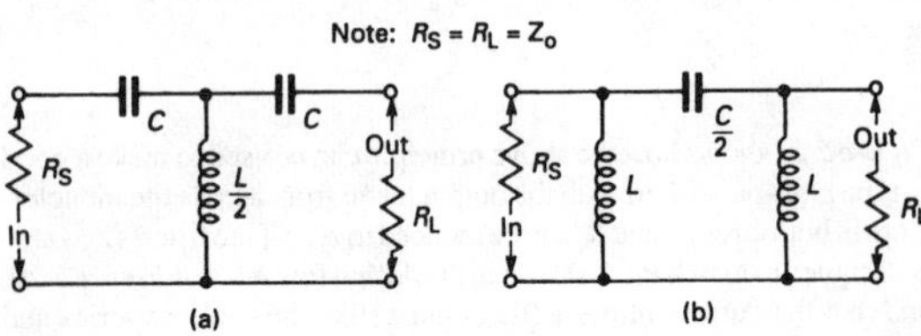

Figure 5.26. *(a)* T-section and *(b)* π-section high-pass filters (see *Figure 5.20* for design formulae).

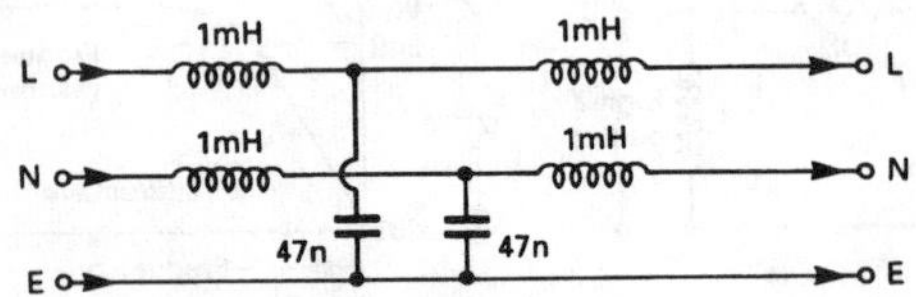

Figure 5.27. T-section mains (power line) input filter rejects line-borne interference up to about 25MHz.

C–R and R–L time constants

Before leaving the subject of passive *C–R* (and R–L) networks, brief mention must be made of their 'time constants', and associated matters.

If a resistor and a discharged capacitor are connected into the circuit of *Figure 5.28(a)*, and S_1 is then switched to 'A', *C* will start to charge up via *R*. Initially, *C*'s charge current equals Vs/*R*; if this current value was maintained the capacitor would charge to the full Vs value in a time, 't', equal to the *R–C* product, as indicated by the dotted lines in the graph of *(b)*, but in practice this current falls off at an exponential rate, and the capacitor's voltage thus rises exponentially, as shown by the solid lines in *(b)*, and reaches 63.2 percent of the Vs value in time 't'. Theoretically, *C*'s voltage never rises to the full Vs value, but in practice it reaches 'virtual' full-charge after about 6 x t. If this fully-charged capacitor is discharged by switching S_1 to position 'B', its voltage falls exponentially as shown in *(c)*, and losses 63.2 percent of its voltage in time 't'. Time 't' is known as the circuit's time constant, and equals the product of the *C* and *R* values; *C*, *R*, and t are most conveniently measured in terms of μF, kΩ, and ms, as shown in *(d)*.

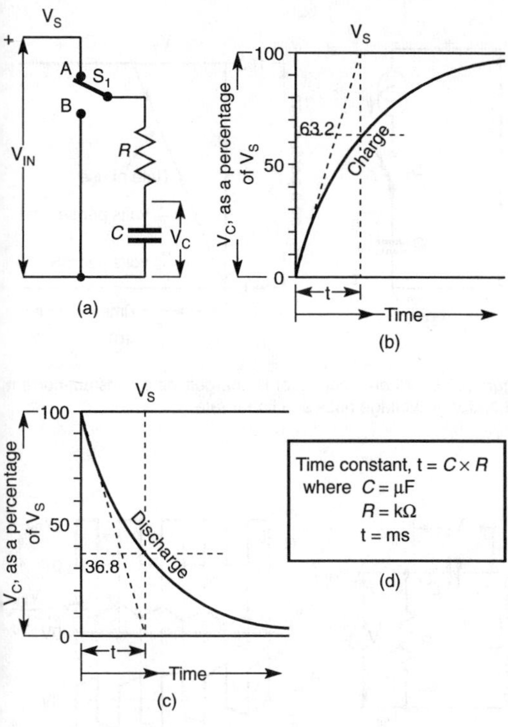

Figure 5.28. Basic *C–R* time constant circuit *(a)*, and its charge *(b)* and discharge *(c)* waveforms and *(d)* formula.

A capacitor can be charged linearly by connecting it to a constant-current generator, as shown in *Figure 5.29*. In this case the capacitor's voltage rises linearly, at a rate of I/*C* volts per second; in practice, these values are most conveniently measured in terms of mA/µF volts per ms, where 'mA' is the charge current value, and µF is the capacitor value.

Variations of the *Figures 5.28* and *5.29* circuits can be used in a variety of component-measurement and waveform generator applications. One common application is as a *C–R* signal coupler, as shown in *Figure 5.30(a)*; if this circuit is fed with a symmetrical square wave with a period equal to the network's 't' value, the waveforms of *(b)* are generated. Note that after a few input cycles '*C*' charges to a mean value of Vpp/2, and the output waveform (which is slightly distorted) then swings symmetrically about the zero-volt line.

If the 't' value of the above circuit is very long relative to the input waveform's period, the output waveform will be undistorted, as shown in *Figure 5.31(a)*, but if 't' is very short the waveform will be severely

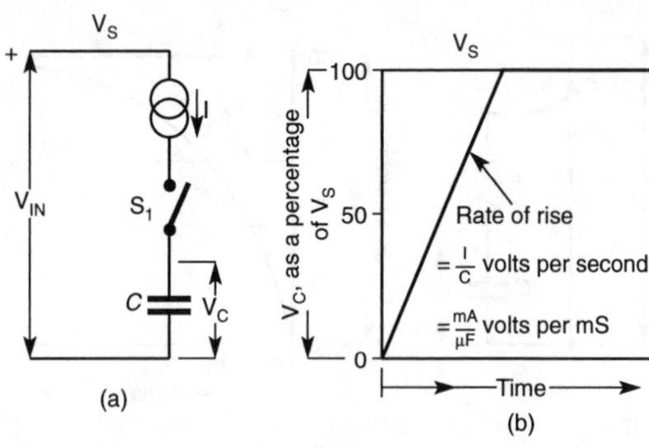

Figure 5.29. When a capacitor is charged via a constant-current generator, its voltage rises at a linear rate.

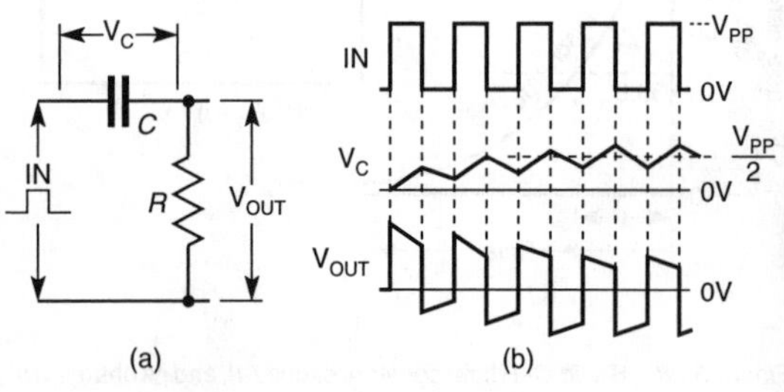

Figure 5.30. Waveforms of a *C–R* coupler circuit fed with a square wave input with a period equal to *C* x *R*.

differentiated, as in *(b)*. If 't' is long, but the input waveform is asymmetrical as shown in *(c)*, the output waveform will not swing symmetrically about the zero volts line, the offset being proportional to the waveform's M/S-ratio.

If an R–L network is connected into the circuit of *Figure 5.32*, and S_1 is set to 'A', a current flows in L and starts at a low value and then rises exponentially and reaches 63.2 percent of its final value in a time, 't', of L/R seconds. The resulting 'current' graph is identical to the voltage graph of *Figure 5.28(b)*, and the voltage graph is as shown in the first half of *Figure 5.32(b)*; it starts at a high value and then decays exponentially to zero. If S_1 is then moved to 'B', L's stored energy leaks away via *R*, and L generates a negative voltage that decays exponentially, as shown.

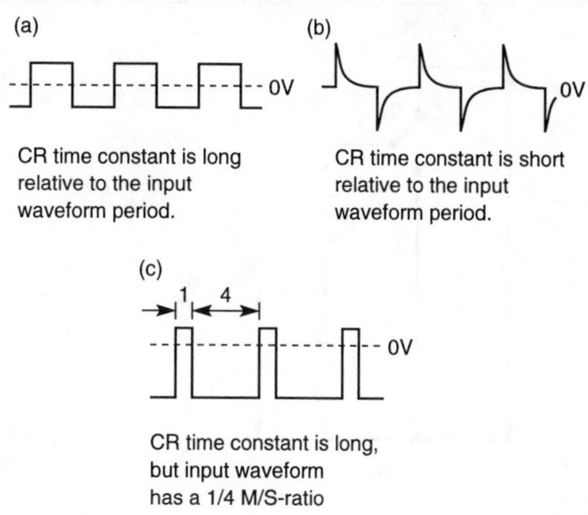

Figure 5.31. Output waveforms of a *C–R* coupler when fed with various types of square wave input.

C–R active filters

An active filter is a circuit that combines passive *C–R* networks and one or more amplifier or op-amp stages, to form a filter that can either outperform normal *C–R* filters or can give a performance that is unobtainable from purely passive networks. A good selection of practical active filters, including multi-order high-pass and low-pass types, have already been described in Volume 1 (Linear ICs) of the *Newnes Electronics Circuits Pocket Book* series, and these will not be repeated here. Volume 1 did not, however, deal with active tone and notch filters, and a selection of these are shown in *Figures 5.33* to *5.37*.

Active tone and notch filters

Excellent active *C–R* tone filters, with very high effective 'Q' values, can be made by using twin-T or Wien networks in the feedback loops of suitable op-amp circuits. A 1kHz twin-T design has already been described in *Figure 5.15*. *Figure 5.33* shows a 1kHz Wien bridge based tone or 'acceptor' filter; the circuit's Q is adjustable via 'R_2', and the circuit becomes an oscillator if R_2 is reduced to less than twice the R_1 value.

The basic twin-T notch filter has a very low Q. The filter's Q, and thus the notch 'sharpness', can be greatly increased by incorporating the twin-T in the feedback network of an active filter. There are two standard ways of doing this. The first way is to use the shunt feedback technique shown in *Figure 5.34*, in which the input signal is fed to the twin-T via R_1, and an amplified and inverted version of the filter's output is fed back to the filter's input via R_2, which has the same value

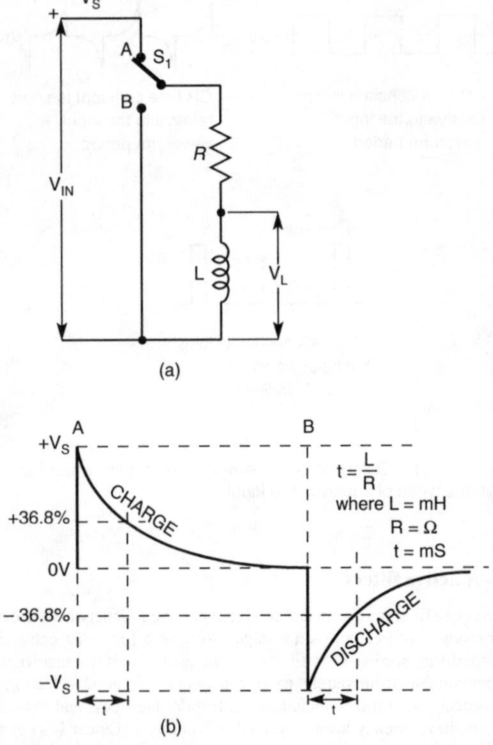

Figure 5.32. Basic R–L time constant circuit *(a)* and its charge and discharge voltage waveforms *(b)*.

as R_1. *Figure 5.35* shows the practical circuit of a 1kHz version of this type of active filter. The network's null point can be adjusted via the 1k0 variable resistor.

The second (and more modern) Q-boosting method is the bootstrapping technique, which has already been described and shown in basic form in *Figure 5.12*. *Figure 5.36* shows a practical 1kHz variable-Q version of the circuit. The twin-T's output is buffered by the upper op-amp voltage follower, and part of the buffered output is tapped off via RV_3 and fed to the bottom of the twin-T (as a bootstrap signal) via the lower op-amp voltage follower. When RV_3's slider is set to the lowest point the circuit acts like a standard twin-T filter with a Q of 0.24. When RV_3's slider is set to the highest point the network has heavy bootstrapping, and the filter has an effective Q of about 8 and provides a very sharp notch. The filter's centre-frequency can be trimmed slightly via RV_1, and the null point can be adjusted via RV_2, which should be a multi-turn type.

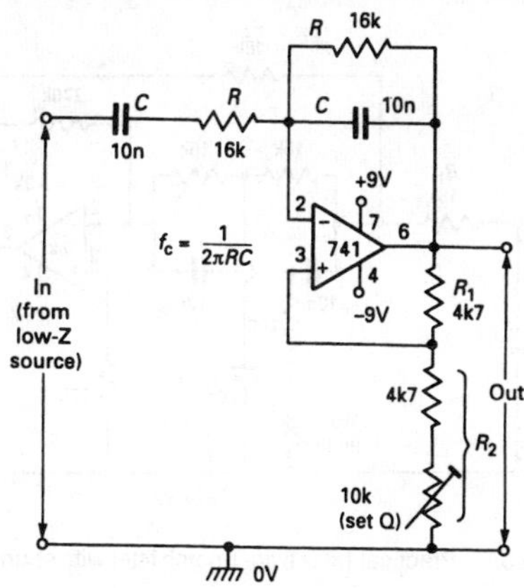

Figure 5.33. Wien bridge based 1kHz high-Q tone filter.

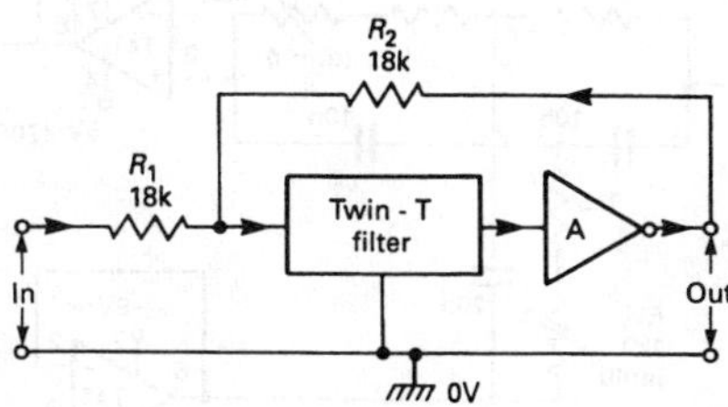

Figure 5.34. Basic twin-T notch filter using shunt feedback.

A THD (distortion) meter

The bootstrapped twin-T notch filter can be used as the basis of an excellent THD or 'distortion' meter. Here, the filter's notch is tuned to the basic frequency of the input test signal, and totally rejects the fundamental frequency of the signal but gives zero attenuation to the signal's unwanted harmonics and mush, which appear at the filter's output; the output signals must be read on a true r.m.s. volt or millivoltmeter. Thus, if the original input signal has a r.m.s. amplitude of 1000mV, and the nulled output has an amplitude of 15mV, the THD (total harmonic distortion) value works out at 1.5%.

Figure 5.37 shows a practical high-performance 1kHz THD meter. This filter's Q is set at a value of 5 via the 820R–10k divider, to give the benefits of easy tuning combined and near-zero second harmonic (2kHz) signal attenuation. The input signal to the filter is variable via RV_3, and the filter's tuning and nulling are variable via RV_1 and RV_2

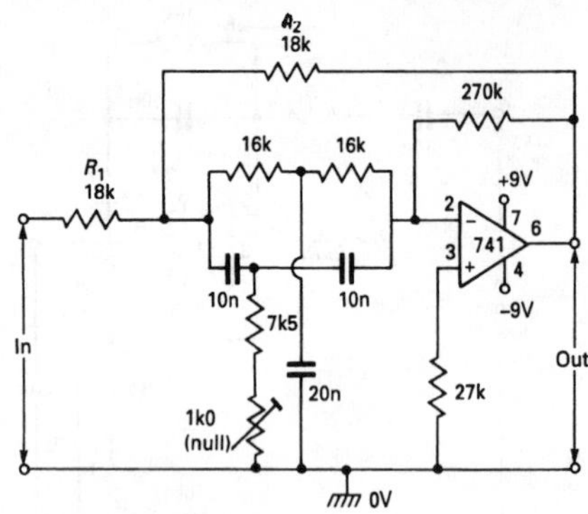

Figure 5.35. Practical 1kHz twin-T notch filter with shunt feedback.

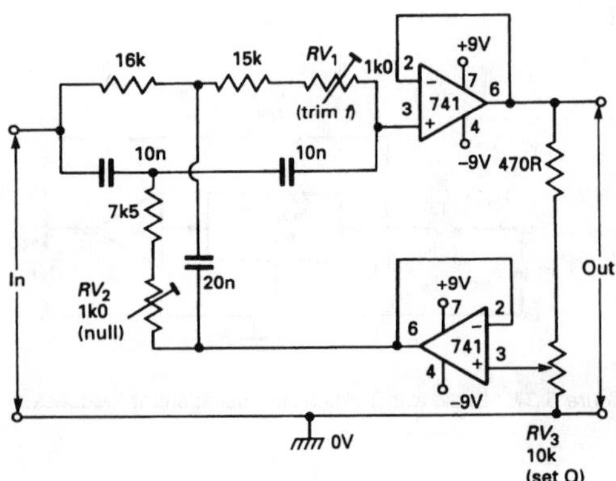

Figure 5.36. 1kHz variable-Q bootstrapped twin-T notch filter.

respectively. SW_1 enables either the filter's input or its distorted output to be fed to an external true r.m.s. meter; note that the meter feed line incorporates a 10kHz low-pass filter, to help reject unwanted 'noise' signals.

To use the *Figure 5.37* THD meter, first set SW_1 to the input position connect the 1kHz input test signal, and adjust RV_3 to set a convenient (say 1 volt) reference level on the true r.m.s. meter. Next, set SW_1 to the distort position, adjust the input frequency for an approximate null then trim RV_1 and RV_2 alternately until the best possible null is obtained. Finally, read the nulled voltage value on the meter and calculate the distortion factor on the basis of:

$$\text{THD (in percent)} = (V_{DIST} \times 100)/V_{IN}.$$

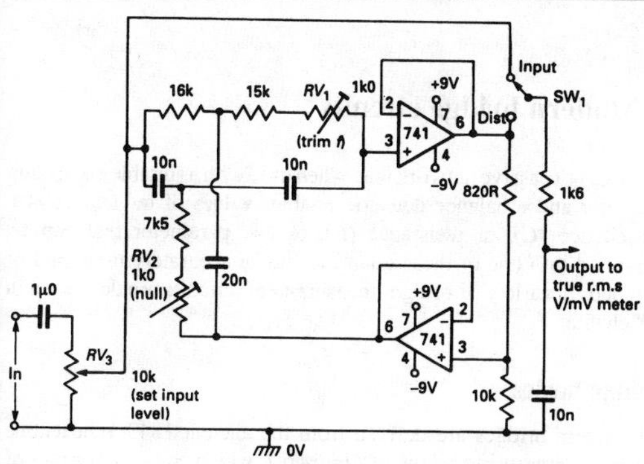

Figure 5.37. 1kHz THD (distortion) meter circuit.

6 Modern bridge circuits

A bridge is a passive network that, when used with a suitable energising generator and a balance detector, enables values of inductance (L), capacitance *(C)*, or resistance *(R)*, or any parameter that can be converted into one of these quantities, to be accurately measured or matched. A variety of bridge 'measurement' circuits are described in this chapter.

Bridge basics

All modern bridges are derived from the ancient 1843 Wheatstone resistance-measuring bridge of *Figure 6.1*, which consists of a pair of d.c.-energised potential dividers (R_2/R_1 and R_y/R_x) with a sensitive meter wired between them. R_2/R_1 have a 1:1 division ratio, and form the bridge's 'ratio arms'; R_x is the 'unknown' resistor, and R_y is a calibrated variable resistor. In use, R_y is adjusted to give a zero or 'null' reading on the meter, at which point R_2/R_1 and R_y/R_x are giving equal output voltages and the bridge is said to be 'balanced' or 'nulled'; under this condition the ratio R_y/R_x equals 1:1, and R_x equals R_y; the bridge's balance is not influenced by variations in energising voltage.

If this bridge is energised from 10V d.c., 5V is developed across all resistors at balance, and a resistance shift of a mere 0.1% will thus give a 5mV reading on the meter; the bridge thus has a very high *null sensitivity*. In practice, this circuit may, when using a fairly simple null-detecting d.c. amplifier, have a null sensitivity factor (i.e., percentage out-of-balance detection value) of about 0.003%. The bridge's major disadvantage is that R_y needs a vast range of values if it is to balance all possible values of R_x. In 1848 Siemens overcame this snag with the *Figure 6.2* modification, in which the R_2/R_1 ratio can be any desired decade multiple or sub-multiple of unity. The following basic truths apply to this circuit:

(i) At balance, $R_2/R_1 = R_y/R_x$.
(ii) At balance, $R_1 \text{ x } R_y = R_2 \text{ x } R_x$.
(iii) At balance, $R_x = Ry(R1/R2)$.

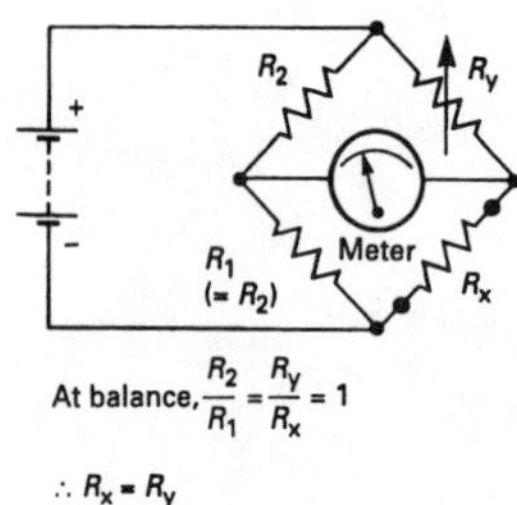

Figure 6.1. Original (1843) version of the basic Wheatstone bridge.

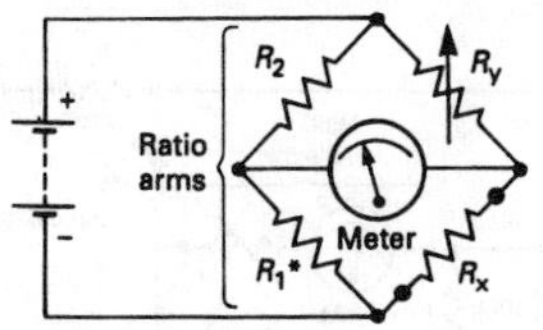

At balance, $\frac{R_2}{R_1} = \frac{R_y}{R_x}$, and $R_1 \times R_y = R_2 \times R_x$

$\therefore R_x = R_y\left(\frac{R_1}{R_2}\right)$

***Note: R_1 is a decade multiple or sub-multiple of R_2**

Figure 6.2. Conventional version of the Wheatstone bridge.

Note from (iii) that, at balance, the R_x value equals R_y multiplied by the R_1/R_2 ratio; one way to vary this ratio is to give R_2 a fixed value and make R_1 switch-selectable, as in the six-range d.c. Wheatstone bridge circuit of *Figure 6.3*, which is based on one used in a high quality '1970 style' laboratory instrument.

The *Figure 6.3* circuit can measure d.c. resistances up to 1 megohm. R_y is a calibrated 10k variable, R_M controls the sensitivity of the balance-detecting centre-zero meter, and R_L limits the bridge current to a few mA. A major weakness of this 1970's bridge is that its null sensitivity degenerates in proportion to the R_1/R_2 ratio's divergence from unity. Thus, the sensitivity is nominally 0.003% on the 10k range, where the R_1/R_2 ratio is 1/1, but degenerates to 0.3% on the 100R and 1M0 ranges, where the R_1/R_2 ratios are 1/100 and 100/1 respectively.

To be of real value the *Figure 6.3* circuit must be used with a sensitive null-balance detector. *Figure 6.4* shows a x10 d.c. differential amplifier that can be used with an external analogue meter to make such a detector, and must have its own 9V battery supply. The external meter can be set to its 2.5V d.c. range for low-sensitivity measurements, or to its 50µA or 100µA range for high-sensitivity ones; in the latter case the circuit must, before use, first be balanced by shorting its input terminals together and trimming the multi-turn set balance control for a 'zero' meter reading.

Wheatstone bridge variations

The *Figure 6.2* Wheatstone bridge circuit can be arranged in three other ways without invalidating its three basic 'balance' truths, as shown in *Figure 6.5*; in each case R_1/R_2 form the 'ratio arms'. The most useful bridge variation is that of *Figure 6.5(a)*, and *Figure 6.6* shows a modern six-range version in which the balance sensitivity (which is proportional to the R_y/R_2 ratio at balance) is very high on all ranges, and varies from 0.003% at R_y's full scale balance value, to 0.03% at one tenth of full scale, and so on. Thus, this circuit can, by confining all measurements to the top 9/10ths of the R_y range, measure all values from 1Ω to 1 megohm with 0.003% to 0.03% sensitivity.

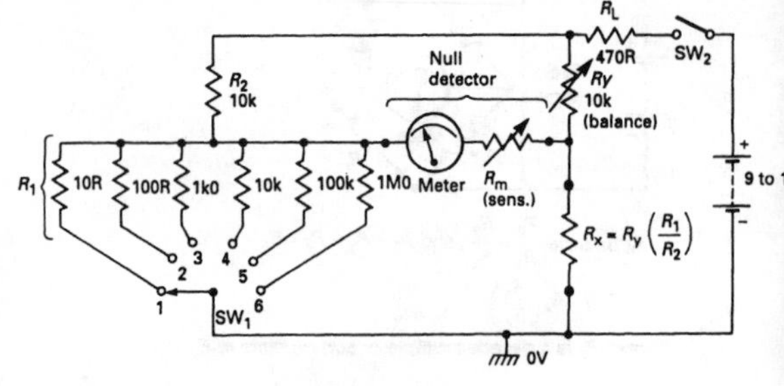

SW_1 range	Bridge range	R_1 value	R_1/R_2 ratio	Bridge null sensitivity (nom.)
1	0 – 10R	10R	1/1000	3.0%
2	0 – 100R	100R	1/100	0.3%
3	0 – 1k0	1k0	1/10	0.03%
4	0 – 10k	10k	1/1	0.003%
5	0 – 100k	100k	10/1	0.03%
6	0 – 1M0	1M0	100/1	0.3%

Figure 6.3. Circuit and tabulated details of a conventional Wheatstone version of a six-range d.c. resistance-measuring bridge.

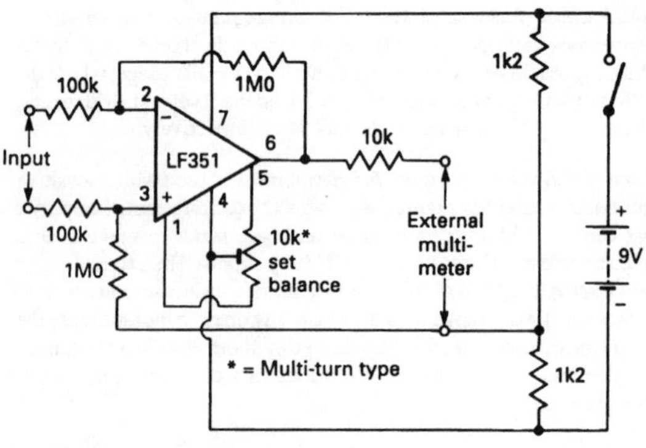

Figure 6.4. D.c. null-point amplifier, for use with an external multimeter.

A Wheatstone bridge can be energised from either an a.c. or d.c. source, without upsetting its balance truths. *Figure 6.7* shows an a.c.-energised version in which the balance condition is obtained via an infinitely-variable pair of 'ratio' arms made up by RV_1, and balance sensitivity is so high that balance-detection can be made via a pair of earphones.

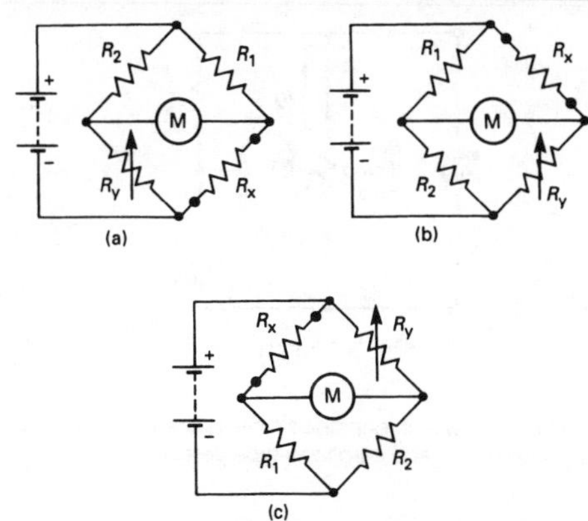

Figure 6.5. Each of these three alternative versions of the Wheatstone bridge has the same balance formulae as the *Figure 6.2* circuit.

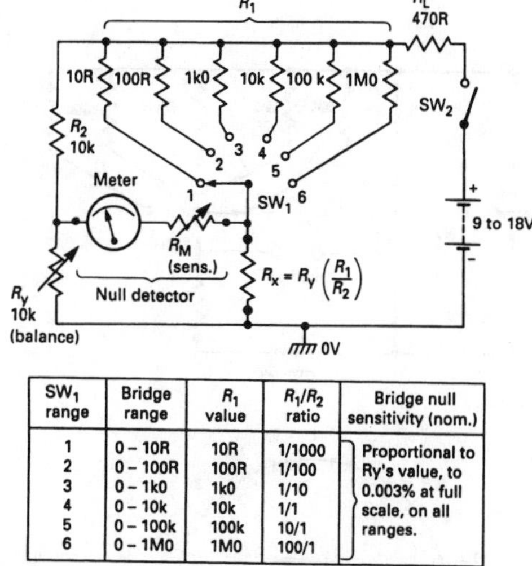

SW1 range	Bridge range	R1 value	R1/R2 ratio	Bridge null sensitivity (nom.)
1	0 – 10R	10R	1/1000	Proportional to Ry's value, to 0.003% at full scale, on all ranges.
2	0 – 100R	100R	1/100	
3	0 – 1k0	1k0	1/10	
4	0 – 10k	10k	1/1	
5	0 – 100k	100k	10/1	
6	0 – 1M0	1M0	100/1	

Figure 6.6. Circuit and tabulated details of a high-sensitivity Wheatstone version of a six-range d.c. resistance-measuring bridge.

Figure 6.8 shows a five-range version of this circuit that spans 10Ω to 10MΩ with good precision. The RV_1 'ratio' equals unity when RV_1's slider is at mid-range; the diagram shows the typical scale markings of this control, which must be hand-calibrated on test. To use the bridge, connect it to a 1kHz sinewave source, fix R_x in place, and adjust SW_1

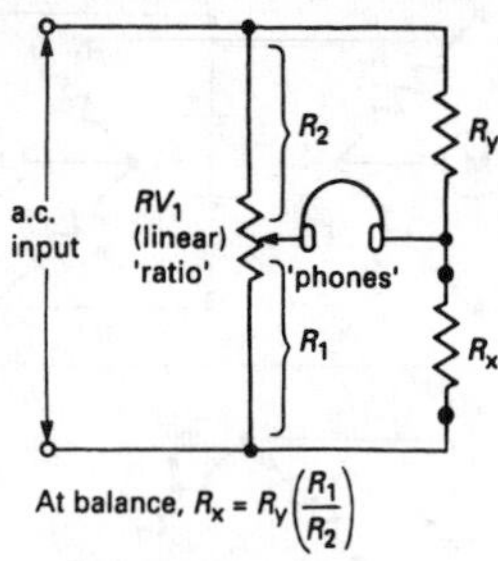

Figure 6.7. Basic a.c.-energised Wheatstone bridge with variable-ratio-arm balancing and earphone-type detection.

and RV_1 until a null is detected on the earphones, at which point R_x equals the SW_1 resistor value multiplied by the RV_1 scale value. In practice, a balance is available on any range, but to get the best precision should occur with a RV_1 scale reading between roughly 0.27 and 3.0.

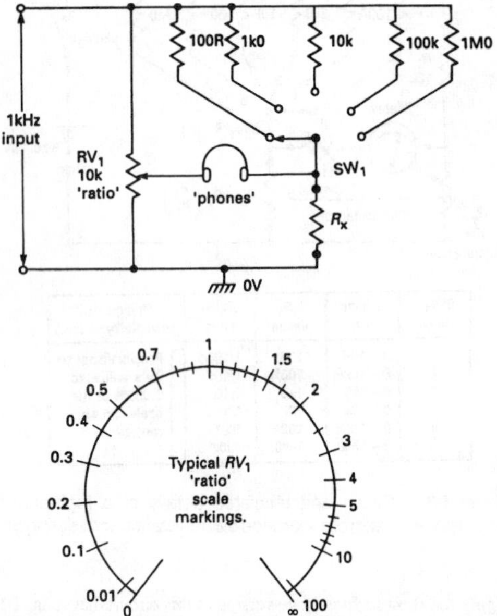

Figure 6.8. Five-range resistance bridge, with typical RV_1 scale markings.

To calibrate the RV_1 scale, fit an accurate 10k resistor in the R_x position and move SW_1 progressively through its 100R, 1k0, 10k, 100k, and 1M0 positions and mark the scale at each sequential balance point as 0.01, 0.1, 1 (mid-scale), 10, and 100. Repeat this process using R_x values that are decade multiples or sub-multiples of 1.5, 2, 3, 4, 5, and so on, until the scale is adequately calibrated, as in the diagram.

Resolution and precision

The three most important features of a bridge are (apart from its measurement range) its *balance-sensitivity* (which has already been described), its *resolution*, and its *precision*. The term *resolution* relates to the sharpness with which the R_x value can be read off on the bridge's controls. Thus, in *Figures 6.3* and *6.6*, R_y gives a resolution of about ±1% of full-scale if it is a hand-calibrated linearly-variable resistor, or of ±0.005% of full-scale if it takes the form of a 4-decade '*R*' box. The *Figure 6.8* circuit's resolution varies from ±1% at a 1 ratio, to ±2% at a 0.3 or 3.0 ratio, to ±5% at a 0.1 or 10 ratio, and so on.

The term *precision* relates to the intrinsic accuracy of the bridge, assuming that it has perfect sensitivity and resolution, and equals the sum of the R_1/R_2 ratio tolerance and the tolerance of the resistance standard (R_y). If the R_1/R_2 ratio is set by using precision resistors, the ratio's precision equals the sum of the R_1 and R_2 tolerances; note, however, that it is possible, using techniques described later in this chapter, to reduce ratio errors to below ±0.01%.

The overall quality of a bridge depends on its balance-sensitivity, its resolution, and its precision. Thus, the *Figure 6.6* circuit has excellent sensitivity and potentially good resolution and precision, so can be used as the basis of a precision laboratory instrument, but the *Figure 6.8* design has intrinsically poor resolution and precision, and can thus only be used as the basis of a cheap-and-simple 'service' instrument.

Service-type *C* and L bridges

The a.c.-energised Wheatstone bridge of *Figure 6.7* can measure reactance as well as resistance, and *Figure 6.9* shows how it can be used to measure *C* or L values by replacing R_x and R_y with reactances of like

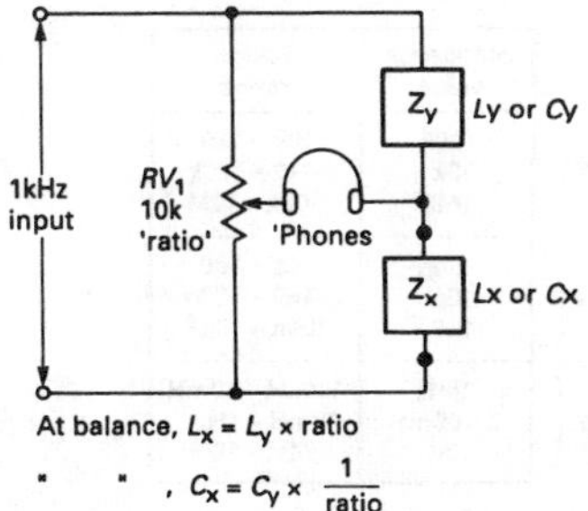

Figure 6.9. A Wheatstone bridge can be used to balance both capacitive and inductive reactances.

types, provided that C_x or L_x have impedances in the range 1Ω to 10 megohms at 1kHz. The problems with trying to measure inductance using this circuit are that accurate inductors (for use in the Z_y position) are hard to get, and that inductive impedances are only 6.28Ω per millihenry at 1kHz. The only capacitance measuring problem is that the C_x value is proportional to the reciprocal of the RV_1 'R' scale markings, so two calibrated sets of RV_1 scales are needed. This snag can be overcome by fitting RV_1 with a reversing switch, as shown in the L–C–R 'service'-type bridge of Figure 6.*10*, so that only a single *Figure 6.8* type scale is needed.

The *Figure 6.10* circuit is quite versatile; SW_2 enables it to use either internal or external L, *C*, or *R* standards. The mid-scale value of each range is equal to the value of standard used on that range. The circuit can be built exactly as shown, or can be built as a self-contained instrument with integral oscillator and detector circuits. In the latter

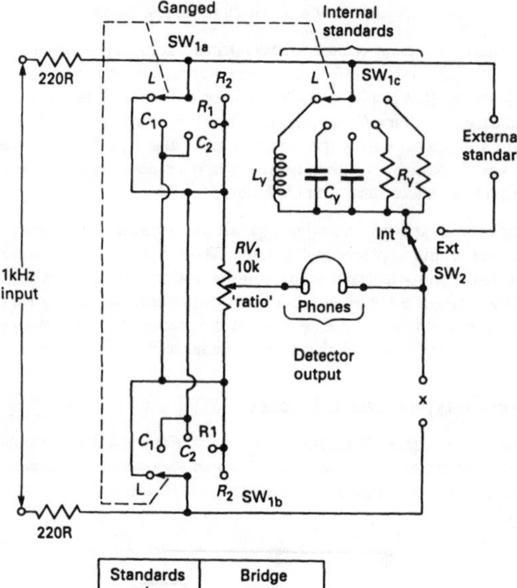

	Standards value	Bridge range
Ry	100R	10R – 1k0
	10k	1k0 – 100k
	1M0	100k – 10M
Cy	100pF	10p – 1n0
	10n	1n0 – 100n
	1μF	100n – 10μF
Ly	1mH	100μH – 10mH
	100mH	10mH – 1H
	10H	1H – 100H

Figure 6.10. Service type L–C–R bridge with earphone-type detector.

case note that, since a bridge can not share common input and output terminals, the oscillator must be effectively 'floating' relative to the balance detector circuitry. The designer has two basic options in this respect, as shown in *Figure 6.11*. One option is to power both circuits from the same supply but to isolate the oscillator by transformer-coupling its output to the bridge, as in *(a)*; the other is to power the oscillator from its own floating supply, as in *(b)*. This second option is highly efficient, and is generally to be preferred.

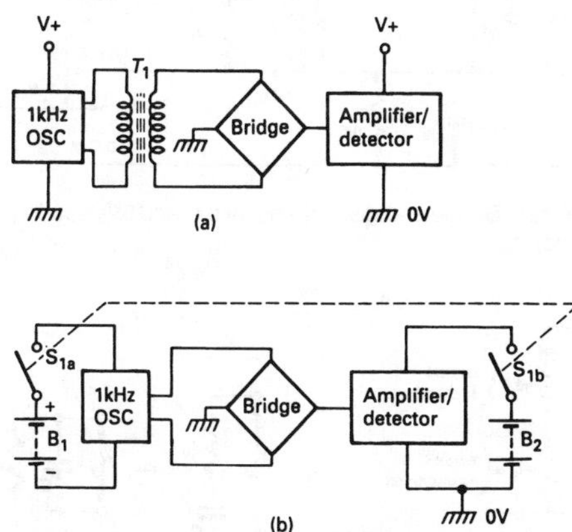

Figure 6.11. Alternative ways of providing a bridge with independent energisation and detection.

Figure 6.12 shows a practical battery-powered 'bridge energiser' that can give either a 9V d.c. output or an excellent 1kHz sinewave output with a peak-to-peak amplitude of 5 volts. The oscillator is a diode-stabilized Wien type, effectively operated from a split supply derived from the battery via R_1 and R_2; it has a low-impedance output, and consumes a quiescent current of less than 4mA. To set up the oscillator, connect its output to an oscilloscope and trim RV_1 to give a reasonably pure sinewave output of about 5V peak-to-peak.

Precision *C* and L Bridges

All 'precision' *C* and L bridges incorporate facilities for balancing both the resistive and the reactive elements of test components. The basic principles of the subject are detailed in *Figures 6.13* to *6.15*.

Any practical capacitor has the equivalent circuit of *Figure 6.13(a)*, in which *C* represents a pure capacitance, R_S represents dielectric losses, R_P the leakage losses, and L_S the inductance of electrode foils, etc. At frequencies below a few kHz, L_S has (except in high-value electrolytics) negligible practical effect, but R_S and R_P cause a finite shift in the capacitor's voltage/current phase relationship. This same phase shift can be emulated by wiring a single 'lumped' resistor in series or in parallel with a pure capacitor, as in *Figures 6.13(b)* and *6.13(c)*.

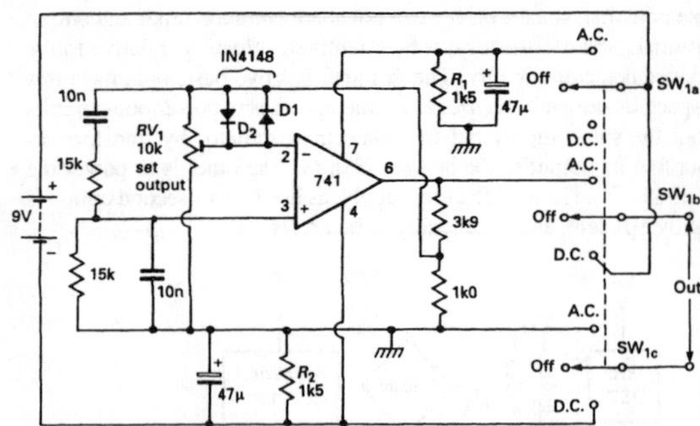

Figure 6.12. Bridge energiser, giving 9V d.c. and 5V peak-to-peak 1kHz outputs.

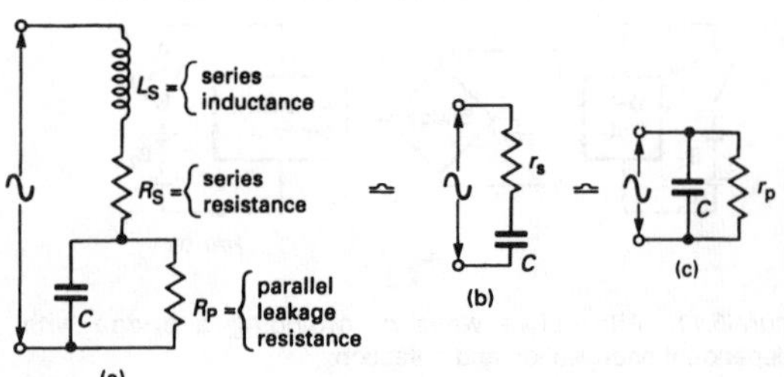

Figure 6.13. Absolute *(a)* and simplified series *(b)* and parallel *(c)* equivalent circuits of a capacitor.

A similar story is true of inductors, which can, in terms of phase shifts, be regarded as 'negative' capacitors. Either device can, at any given frequency, be treated as a pure reactance, X, that is either in series or parallel with a single 'loss' resistor (r_S or r_P), as shown in *Figures 6.14(a)* or *6.14(b)*. The ratio between *r* and X is normally called 'Q' in an inductor, or 'D' or the 'loss factor' in a capacitor; one of its major effects is to shift the device's voltage-to-current phase relationship, ø (pronounced phi), from the ideal value of 90° to some value between zero and 90°; the difference between this and the ideal value is known as the loss angle, ∂ (pronounced delta) of the phasor diagram. Another major effect of Q or D is to shift the component's impedance (Z) away from its pure reactance (X) value. All the formulae relevant to the subject are shown in *Figure 6.14*.

Figure 6.15 lists the relationships between phase and loss angles and the Z/X-ratios of both series and parallel equivalent circuits at various decade-related Q and D values. Note that components with Q values of 10 or more, or D values of 0.1 or less, have similar Z and X values. Thus, a coil with a Q of 10 and an X of 1000Ω at a given frequency can be said

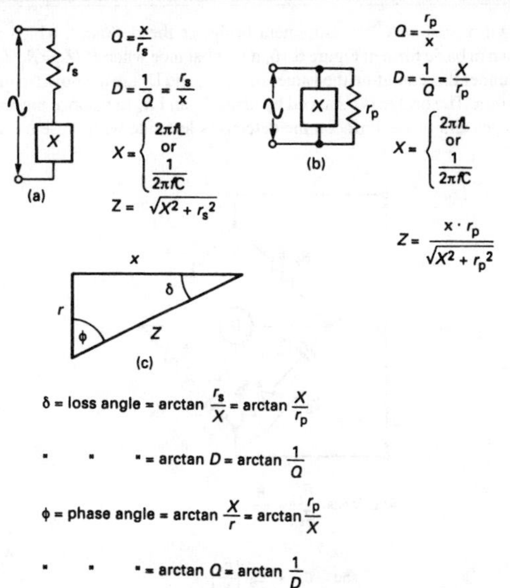

$Q = \frac{X}{r_s}$
$D = \frac{1}{Q} = \frac{r_s}{X}$
$X = \begin{cases} 2\pi fL \\ \text{or} \\ \frac{1}{2\pi fC} \end{cases}$
$Z = \sqrt{X^2 + r_s^2}$
(a)
$Q = \frac{r_p}{X}$
$D = \frac{1}{Q} = \frac{X}{r_p}$
$X = \begin{cases} 2\pi fL \\ \text{or} \\ \frac{1}{2\pi fC} \end{cases}$
$Z = \frac{X \cdot r_p}{\sqrt{X^2 + r_p^2}}$
(b)
X
r
Z
δ
φ
(c)
$\delta = \text{loss angle} = \arctan \frac{r_s}{X} = \arctan \frac{X}{r_p}$
$= \arctan D = \arctan \frac{1}{Q}$
$\phi = \text{phase angle} = \arctan \frac{X}{r} = \arctan \frac{r_p}{X}$
$= \arctan Q = \arctan \frac{1}{D}$

Figure 6.14. Any capacitor or inductor can be represented by a series *(a)* or parallel *(b)* equivalent circuit, which determines its phasor diagram *(c)*.

to have either a 100Ω series (r_S) or a 10000Ω parallel (r_P) resistance; in either case, it gives a loss angle of 5.71 degrees and an impedance that is within 0.5% of 1000Ω at that frequency. Also note that series and parallel 'equivalent' circuits produce significantly different Z/X-ratios at low values of Q.

The relevance of all this is that, at any given frequency, the true X, Q and D values of a capacitor or inductor can be deduced by measuring the device's impedance and loss angle, and it is this principle that forms the basis of most precision '*C*' or 'L' measurement bridges. The best-

Q	D	Phase angle φ	Loss angle δ	Z/X – ratio	
				Series circuit	Parallel circuit
∞	0	90°	0°	1.000	1.000
1000	0.001	89.94°	0.06°	1.000	1.000
100	0.01	89.43°	0.57°	1.000	1.000
10	0.1	84.29°	5.71°	1.005	0.995
1.0	1.0	45°	45°	1.414	0.7071
0.1	10	5.71°	84.29°	10.05	0.0995
0.01	100	0.57°	89.43°	100.0	0.0100

Figure 6.15. Relationship between ø, ∂, and Z/X-ratio in series and parallel equivalent circuits at various values of Q and D.

known precision '*C*' measurement bridge is the de Sauty, which is shown in basic form in Figure 6.16; it is at balance when $R_y/Z_S = R_1/Z_x$, and under this condition the values of C_x, r_x, and D_x are as shown in the diagram. The bridge is balanced by using R_y and R_S to balance the a.c. voltages and phase shifts on the detector's left side with those on its right.

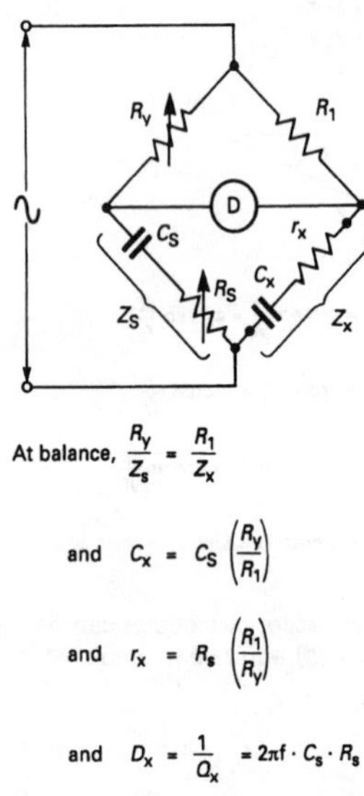

R_y
R_1
D
C_S
r_x
C_x
R_S
Z_S
Z_x
At balance, $\frac{R_y}{Z_s} = \frac{R_1}{Z_x}$
and $C_x = C_S\left(\frac{R_y}{R_1}\right)$
and $r_x = R_s\left(\frac{R_1}{R_y}\right)$
and $D_x = \frac{1}{Q_x} = 2\pi f \cdot C_s \cdot R_s$

Figure 6.16. Basic de Sauty capacitance bridge.

Figure 6.17 shows a practical de Sauty bridge that spans 1pF to 10μF in six decade ranges. This is a very sensitive design in which roughly half of the a.c. energising voltage appears at each end of the detector at balance; the detector can thus take a very simple form (such as earphones). R_S enables nulling to be obtained with C_x capacitors with D values as high as 0.138 (equal to a Q of 7.2). If desired, R_S can be calibrated directly in D values, since (in this design, at 1kHz) D = 0.001 per 15.9Ω of R_S value.

The two best known 'L' measurement bridges are the Hay and the Maxwell types of *Figures 6.18* and *6.19*. These work by balancing the inductive phase shift of L_x against a capacitive shift of the same magnitude in the diametrically opposite arm of the bridge. The Hay bridge uses a series equivalent (C_S–R_S) balancing network, and is useful for measuring high-Q coils (it can be very inaccurate at Q values below 10, as implied by the formulae of *Figure 6.18*). The Maxwell bridge uses a parallel equivalent (C_S–R_S) balancing network, and is useful for measuring coils with Q values below 10. Note that the Q values referred to here are those occuring at the test frequency of (usually) 1kHz; a coil that has a high-frequency Q of 100 may have a Q of 1 or less at 1kHz.

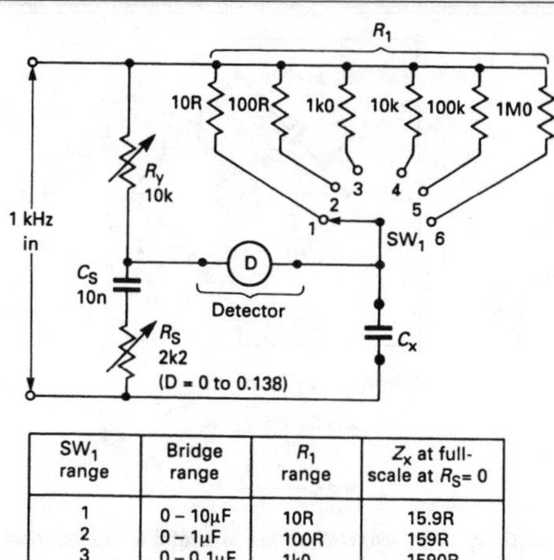

SW_1 range	Bridge range	R_1 range	Z_x at full-scale at R_S= 0
1	0 – 10μF	10R	15.9R
2	0 – 1μF	100R	159R
3	0 – 0.1μF	1k0	1590R
4	0 – 10nF	10k	15.9k
5	0 – 1nF	100k	159k
6	0 – 100pF	1M0	1.59M

Figure 6.17. Sensitive six-range de Sauty capacitance bridge.

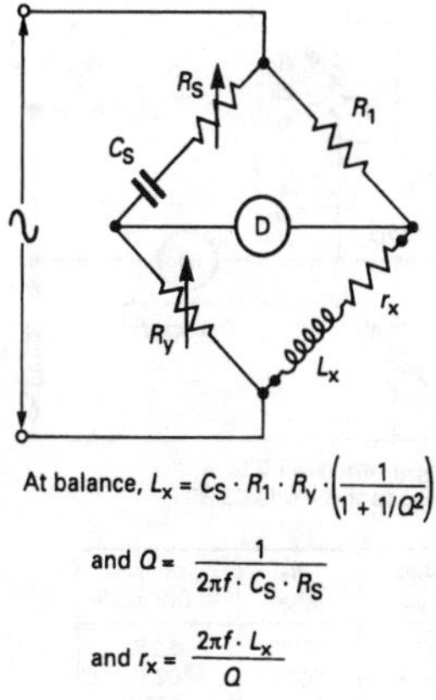

Rs
R1
Cs
~
D
rx
Ry
Lx
At balance, $L_x = C_S \cdot R_1 \cdot R_y \cdot \left(\frac{1}{1 + 1/Q^2}\right)$
and $Q = \frac{1}{2\pi f \cdot C_S \cdot R_S}$
and $r_x = \frac{2\pi f \cdot L_x}{Q}$

Figure 6.18. The Hay inductance bridge is useful for measuring high-Q coils.

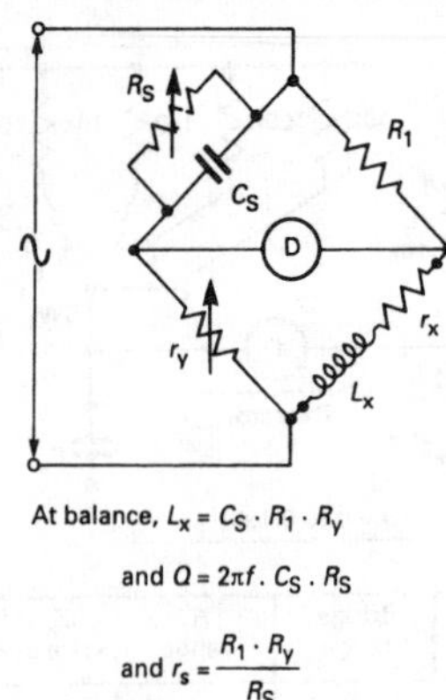

Figure 6.19. The Maxwell bridge is useful for measuring low-Q coils.

Figure 6.20 shows a practical inductance bridge that spans 10µH to 100Henrys in six ranges; it uses a Hay configuration for high-Q measurements (with SW_2 in the 'H' position) and a Maxwell layout for low-Q ones (with SW_2 set to 'L'). This is another sensitive design, in which the a.c. voltages at either end of the detector are close to the half-supply value at balance, and can use a very simple type of detector.

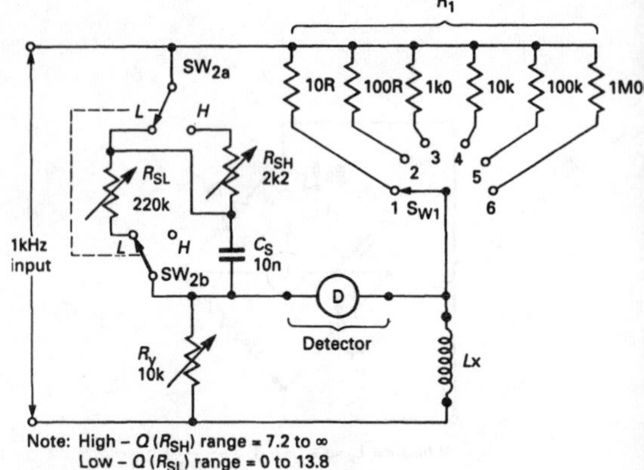

SW_1 range	Bridge range	R_1 value	Z_x at 1kHz at full scale
1	0 – 1mH	10R	6.2R
2	0 – 10mH	100R	62R
3	0 – 100mH	1k0	620R
4	0 – 1H	10k	6.2k
5	0 – 10H	100k	62k
6	0 – 100H	1M0	620k

Figure 6.20. Six-range Hay/Maxwell inductance bridge.

A precision L–C–R bridge

Figure 6.21 shows a precision 18-range L–C–R bridge that combines the circuits of *Figures 6.20, 6.17,* and *6.6*, to make a highly sensitive design that can use very simple types of balance detector (such as a multimeter on the d.c.-driven *'R'* ranges, or earphones on the a.c.-energised *'C'* and 'L' ranges). This circuit's 'y' null-balance network is modified by the addition of resistor R_{yx} and switch S_y, which enable the coverage of each range to be extended by ten percent.

When this bridge is used on its 0–100pF range, considerable errors occur due to the effects of stray capacitance, and measurements should thus be made by using the 'incremental' method, as follows: first, with no component in place across the 'x' terminals, null the bridge via R_y and note the resultant 'residual' null reading (typically about 15pF); now fit the test component in place, obtain a balance reading (say 83pF), and then subtract the residual value (15pF) to obtain the true test capacitor value (68pF).

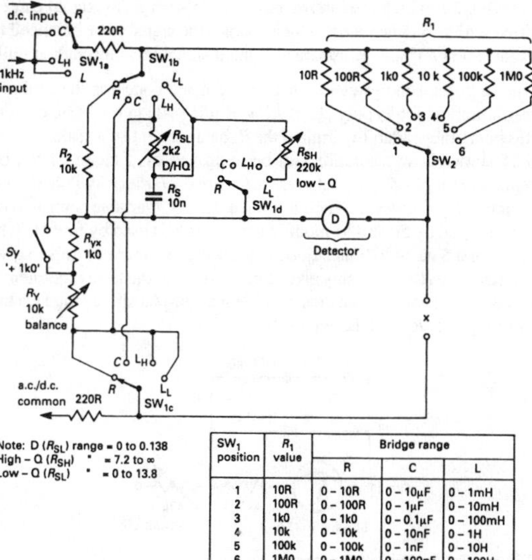

SW1 position	R1 value	Bridge range R	C	L
1	10R	0 - 10R	0 - 10µF	0 - 1mH
2	100R	0 - 100R	0 - 1µF	0 - 10mH
3	1k0	0 - 1k0	0 - 0.1µF	0 - 100mH
4	10k	0 - 10k	0 - 10nF	0 - 1H
5	100k	0 - 100k	0 - 1nF	0 - 10H
6	1M0	0 - 1M0	0 - 100pF	0 - 100H

Figure 6.21. Sensitive eighteen-range laboratory-standard L–C–R bridge.

The *Figure 6.21* circuit can either be built exactly as shown and used with external energising and null-detection circuitry, or can be modified in various ways to suit individual tastes. *Figure 6.22*, for example, shows how an extra wafer (e) can be added to SW_1 to facilitate the use of a built-in d.c./a.c. energiser (which can use the *Figure 6.12* design). Similarly, d.c. and a.c. null-balance detectors can easily be built in; the d.c. detector can take the form of a 50µA–0–50µA meter, with overload

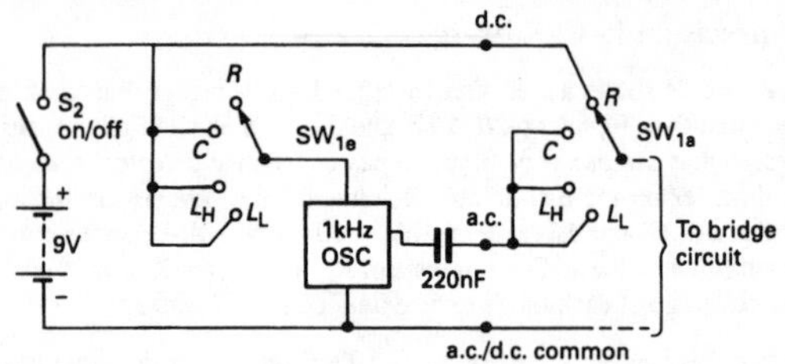

Figure 6.22. Built-in energiser for the L–C–R bridge.

protection given via a couple of silicon diodes and sensitivity adjustable via a series resistor, as shown in *Figure 6.23*. The a.c. detector can take the form of any single-ended a.c. analogue millivoltmeter circuit; in this case the 'low' input of the detector and the left-hand 'detector' junction of the bridge can both be grounded to chassis, as shown in *Figure 6.24*, which also shows how the a.c. energising signal can be fitted with a Wagner earth, which enables the signal to be balanced to ground (via RV_1), to eliminate unwanted signal breakthrough at null.

One important modification that can be made concerns the bridge's resolution, which is only about ±1% of full-scale in the basic design, this being the readability limit of the R_y balance control's scale. *Figure 6.25* shows how resolution can be improved by a factor of ten by replacing the S_y–R_{yx}–R_y network of *Figure 6.21* with a switched-and-variable 'R_y' network; switch S_y enables the RV_y variable control to be over-ranged by 50%. The modified bridge is best used by first switching S_y and SW_y to '0' and adjusting the bridge's range controls to give a balance on RV_y only; this gives a good guide to the test component's value; a final balance can then be read on a more sensitive range via the full range of 'R_y' balance controls.

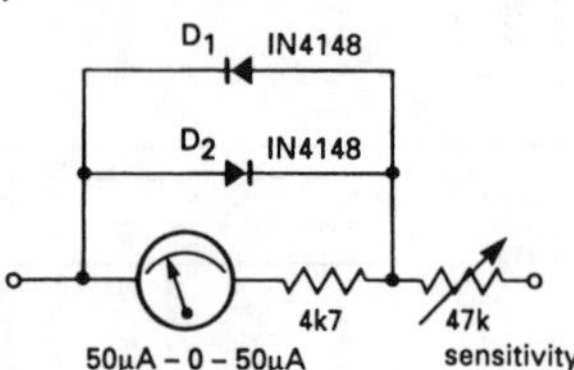

Figure 6.23. D.c. null detector.

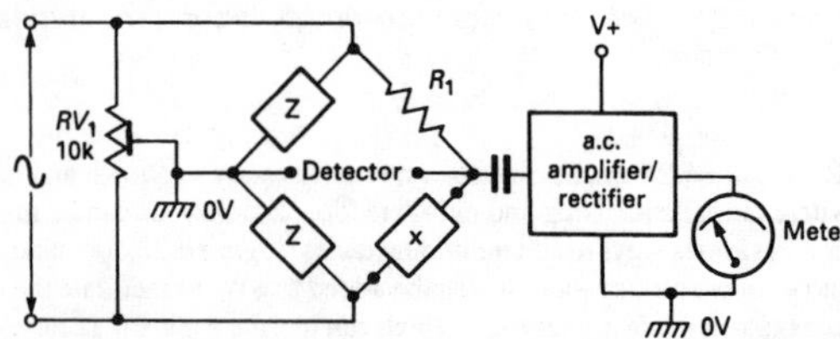

Figure 6.24. Provision of a Wagner earth and single-ended a.c. null detector.

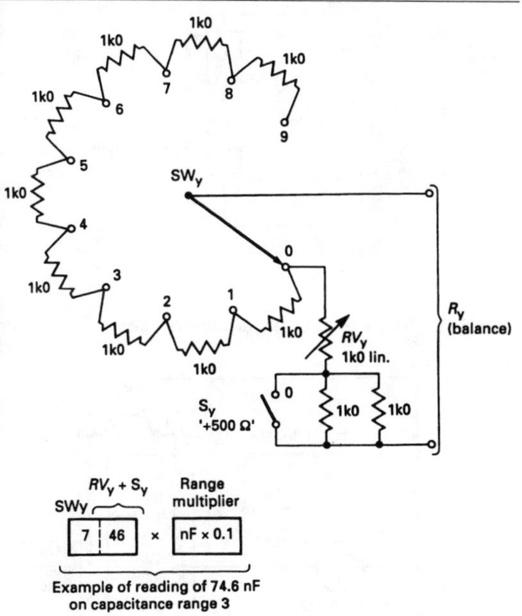

Figure 6.25. This circuit can be used to raise the null resolution (readability) of the L–C–R bridge to ±0.1% of full scale.

Special bridge circuits

In addition to the types of bridge already described, two others are of special value. One of these is the transformer ratio-arm bridge, shown in basic form in *Figure 6.26*. The ratio arm values (shown switch-selectable at 0.1/1, 1/1, or 10/1 in this example) equal the transformer's turns-ratios, and can easily be wound with a precision better than 0.01%. The value of an unknown (x) impedance can be balanced against that of a standard either by varying the value of the standard or the value of the ratio arms. Resistors or capacitors can be balanced against each other by placing them on opposite sides of the bridge as shown, or a capacitor can be balanced against an inductance by placing both components on the same side of the bridge.

The other important bridge is the d.c. resistance-matching type; *Figure 6.27* shows a simple version that enables resistors to be matched to within ±0.1% or better. The basic principles involved here are, first, that if R_A and R_B are equal, R_{MATCH} will equal R_S at balance; and second, that if R_A and R_B are equal they will give exactly the same output voltage whichever way they are connected to the supply. With this second principle in mind, R_A and R_B are joined by 500R multi-turn pot RV_1 and are connected to the d.c. supply via biased polarity-reversal switch SW_1.

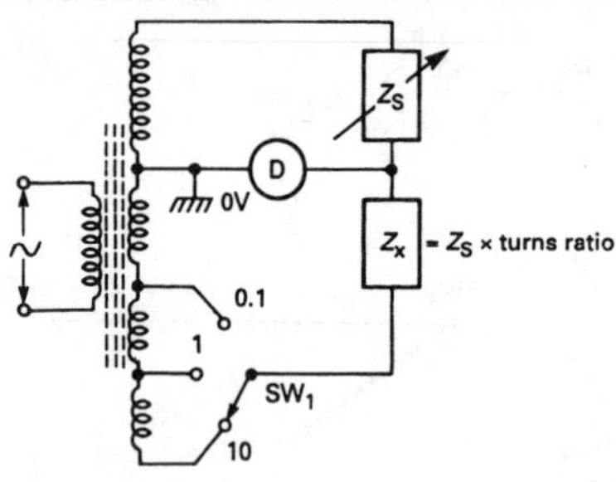

Figure 6.26. Simple transformer ratio-arm bridge.

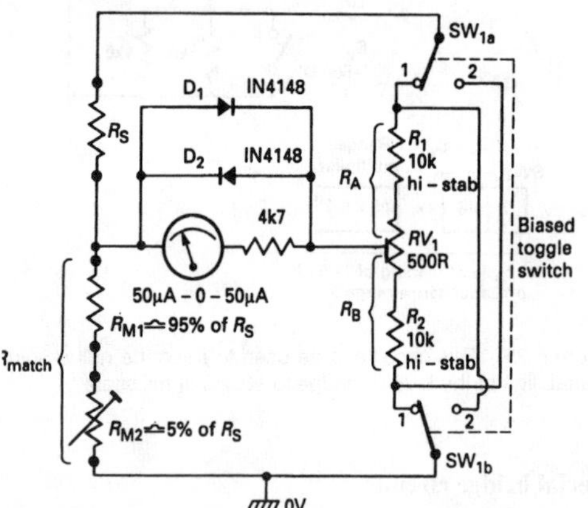

Figure 6.27. Simple resistance-matching bridge, gives matching within ±0.1%.

To use this bridge, first fit R_S and R_{MATCH} in place, noting that R_{MATCH} is, for simplicity, shown as being made up of a fixed (R_{M1}) and a variable (R_{M2}) element. With the d.c. supply connected, trim R_{M2} and RV_1 to bring the meter reading to a sensible level; now repeatedly toggle SW_1 and trim RV_1 until identical meter readings are given in either toggle position. That completes the R_A–R_B adjustment. The value of R_{MATCH} should now be trimmed (via R_{M2}) to bring the meter reading to zero, at which point R_S and R_{MATCH} are matched. Note that, after RV_1 has been initially set, it should only rarely need readjustment, and that the circuit's 'matching' fidelity is limited to ±0.1% only by the balance-detection meter's sensitivity.

When building a resistance-matching bridge or when matching resistors, use only low-temperature-coefficient components, and never physically touch them during balancing/matching operations. *Figure*

Resistor type	Typical temperature coefficient	
	± ppm/°C	± %/°C
Carbon film	300 – 1000	0.03% – 0.1%
Thick film metal	100 – 300	0.01% – 0.03%
Metal film	50 – 100	0.005% – 0.01%
Precision metal film	15 – 50	0.0015% – 0.005%
Vitreous wire wound	75	0.0075%
Precision wire wound	5 – 15	0.0005% – 0.0015%

Figure 6.28. Typical temperature coefficients of modern resistors.

6.28 helps clarify this point by listing the temperature coefficients of various types of resistor. Note when using the bridge that the R_{MATCH} value can be trimmed (to make it equal R_S) by using series resistance to increase its value or shunt resistance to reduce it.

A precision resistance-matching bridge

Figure 6.29 shows a precision resistance-matching bridge that incorporates a meter-driving d.c. differential amplifier that gives such high balance-detection sensitivity that resistors can be matched to within ±0.003%. This bridge also has a facility for indicating, on RV_1's calibrated scale, the percentage out-of-match error of R_{MATCH}; this scale spans ±0.5%, ±0.05%, and ±0.005% in three switch-selected ranges.

To initially set up this bridge, fit R_S and R_{MATCH} in place, connect the op-amp's output to an external meter, and switch on via SW_4. Now close SW_3 and trim the op-amp's set balance control for zero reading on the meter's most sensitive d.c. current range; release SW_3. Now set RV_1 to mid-scale and, with SW_2 initially set to its 'x1' scale, start toggling SW_1 and trim RV_2 (and if necessary, R_{M2}) to find a setting where identical meter readings are obtained in both toggle positions; as RV_2 nears the balance point, increase balance sensitivity via SW_2, until eventually a perfect balance is obtained on the 'x0.01' range. That completes the initial setting up procedure, and R_{MATCH} can then be matched to R_S by trimming R_{M2} for a zero reading on the meter. Once the circuit has been initially set up as described, RV_2 and the op-amp's set balance control should only rarely need readjustment, and in all further 'matching' operations the following procedure can (after making a brief initial check that the meter and toggle balances are correct) be used.

Fit R_S, R_{MATCH}, and the external meter in place. Turn SW_2 to the 'x1' position, and switch the bridge on via SW_4. If R_{MATCH} is within ±0.5% of the R_S value it should now be possible to set the bridge to a null via RV_1; if necessary, trim the R_{MATCH} value until a null can be obtained. At null, read off the R_{MATCH} error on the RV_1 scale (see *Figure 6.29*), and then make the appropriate error correction; to increase the R_{MATCH} value by a fixed percentage, add a series resistor with a value of:

$$R_{SERIES} = (R_{MATCH}/100) \text{ x \% error.}$$

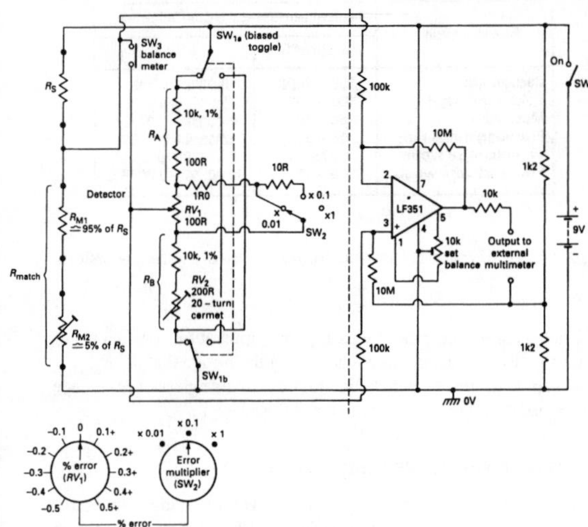

Figure 6.29. High-precision resistance-matching bridge, gives matching to within ±0.003%.

To reduce the R_{MATCH} value by a fixed percentage, add a shunt resistor with a value of:

$$R_{SHUNT} = (R_{MATCH} \times 100)/\%\ \text{error}.$$

Thus, a 1000Ω (nominal) resistor can be increased by 0.3% by adding a 3R0 series resistor, or reduced by 0.3% by adding a 330k shunt resistor. Once a good match has been obtained on SW_2's 'x1' range, repeat the process on ranges 'x0.1' and 'x0.01', until the match is adequate. An alternative to this technique is to simply leave RV_1 in its mid-scale position and trim R_{MATCH}, via R_{M2}, to obtain a null on all ranges of SW_2.

Resistor-matching bridge applications

A precision resistor-matching bridge has several useful applications in the electronics laboratory. One of these is in the duplication of precision resistor values, and *Figure 6.30* shows an example of how this facility can be put to good use. Here, ten duplicates of a precision 1k0 resistor are so wired that they can easily be used in series, to act as a resistance that increases in 1k0 steps up to a maximum of 10k, or in parallel, to act as a resistance that decreases in steps down to 100Ω. Note that as more and more resistors are wired in parallel or series, their ±0.003% duplication errors average out and diminish, so that the precision of the final 10k series or 100R parallel resistance is equal, for all practical purposes, to that of the original 'master' resistor. The actual value of 'summed' duplication error is equal to the original error divided by the square root of the number of summed resistors (10), and equals 0.001% in this case.

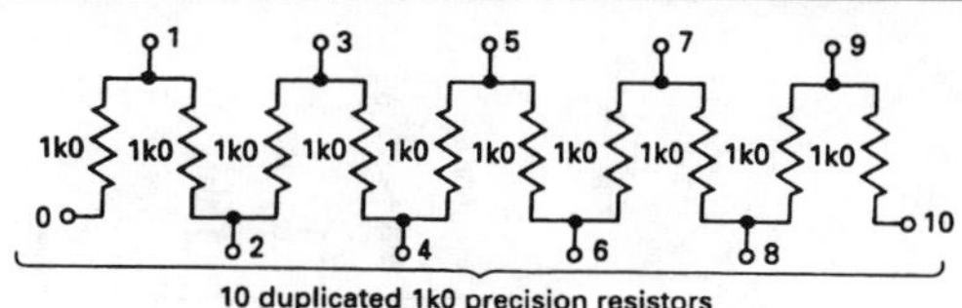

Figure 6.30. This matrix of 1k0 resistors totals 10k when series connected, or 100R when parallel connected.

Another important resistor-matching bridge application is in the creation of a precision ratio-matching bridge of the type shown in *Figure 6.31*. Here, 14 resistors are duplicated from a 1k0 (nominal) master and wired together to create a three-ratio (10/1, 1/1, and 1/10) divider. The fact that these resistors are all precision-matched to within ±0.003% ensures that the ratios are intrinsically defined with great precision, the actual precision being ±0.002% on the 1/1 range and ±0.005% on the 10/1 and 1/10 ranges. This bridge can itself be used to produce direct or decade multiple or sub-multiple duplicates of a master resistor. If the bridge is used with a sensitive null detector that gives a duplication precision of 0.003%, the overall precision of duplication is ±0.005% on the 1/1 range, and ±0.008% on the 10/1 and 1/10 ranges.

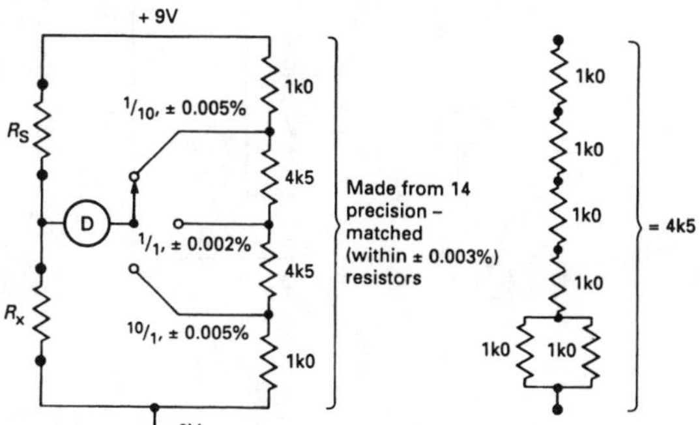

Figure 6.31. Precision ratio-matching bridge.

One obvious application of the *Figure 6.31* circuit is in matching the range and ratio arms of conventional bridges, to enhance bridge precision. Another is in generating high-precision resistors for use in decade *'R'* boxes. *Figure 6.32* shows a 3-decade *'R'* box that spans 0 to 99.9k in 100Ω steps, with 100k over-ranging available via SW_4. This type of circuit can be generated from a single precision reference (1k0 in this case), and has a multitude of applications in the laboratory, including those of calibrating bridge scales and finding resistor values by substitution.

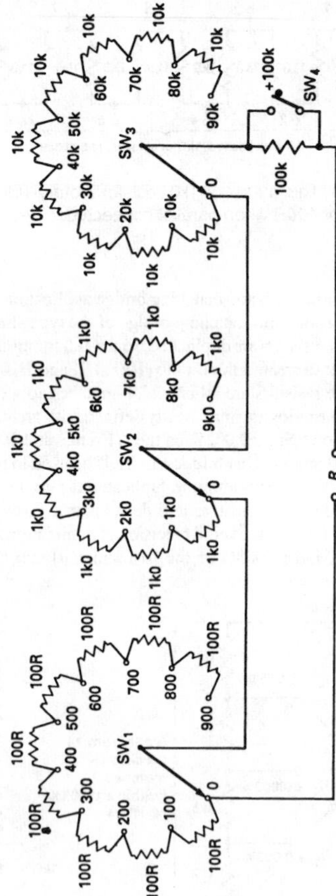

Figure 6.32. Three-decade *R* box spans 0 to 99.9k in 100R steps.

7 Basic diode circuits

The discrete solid-state diode is the most fundamental element used in modern electronics. It is available in a variety of forms, including those of signal detector, rectifier, zener 'voltage reference', noise-generator, varicap 'variable capacitor', light-sensitive diode, and light-emitting diode (LED). This chapter looks at the basic characteristics of these devices, and shows various ways of using standard diodes and rectifiers.

Basic diode characteristics

Most modern diodes are of the 'junction' type, and use the basic structure (and symbol) shown in *Figure 7.1*. They are made from a single p–n junction; the 'p' terminal is the anode, and the 'n' terminal the cathode.

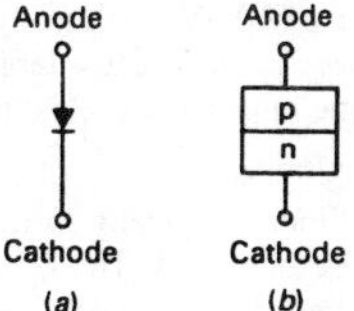

Figure 7.1. Symbol *(a)* and structure *(b)* of solid-state diode.

Figure 7.2 illustrates the basic characteristics of the diode. When forward biased (with the anode positive to the cathode) it has a low resistance and readily passes current, but when reverse biased it has a high resistance and blocks current: this action is implied by the diode symbol, which resembles an arrow pointing in the direction of easy current conduction.

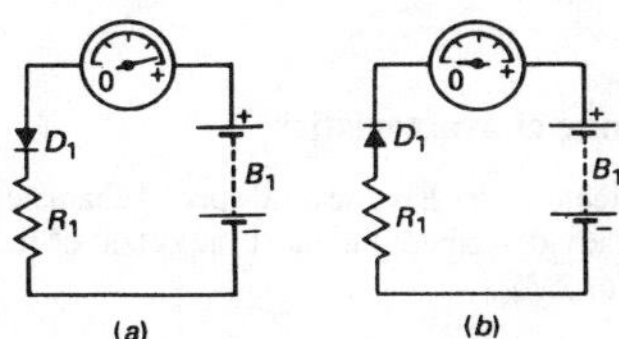

Figure 7.2. Diode conduction when *(a)* forward and *(b)* reverse biased.

Most junction diodes are made from either germanium or silicon materials. *Figure 7.3* compares the basic characteristics of the two types of device when operated at a normal room temperature of 20°C; note the following important points.

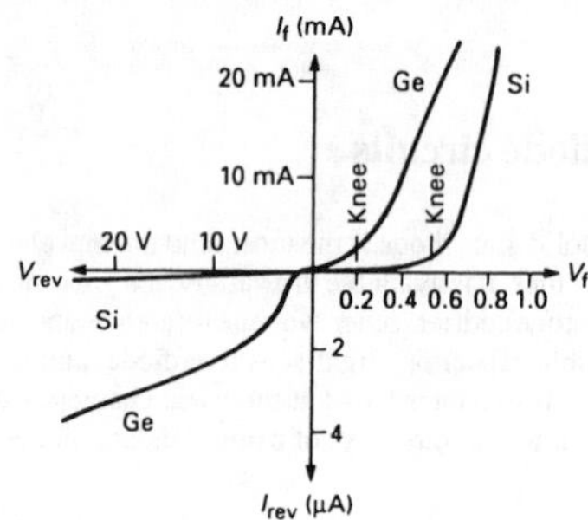

Figure 7.3. Basic characteristics of germanium (Ge) and silicon (Si) junction diodes (at 20°C).

(1) A forward biased diode passes little current (I_f) until the applied voltage (V_f) exceeds a certain 'knee' value (typically 150 to 200mV in germanium diodes, 550 to 600mV in silicon types). When V_f exceeds the knee value, small increases in V_f cause large increases in I_f, and the diode's forward dynamic impedance (Z_f) is inversely proportional to applied voltage.

(2) The Z_f of a silicon diode is typically 25/I, where I is in mA; i.e., Z_f = 25Ω at 1mA, or 0.25Ω at 100mA. The Z_f of a germanium diode is greater than that of a silicon type, and its V_f usually exceeds that of a silicon type at I_f values above a few tens of mA.

(3) When a diode is reverse biased by more than 1V or so it passes a reverse leakage current (I_r) that is proportional to the reverse voltage (V_r) value. At normal room temperatures I_r values are measured in microamps in germanium diodes, and in nanoamps in silicon ones. I_r typically doubles with each 8°C increase in junction temperature.

Because of their low knee voltage values, germanium diodes are used almost exclusively in low-level 'signal detection' applications. Silicon types can be used in many general-purpose applications. Diodes that have high voltage and current ratings are, by convention, usually called 'rectifiers'.

Special diode characteristics

Ordinary silicon diodes have several special characteristics additional to those already described; the most important of these are shown in *Figures 7.4* to *7.7*.

If a silicon diode is increasingly reverse biased a point is eventually reached where its reverse current suddenly starts to increase, and any further increase in V_r causes a sharp rise in I_r, as shown in *Figure 7.4*. The voltage at which this action occurs is known as the avalanche or 'zener' value of the device. 'Zener diodes' are specially made to exploit this effect, and are widely used as 'reference voltage' generators. Note, however, that when zener diodes are operated at low currents their impedances fluctuate in a rapid and random manner, and they can thus be used as excellent 'white-noise' generators.

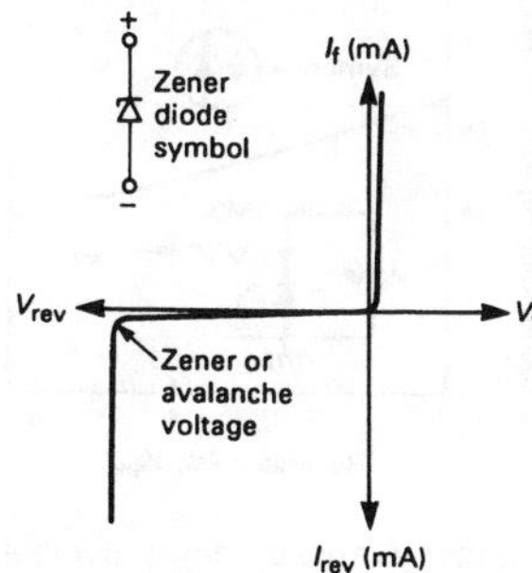

Figure 7.4. Zener diode symbol and characteristics.

If a silicon diode is forward biased at a constant current, its V_f value varies with junction temperature at a rate of –2mV/°C, as shown in *Figure 7.5*. Thus, if V_f = 600mV at +20°C, it falls to 440mV at 100°C or rises to 740mV at –50°C. Silicon diodes can thus be used as temperature-to-voltage converters.

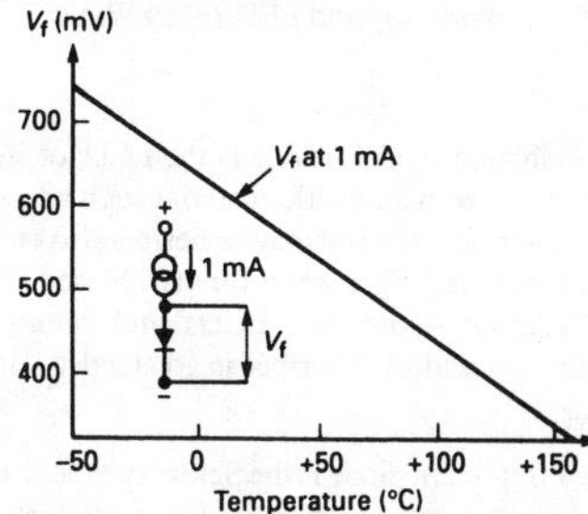

Figure 7.5. Thermal characteristics of a silicon diode at I_f = 1mA.

If a silicon diode is reverse biased from a high-impedance source (as shown in *Figure 7.6*), its junction capacitance decreases (from perhaps 17pF at –1 volt to maybe 10pF at –8 volts) as the reverse bias is increased. 'Varicap' (or Varactor) diodes are specially made to exploit this 'voltage-variable-capacitor' effect; they use the circuit symbol shown in the diagram.

When p–n junctions are reverse biased their leakage currents and impedances are inherently optosensitive; they act as very high impedances under dark conditions and as low impedances under bright ones. Normal diodes are shrouded in opaque material to stop this unwanted effect, but 'photo-diodes' are specially made to exploit it; they use the *Figure 7.7(a)* symbol. Some photo-diodes are designed to respond to visible light, and some to infra-red (IR) light.

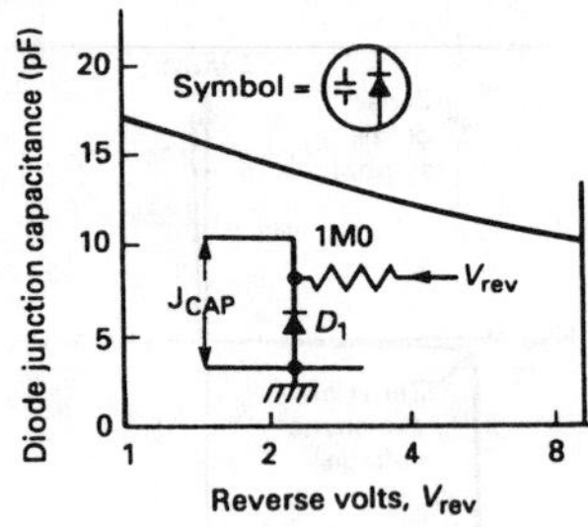

Figure 7.6. Varactor (varicap) diode symbol and typical characteristics.

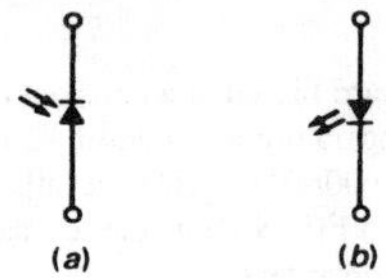

Figure 7.7. Photodiode *(a)* and LED *(b)* symbols.

Another useful 'junction diode' device is the LED, or light emitting diode, which is made from an exotic material such as gallium phosphide or gallium arsenide, etc., and may be designed to emit either red, green, yellow, or infra-red light when forward biased. They use the *Figure 7.7(b)* symbol. Note that LEDs and photo-diodes are optoelectronic devices, and are described in greater detail in Chapter 13 of this Volume.

Finally, one other important diode is the Schottky type, which uses the standard diode symbol but has a very fast switching action and develops only half as much forward voltage as a conventional silicon diode. They can be used to replace germanium diodes in many 'signal detector' applications, and can operate at frequencies up to tens or hundreds of Gigahertz.

Half-wave rectifier circuits

The simplest applications of a diode is as a half-wave rectifier, and *Figure 7.8* shows a transformer-driven circuit of this type (with the diode's 'V_{in}' value specified in volts r.m.s.), together with relevant output waveforms. If this circuit has a purely capacitive load it acts as a peak voltage detector, and the output (V_{pk}) equals 1.41 x V_{in}; if it has a purely resistive load it acts as a simple rectifier and gives an r.m.s output of 0.5 x V_{in}; if it has a resistively-loaded capacitive load (as in most power supply units) the output is 'rippled' and has an r.m.s. value somewhere between these two extremes. In capacitively-loaded circuits D_1 needs a peak reverse-voltage rating of at least 2.82 x V_{in}; if purely resistive loading is used, the rating can be reduced to 1.41 x V_{in}.

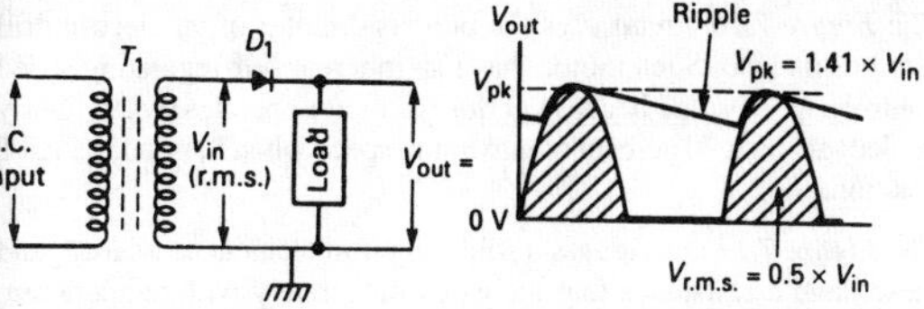

Figure 7.8. Circuit and waveforms of transformer-driven half-wave rectifier.

If a half-wave rectifier circuit is used to power purely resistive loads, they consume only a quarter of 'maximum' power, since power is proportional to the square of applied r.m.s. voltage. Some loads are not purely resistive, and *Figures 7.9* to *7.11* show how the basic half-wave rectifier circuit can be adapted to give 2-level power control of lamps, electric drills, and soldering irons that are operated from the a.c. power lines. Note in each of these circuits that the r.m.s. voltage fed to the load equals V_{in} when S_1 is in position 3, or 0.5 x V_{in} when S_1 is in position 2.

The *Figure 7.9* circuit uses a lamp load, which has a resistance roughly proportional to its filament temperature; when it is operated at half of maximum voltage its resistance is only half of maximum, so the lamp operates at about half of maximum power and thus burns at half-brilliance with S_1 in the DIM position.

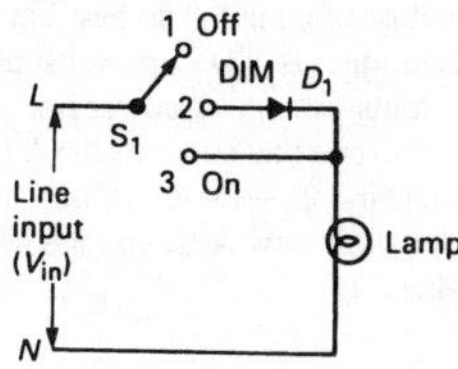

Figure 7.9. Lamp burns at half brightness in DIM position

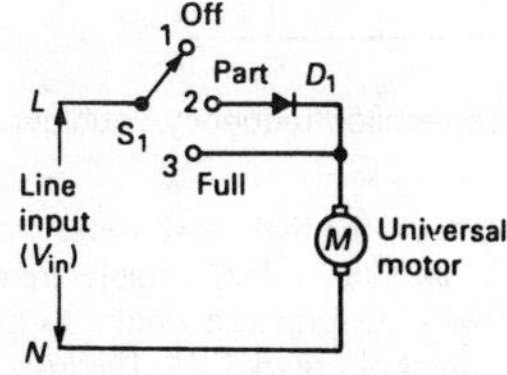

Figure 7.10. Drill motor runs at 70% of maximum speed in PART position

The *Figure 7.10* circuit uses the universal motor of an electric drill (etc.) as its load. Such motors have an inherent self-regulating speed control capacity, and because of this the motor operates (when lightly loaded) at about 70 percent of maximum speed when S_1 is in the PART position.

The *Figure 7.11* circuit uses a soldering iron element as its load, and these have a resistance that increases moderately with temperature; thus, when the iron is operated at half voltage its resistance is slightly reduced, the net effect being that the iron operates at about one third of maximum power when S_1 is in the SIMMER position, thus keeping the iron heated but not to such a degree that its bit deteriorates.

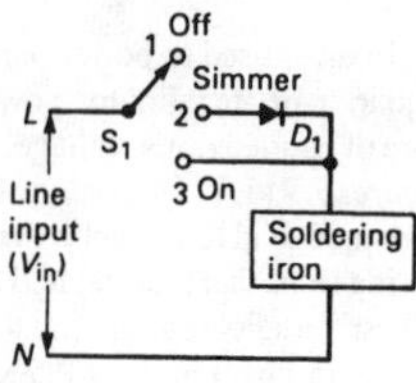

Figure 7.11. Soldering iron operates at 1/3rd power in SIMMER position.

Full-wave rectifier circuits

Figure 7.12 shows four diodes connected in 'bridge' form and used to give full-wave rectification from a single-ended input signal; the output frequency is double that of the input. The best known application of full-wave rectifying techniques is in D.C. power supply circuits, which provide D.C. power outputs from A.C. power line inputs, and consist of little more than a transformer that converts the A.C. line voltage into an electrically isolated and more useful A.C. value, and a rectifier-filter combination that converts this new A.C. voltage into smooth D.C. of the desired voltage value.

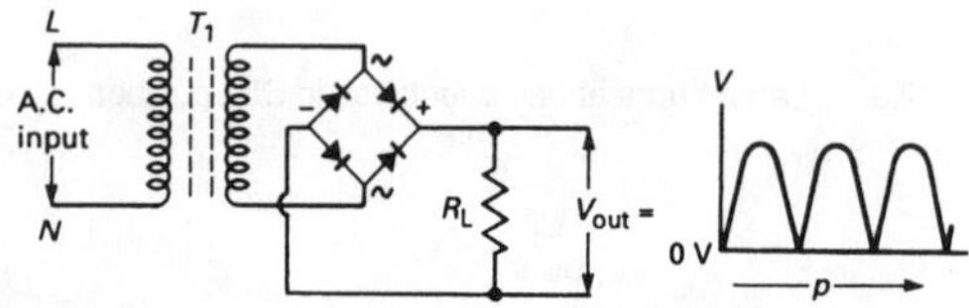

Figure 7.12. Bridge rectifier/frequency-doubler circuit.

Figures 7.13 to *7.16* show the four most useful basic power supply circuits. *Figure 7.13* provides a D.C. supply from a single-ended transformer, and gives a performance similar to that of the centre-tapped transformer circuit of *Figure 7.14*. The *Figures 7.15* and *7.16* circuits each provide split or dual D.C. supplies with nearly identical performances. The rules for designing these circuits are quite simple, as follows.

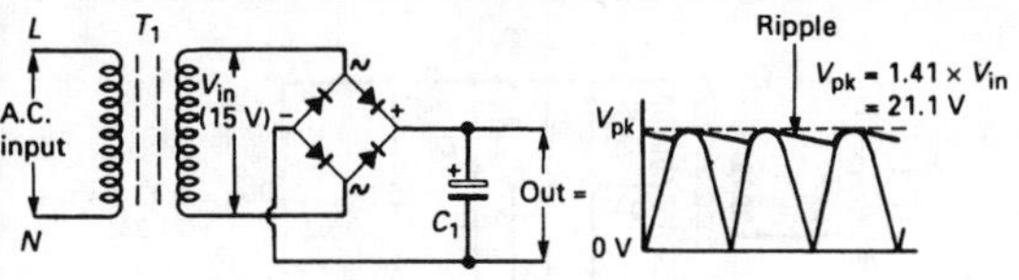

Figure 7.13. Basic single-ended PSU using bridge rectifier.

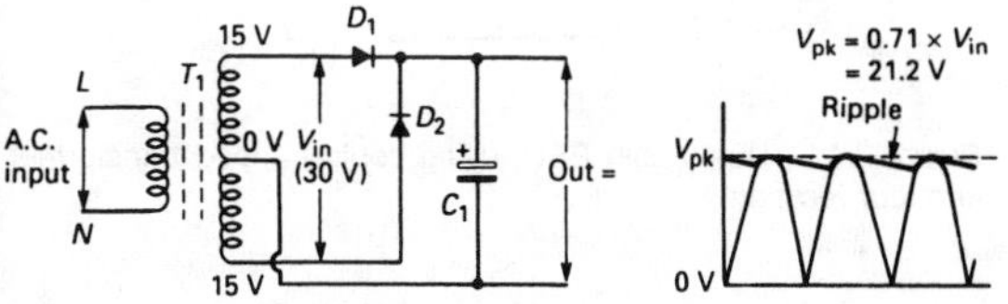

Figure 7.14. Basic single-ended PSU using centre-tapped transformer.

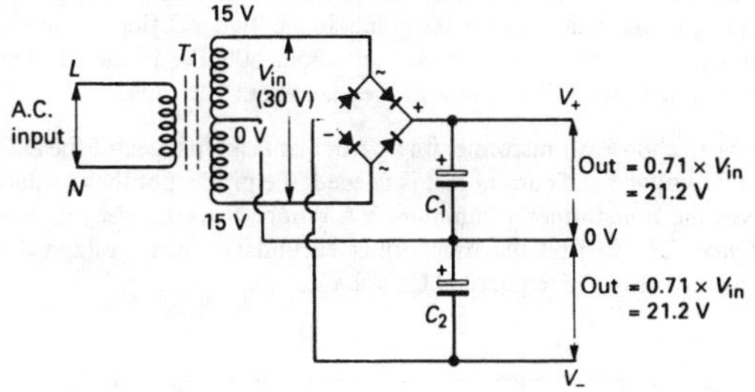

Figure 7.15. Dual (split) PSU, using centre-tapped transformer and bridge rectifier.

Transformer-rectifier selection

The three most important parameters of a transformer are its secondary voltage, its power rating, and its regulation factor. The secondary voltage is always quoted in r.m.s. terms at full rated power load, and the power load is quoted in terms of volt-amps or watts. Thus, a 15V 20VA transformer gives a secondary voltage of 15V r.m.s. when its output is loaded by 20W. When the load is reduced to zero the secondary voltage rises by an amount implied by the *regulation factor*. Thus, the output of a 15V transformer with a 10% regulation factor (a typical value) rises to 16.5V at zero load.

Note that the r.m.s. output voltage of the transformer secondary is *not* the same as the D.C. output voltage of the complete full-wave rectified

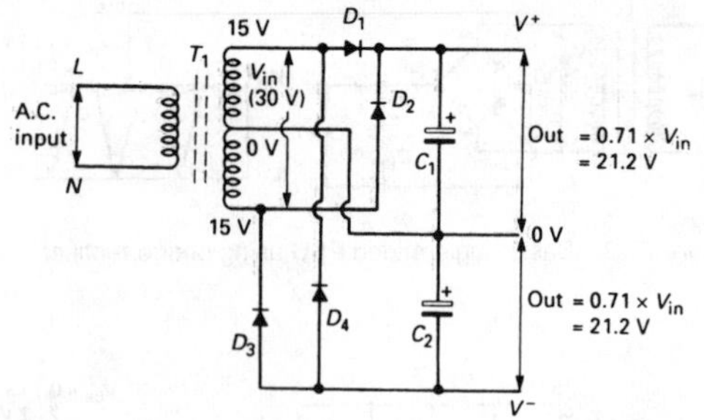

Figure 7.16. Dual (split) PSU, using centre-tapped transformer and four rectifiers.

power supply which, as shown in *Figure 7.17*, is 1.41 times greater than that of a single-ended transformer, or 0.71 times that of a centre-tapped transformer (ignoring rectifier losses). Thus, a single-ended 15V r.m.s. transformer with 10% regulation gives an output of about 21V at full rated load (just under 1amp at 20VA rating) and 23.1 volts at zero load. When rectifier losses are taken into account the output voltages are slightly lower than shown in the graph. In the 'two rectifier' circuits of *Figures 7.14* and *7.16* the losses are about 600mV; in the 'bridge' circuits of *Figures 7.13* and *7.15* they are about 1.2 volts.

Thus, to choose a transformer for a particular task, first decide the D.C. output voltage and current that is needed; the product of these values gives the transformer's minimum VA rating. Now use the graph of *Figure 7.17* to find the transformer secondary r.m.s. voltage that corresponds to the required D.C.. voltage.

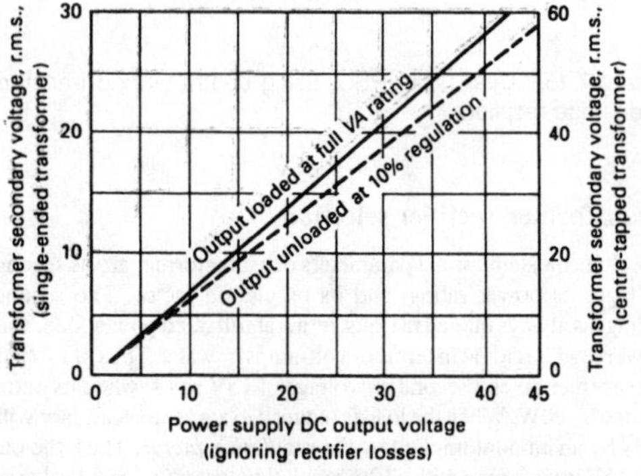

Figure 7.17. Transformer selection chart. To use, decide on the required loaded D.C. output voltage (say 21V), then read across to find the corresponding secondary voltage (15V single-ended or 30V centre-tapped).

The filter capacitor

The filter capacitor's task is to convert the full-wave output of the rectifier into a smooth D.C. output voltage; its two most important parameters are its working voltage, which must be greater than the off-load output value of the power supply, and its capacitance value, which determines the amount of ripple that will appear on the D.C. output when current is drawn from the circuit.

As a rule of thumb, in a full-wave rectified power supply operating from a 50 – 60Hz power line, an output load current of 100mA causes a ripple waveform of about 700mV peak-to-peak to be developed on a 1000µF filter capacitor; the ripple magnitude is proportional to load current and inversely proportional to the capacitance value, as shown in the design guide of *Figure 7.18*. In most practical applications, ripple should be kept below 1.5 volts peak-to-peak at full load.

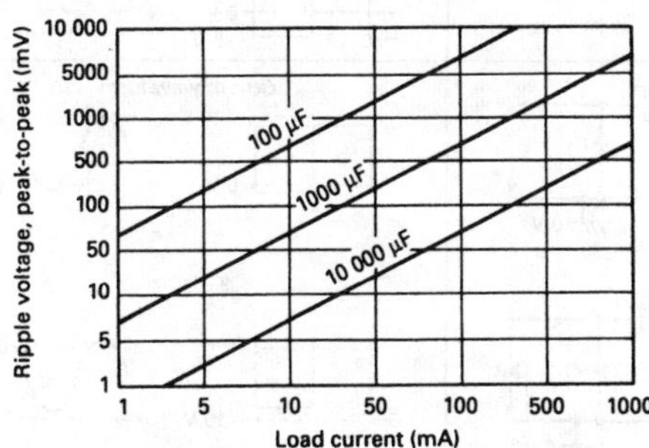

Figure 7.18. Filter capacitor selection chart, relating capacitor size to ripple voltage and load current in a full-wave rectified 50–60Hz powered circuit.

Rectifier ratings

Figure 7.19 summarises the characteristics of the three basic types of rectifier circuit and gives the minimum PIV (peak inverse voltage) and current ratings of the individual rectifiers. Thus, the full-wave circuit (using a centre-tapped transformer) and the bridge circuit (using a single-ended transformer) each give a typical full-load output voltage of about 1.2 x E and need diodes with minimum current ratings of 0.5 x I (where I is the load current value), but the bridge circuit's PIV requirement is only half as great as that of the full wave circuit.

Clamping diode circuits

A clamping diode circuit is one that takes an input waveform and provides an output that is a faithful replica of its shape but has one edge tightly clamped to the zero-voltage reference point. *Figure 7.20(a)* shows a version which clamps the waveform's negative edge to zero and gives a purely 'positive' output, and *Figure 7.20(b)* shows a version which clamps the positive edge to zero and gives a purely 'negative' output.

Circuit	V_{in} (r.m.s.)	No-load output	Full-load output	Rectifier rating	
				PIV	Current
Half wave	E	1.41 × E	E	2.82 × E	I
Full wave	E + E	1.41 × E	1.2 × E	2.82 × E	0.5 × I
Bridge	E	1.41 × E	1.2 × E	1.41 × E	0.5 × I

Figure 7.19. Rectifier circuit characteristics.

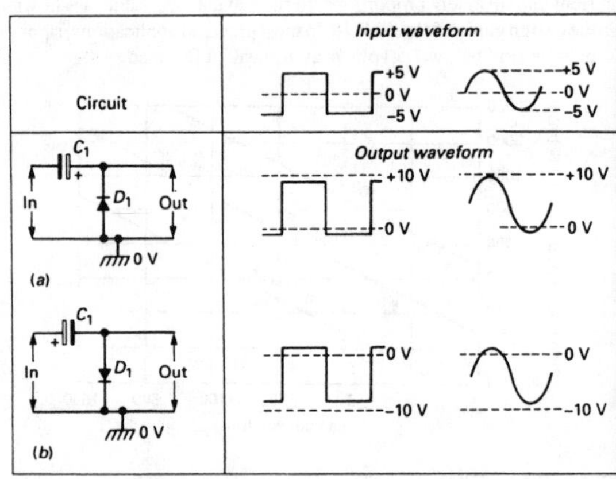

Figure 7.20. Clamping diode circuits.

Two important points should be noted about these apparently-simple circuits. First, their peak output is (ideally) equal to the peak-to-peak value of the input waveform; thus, if the input swings symmetrically about the zero voltage point (as shown), the peak output value is double that of the input. The second point is that the circuits fall short of the ideal in that the output is in fact clamped to a point that is offset from zero by an amount equal to the diode's V_f value (about 600mV in silicon types), as illustrated in these and many other diagrams in this chapter.

Figure 7.21 shows what happens to these circuits when a 10k resistor is wired across D_1, and the inputs are fed with a good 1kHz (= 1mS period) square wave from a low impedance source. In this case C_1–R_1 form a differentiator network, with a time constant equal to the *C–R* product; if this product is very long (100mS) relative to the 1mS input period, the circuits act like simple clamping diode types, as shown in *(a)* and *(c)*. If the *C–R* product is very short (10µS) relative to the 1mS input period, the *C–R* network converts the square wave's rising and falling edges into positive and negative 'spikes' (each with a peak amplitude equal to the peak-to-peak input value) respectively, and D_1 then eliminates (discriminates against) one or other of these spikes, as

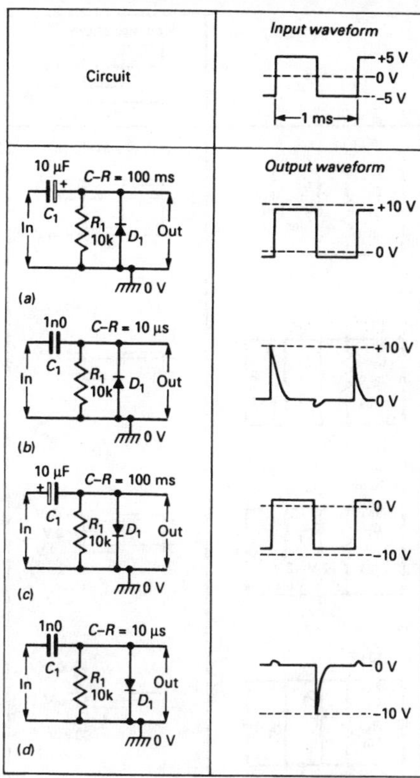

Figure 7.21. Differentiator/discriminator-diode circuits.

shown in *(b)* and *(d)*, which are known as differentiator/discriminator diode circuits.

In an ordinary clamping circuit the diode clamps one edge of the waveform to the zero-voltage reference point. The basic circuit can be used to clamp the waveform edge to a voltage other than zero by simply tying the 'low' side of the diode to a suitable bias voltage; such circuits are known as biased clamping diode types, and a variety of these (with very long *C–R* products) are shown in *Figure 7.22*.

The *Figure 7.22(a)* circuit uses a +2V clamping point and a 'negative output' diode (as in *Figure 7.20(b)*), so its output swings (ideally) from +2V to –8V when fed from a 10V peak-to-peak input.

Figures 7.22(b) to *(d)* show circuits using pairs of clamping diodes. Obviously, a waveform can not be clamped to two different voltages at the same time, so in these circuits one diode acts as a clamp and the other as a waveform clipper. Thus, in *(b)* the output is clamped to zero volts and clipped at +2V, and in *(c)* it is effectively clamped at –2V and

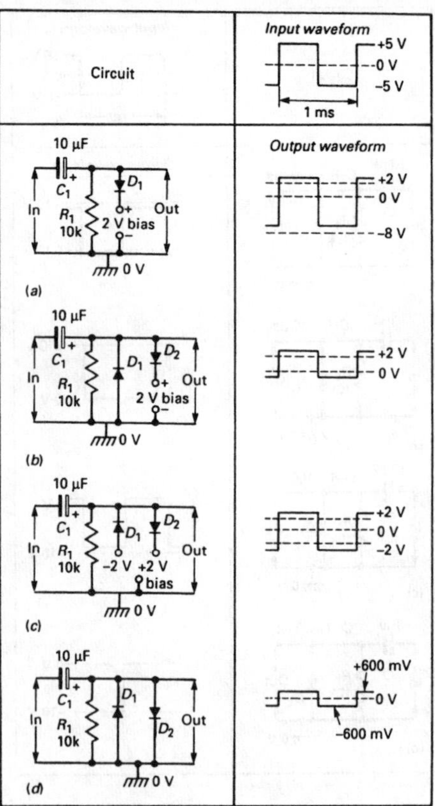

Figure 7.22. Biased clamping diode circuits.

clipped at +2V. Finally, the *(d)* circuit uses a pair of zero-voltage reference points and ideally gives zero output, but, because of the 'offsetting' effects of the D_1 and D_2 V_f voltages (about 600mV each) in fact gives output clipping at +600mV and –600mV.

Diode 'rectifier' circuits

Figure 7.23 shows four different ways of using a single diode as a half-wave rectifier; in all cases it is assumed that the input comes from a low impedance source, the output feeds a high impedance, and the output waveform is 'idealised' (it ignores the effects of diode offset). Thus, the *(a)* and *(d)* circuits give positive outputs only, and *(b)* and *(c)* give negative outputs only. Note, however, that *(a)* and *(b)* give low impedance outputs (roughly equal to the input signal source impedance), but that *(c)* and *(d)* have high-impedance outputs (roughly equal to the R_1 value).

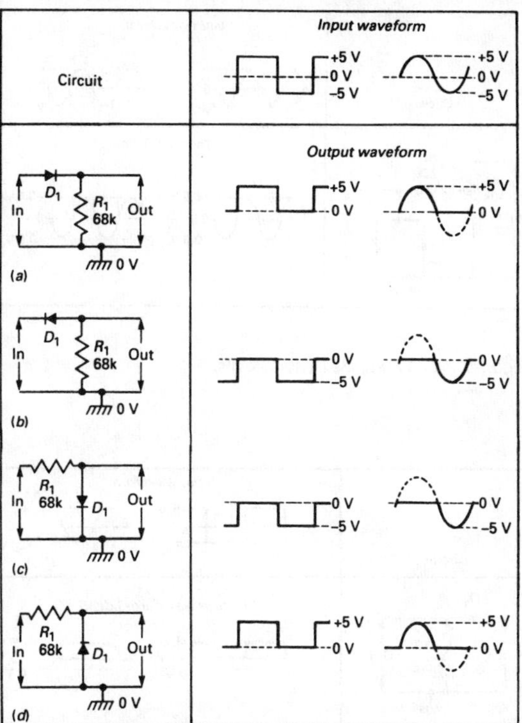

Figure 7.23. Single-diode rectifier circuits.

Figure 7.24 shows a signal 'limiter' that gives an output that is amplitude-limited at ±600mV via silicon diodes D_1 and D_2. It can be used as a triangle-to-sine waveform converter by adjusting RV_1 to give gentle clipping of the triangle peaks (generated sinewave distortion is typically about 2%), or as an audio signal noise limiter by adjusting RV_1 to give clipping of the worst of the noise bursts, as shown.

Figure 7.25 shows how the *Figures 7.23(a)* and *(b)* circuits can be modified to give outputs that are above or below a selected 'bias' or reference level. Thus, *(a)* produces outputs of only +2 volts or greater, *(b)* gives outputs of +2 volts or less, *(c)* of –2 volts or greater, and *(d)* of –2 volts or less. In all cases, the output load impedance is assumed to be small relative to the R_1 value.

Voltage multiplier circuits

Figures 7.26 to *7.28* show various ways of connecting diodes and capacitors to make A.C. 'voltage multipliers' that give a D.C. output equal to some multiple of the *peak* value of an A.C. input voltage. The

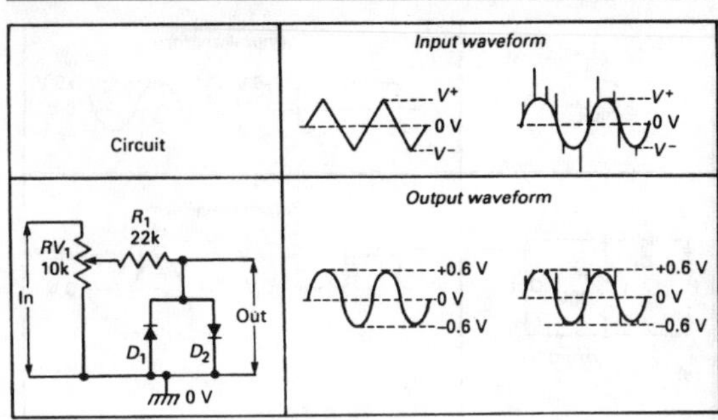

Figure 7.24. Two-diode limiter circuits.

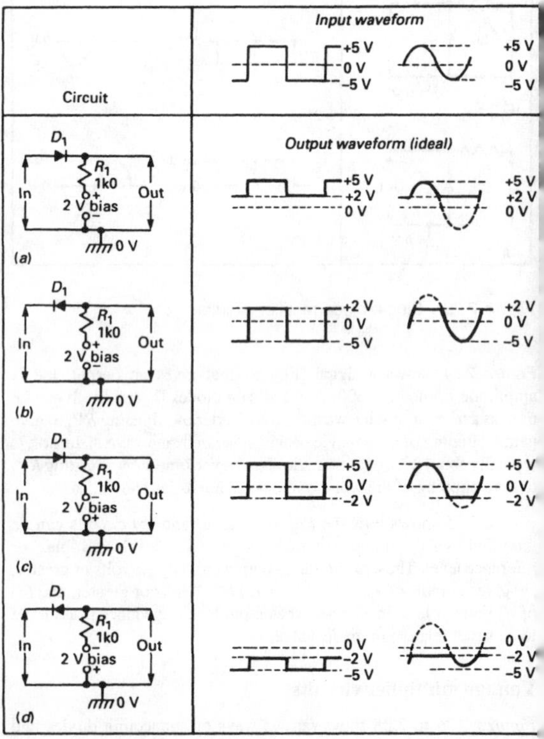

Figure 7.25. Biased single-diode rectifier circuits.

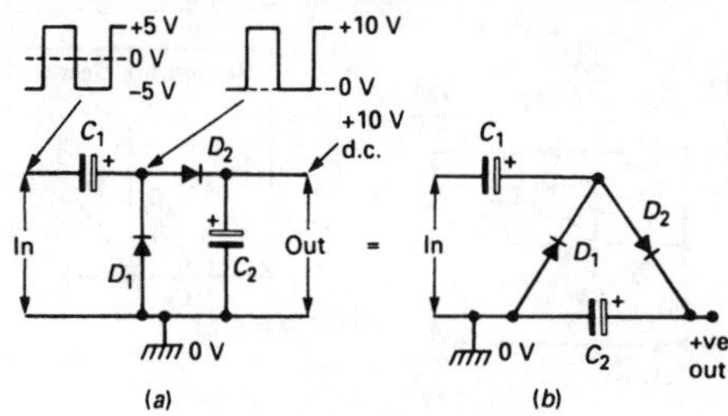

Figure 7.26. Voltage doubler circuit.

'voltage doubler' of *Figure 7.26* consists of a simple C_1–D_1 clamping diode network (like *Figure 7.20(a)*), which gives an A.C. output with a peak value equal to the peak-to-peak value of the input, followed by a peak voltage detector (D_2–C_2) that gives a D.C. output equal to the peak values of D_2's input voltage. *Figure 7.26(a)* shows the conventional diagram of this circuit, and *(b)* shows it redrawn as a 'standard' voltage-doubler section.

Figure 7.27 shows a 'voltage tripler' circuit. In this case (as can be seen from (a)) D_3–C_3 acts as a peak voltage detector that generates +5V on the D_3–C_3 junction, and C_1–D_1–D_2–C_2 acts as a voltage doubler section (identical to *Figure 7.26*) that generates a 'voltage doubled' output on top of this '+5V' potential, thus giving a final 'tripled' output of +15V. This circuit thus consists of a D_3–C_3 'half section' plus a full C_1–D1–D2–C2 'doubler' section, as shown in *(b)*.

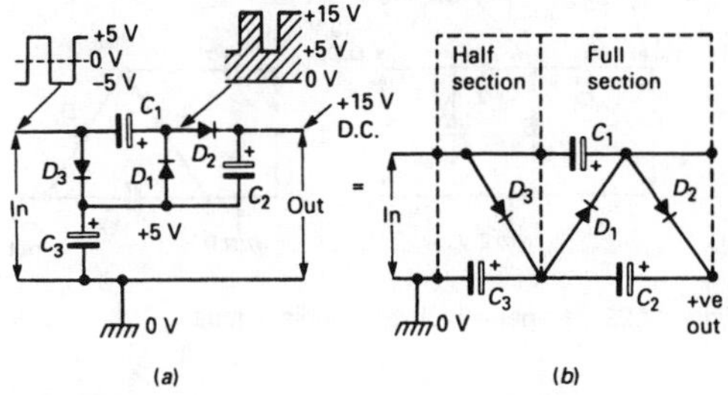

Figure 7.27. Voltage tripler circuit.

Figure 7.28 shows a 'voltage quadrupler' that gives a D.C. output equal to four-times the *peak* voltage value of a symmetrical A.C. input signal. In this case *C*1–D1–D2–*C*2 act as a voltage doubler section that generates +10V on the D2–*C*2 junction, and *C*3–D3–D4–*C*4 act as another voltage doubler section that generates another +10V between the D_2–*C*2 junction and the D_4–C_4 junction, to give a final +20V of output between the D_4–C_4 junction and ground.

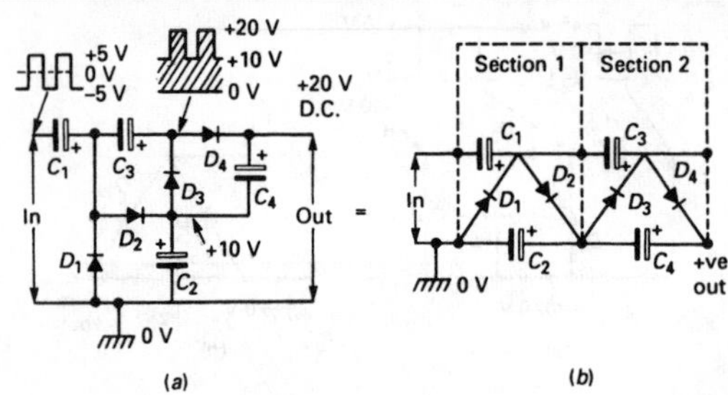

Figure 7.28. Voltage quadrupler circuit.

Note from *Figures 7.27(b)* and *7.28(b)* that any desired amount of voltage multiplication can be obtained by wiring appropriate numbers of full and half 'multiplier' sections in series. Thus, seven-times multiplication can be obtained by wiring three full sections in series with a single 'half' input section. In all cases, all multiplier diodes and capacitors need minimum ratings of twice the peak input voltage value.

The *Figures 7.26* to *7.28* circuits are all designed to give positive output voltages; they can be made to give negative output voltages by simply reversing the polarities of all multiplier diodes and capacitors, as in the negative voltage doubler of *Figure 7.29*.

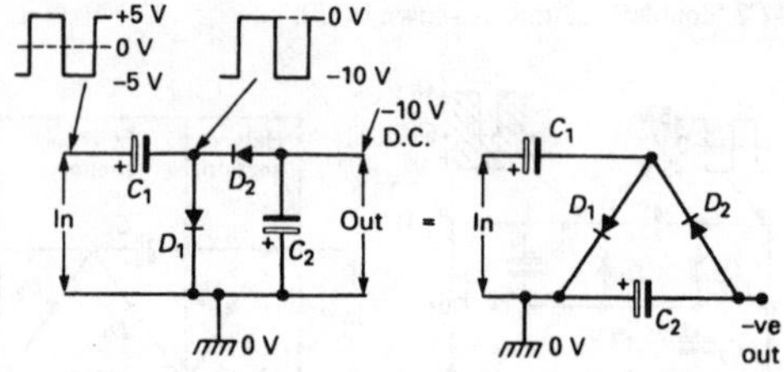

Figure 7.29. Negative voltage doubler circuit.

Damping diode circuits

If an inductive device's operating current is suddenly interrupted it generates a high switch-off back-emf, which may damage associated electronic components, etc. This danger can be eliminated by wiring a damping diode across the inductor, as in the relay circuit of *Figure 7.30*. Here, D_1 stops the RLA-SW_1 junction from swinging more than 600mV above the positive supply. Alternatively, D_2 (shown dotted) can be used to prevent the junction from swinging more than 600mV below the negative supply rail. In critical applications, both diodes can be used, to give maximum protection.

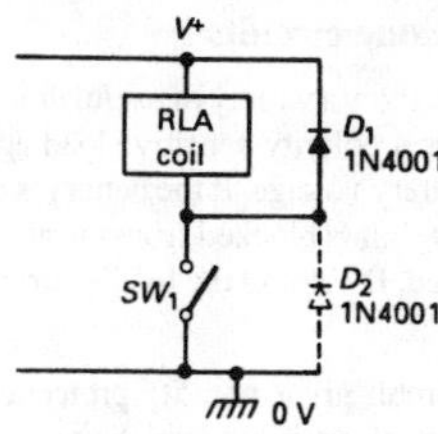

Figure 7.30. Damping diode circuit limits relay coil back-emfs to safe values.

Diode gate circuits

Figure 7.31 shows how a few diodes and a resistor can be used to make an OR gate that gives a high (logic-1) output when any one of its inputs is high, and also shows the truth table of the circuit when it is wired in the 2-input mode. The circuit can be given any desired number of inputs by simply adding extra diodes, as shown dotted by D_3 and D_x.

Figure 7.32 shows an AND type of diode gate; it gives a high output only when all inputs are high, and can have any desired number of diode inputs.

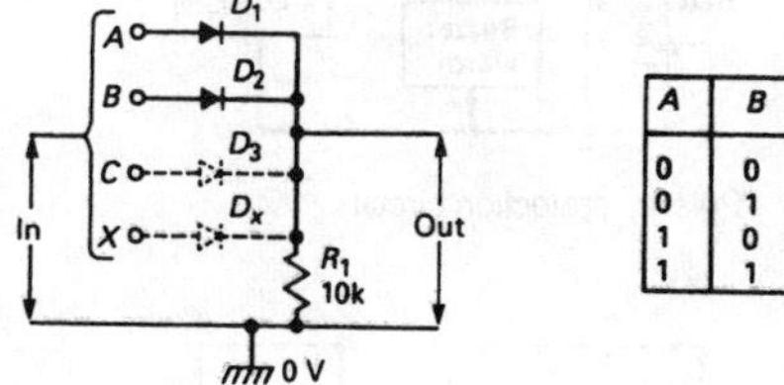

A	B	Out
0	0	0
0	1	1
1	0	1
1	1	1

Figure 7.31. Diode OR gate circuit, with 2-input truth table.

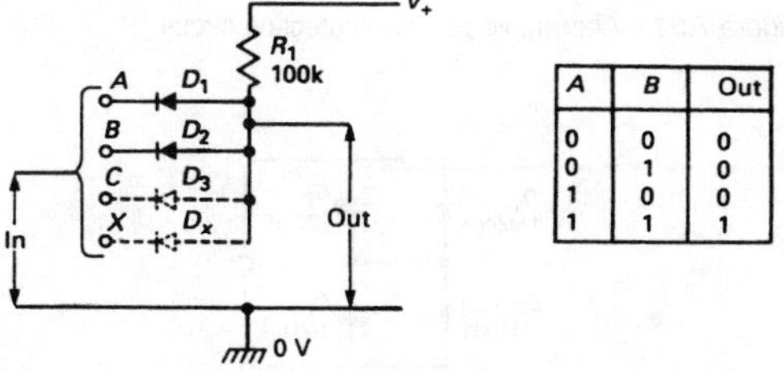

A	B	Out
0	0	0
0	1	0
1	0	0
1	1	1

Figure 7.32. Diode AND gate circuit, with 2-input truth table.

Miscellaneous diode circuits

Figures 7.33 to *7.38* show a variety of useful diode circuits. The *Figure 7.33* design protects a polarity-sensitive load against damage from a wrongly applied battery voltage. If the battery is correctly connected it feeds the load via D_1 but is blocked from the alarm buzzer via D_2; if it is wrongly connected, D_1 blocks the load's current and D_2 enables the alarm buzzer.

The *Figure 7.34* circuit gives polarity protection to the load via the bridge-connected set of rectifiers, which ensure correct load polarity irrespective of the polarity of the supply battery.

Figure 7.35 shows how to make a high-value non-polarised capacitor from two electrolytics and two diodes; each diode effectively shorts out its capacitor if connected to the 'wrong' polarity. The circuit has an effective capacitance equal to the C_1 or C_2 value.

Figure 7.36 shows a pair of silicon diodes used to protect a moving-coil current meter against overload damage. Such meters can withstand 2–3 times full scale deflection (f.s.d.) without damage, and in this circuit R_x must be chosen so that about 300mV is developed across the diodes at f.s.d.; under this condition the diodes pass zero current, but they start to conduct and shunt the meter current at readings of twice f.s.d. or greater.

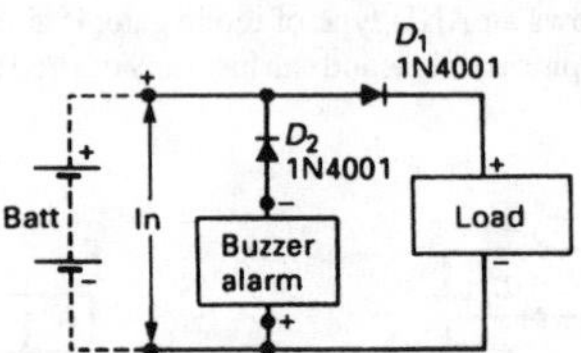

Figure 7.33. Polarity protection circuit.

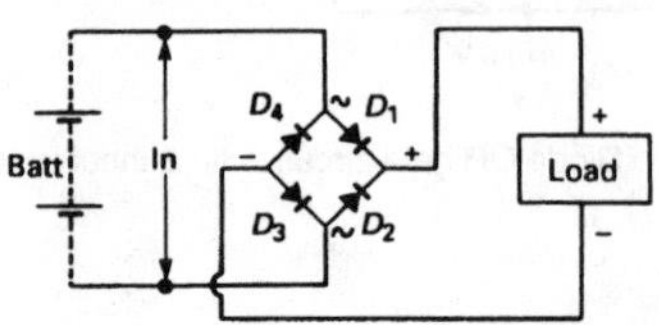

Figure 7.34. Alternative polarity protection circuit.

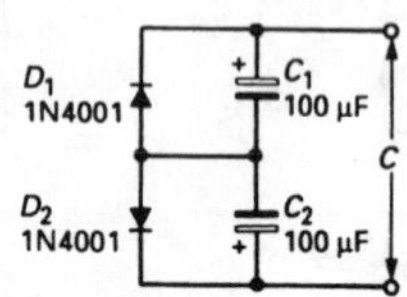

Figure 7.35. Making a high-value (100µF) non-polarised capacitor.

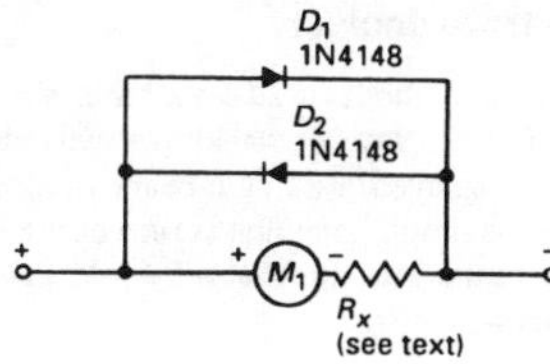

Figure 7.36. D.c.-meter overload protection.

Figure 7.37 shows how two 6 volt relays can be independently controlled via a 12 volt a.c. 2-wire link. The relay coils are wired in series, but are each shunted by a diode so that RLA is turned on only by positive half-cycles and RLB only by negative half-cycles. When SW_1 is set to position 1, both relays are off; in position 2 only positive half-waves are fed to the relays, so RLA turns on; in position 3 only negative half-waves are fed to the relays, so RLB turns on; finally, in position 4 full-wave a.c. is fed to the relays, and RLA and RLB both turn on.

Figure 7.38 shows a modified version of the above circuit, in which each relay can be independently controlled via its own on/off switch. The circuit operates in the same basic way as described above.

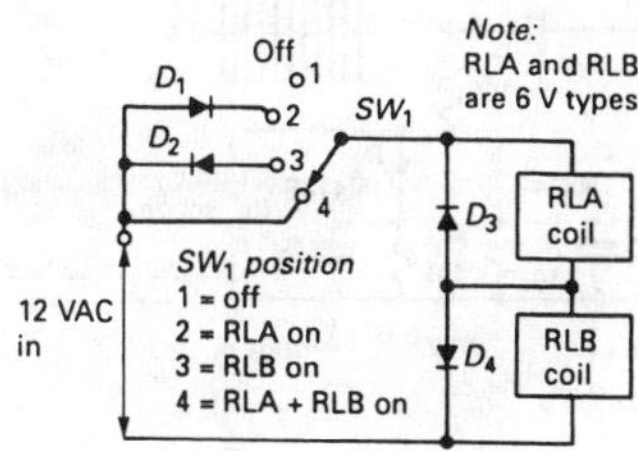

Figure 7.37. Dual relay control.

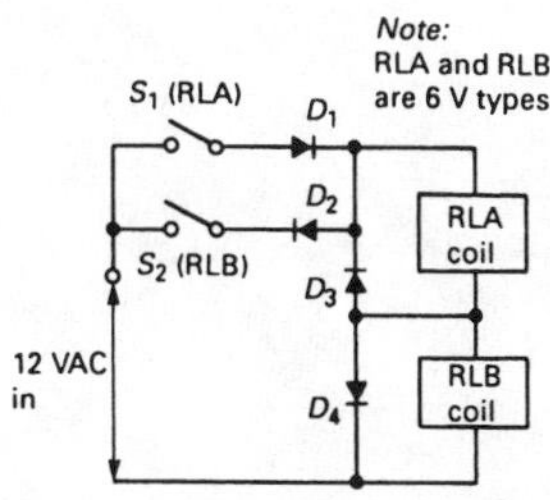

Figure 7.38. Alternative dual relay control.

An oscilloscope trace doubler

Figure 7.39 shows a pair of diodes used as the basis of a simple but very effective oscilloscope trace doubler that allows two individual signals to be simultaneously displayed on a single-beam oscilloscope. The two diodes are connected as simple gates that are driven via a 10 volt square wave input; C_1 causes the gate signal at the C_1–R_1 junction to switch between +5 volts and –5 volts.

When the C_1–R_1 junction is at +5V, D_2 is reverse biased and R_2–C_2 are effectively disconnected from the circuit, but D_1 is forward biased and R_1 and R_3 are effectively shorted together, thus presenting a mean +2.5V, on which the input-1 signal is superimposed, at the output. When the C_1–R_1 junction is at –5V the reverse action is obtained, and D_1 is effectively open circuit and D_2 is short circuit, thus presenting a mean –2.5 volts, on which the input-2 signal is superimposed, at the output. When this complex output signal is fed to the input of a single beam oscilloscope, the vertical switching transitions disappear, and the tube displays input-1 vertically displaced above input-2; the trace separation can be varied by altering the amplitude of the square-wave gate drive signal.

The gate drive square wave frequency can either be made high relative to that of the oscilloscope's time base, or can be exactly half the time base frequency (via a binary divider); in the latter case, the oscilloscope displays the input-1 and input-2 signals on alternate sweeps.

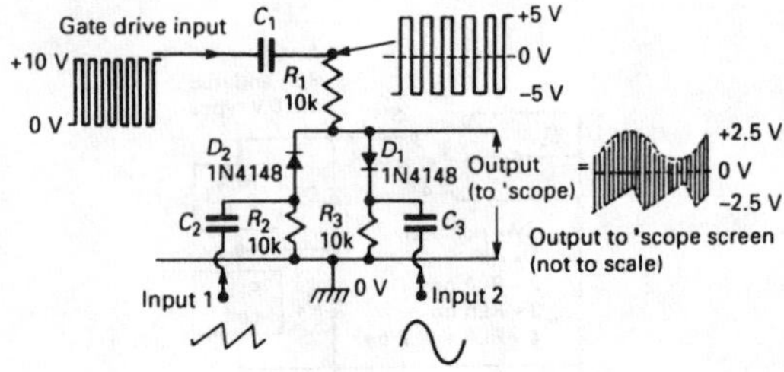

Figure 7.39. Diode gate circuit used as an oscilloscope trace doubler.

8 Special diode circuits

The last chapter looked at the basic characteristics of the junction diode and associated devices and then showed a whole range of practical applications of ordinary diodes and rectifiers. The present chapter continues the theme by showing practical applications of zener, 'signal', and varactor diodes.

Zener diode circuits

Zener diodes are used mainly as 'reference voltage' generating devices. *Figure 8.1* shows how a zener diode can be used to generate a fixed voltage by passing a modest current through it via limiting resistor R_1. The zener output voltage is not greatly influenced by sensible variations in the zener current value (caused by variations in supply voltage or R_1 value, etc), and the output thus acts as a stable reference voltage. The R_1 value is found from:

$$R_1 = (V_{IN} - V_Z)/I_Z.$$

In most applications, I_Z should be set at about 5mA.

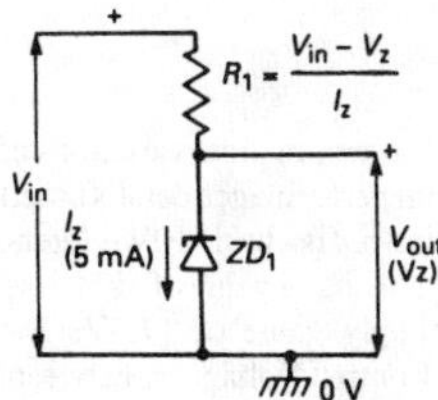

Figure 8.1. Basic zener voltage reference circuit.

Figure 8.2 shows the zener circuit modified to act as a simple voltage regulator that can supply output currents of a few tens of mA to an external load. The R_1 value is selected so that it passes the maximum desired output current plus 5mA; thus, when the specified maximum output load current is drawn the zener passes only 5mA, but when zero load current is drawn it passes all of the R_1 current and dissipates maximum power. The zener's power rating must not be exceeded under this 'no load' condition.

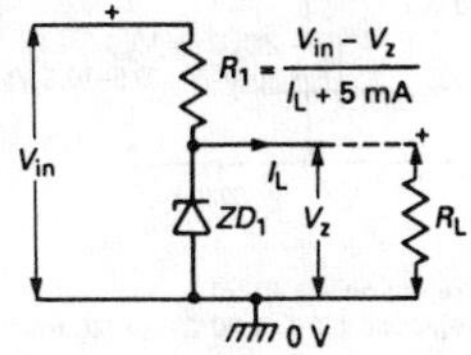

Figure 8.2. Basic zener voltage regulator circuit.

Practical zener diodes are available in a variety of voltage values (ranging in steps from 2.7V to about 100V) and power ratings (typically 500mW, 1.3W, 5W, and 20W), and usually have a basic voltage tolerance within ±5% of their specified value. Other important parameters of a zener diode are its temperature coefficient ('t.c.') and its dynamic impedance. *Figure 8.3* lists the typical parameter values of 500mW zeners with standard voltages in the 2.7V to 16V range.

Vz (volts)	Temp. coeff. (mV/°C)	Dynamic impedance (ohms)	Vz (volts)	Temp. coeff. (mV/°C)	Dynamic impedance (ohms)
2.7	−1.8	120	6.8	+2.7	15
3.0	−1.8	120	7.5	+3.7	15
3.3	−1.8	110	8.2	+4.5	20
3.6	−1.8	105	9.1	+6.0	25
3.9	−1.4	100	10.0	+7.0	25
4.3	−1.0	90	11	+8.0	35
4.7	+0.3	85	12	+9.0	35
5.1	+1.0	75	13	+10.5	35
5.6	+1.5	55	15	+12.5	40
6.2	+2.0	27	16	+14.0	40

Figure 8.3. Typical parameter values of 500mW, 2.7V to 16V zener diodes.

Figures 8.4 to *8.7* show some practical variations of the zener voltage reference circuit, with full performance details based on the parameters listed in *Figure 8.3*. *Figure 8.4* is a basic 10V reference circuit, powered from a 15V to 20V input. R_1 has a value of 1k5, to set the zener current at 5mA at a 'mean' supply value of 17.5V. The zener has a 5% tolerance, so the actual output voltage is between 9.5V and 10.5V. Supply voltage variations (between 15V and 20V) cause the zener current to vary by ±1.6mA and, since the 10V zener has a dynamic impedance of 25Ω, makes the zener output voltage vary by an additional ±40mV (the supply rejection factor). Also, since this zener has a t.c. of +7mV/°C, the output varies by an additional ±140mV when the temperature is varied by ±20°C about a mean +20°C 'room temperature' value.

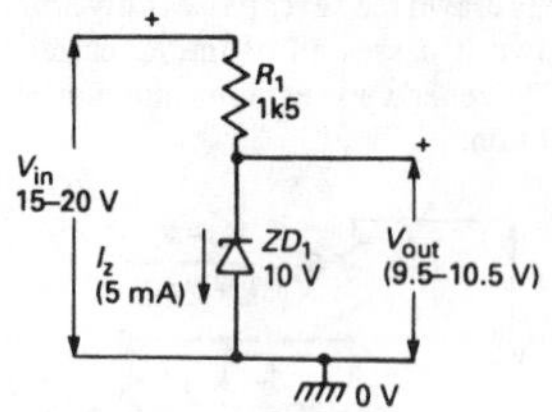

Supply rejection = ± 40 mV
Temp. rejection (20°C ± 20°C) = ±140 mV

Figure 8.4. Simple 10 volt zener reference circuit.

Figure 8.5 shows how the above circuit's thermal regulation can be improved by using two series-connected zeners, with opposing t.c.s, to act as a 'composite' zener with a nominal value of 10.1V (giving an actual voltage in the range 9.4V to 10.6V) and a t.c. of only 0.6mV/°C (giving a variation of only ±12mV over the +20°C±20°C range). Note, however, that this 'zener' has a dynamic impedance of 127Ω, and thus gives a supply rejection factor of ±203mV.

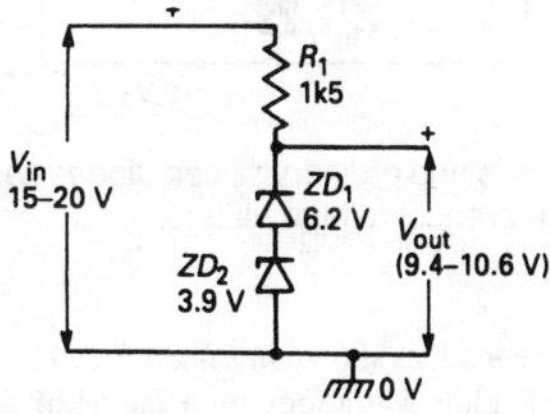

Figure 8.5. Thermally-compensated 10 volt zener reference circuit.

Figure 8.6 shows how the above circuit's performance can be improved by adding a pre-regulating zener stage (ZD_1), which holds the R_1–R_2 junction within ±265mV of a nominal 13V over the full span of supply-voltage variations, thus giving a final supply rejection factor of ±53mV and a thermal rejection factor of ±12mV over the full thermal range. R_2 gives a '10V zener' current of about 5mA, and R_1's value is chosen to supply a current greater than this when V_{IN} is at 15V, so that ZD_1 does not cut off under this condition.

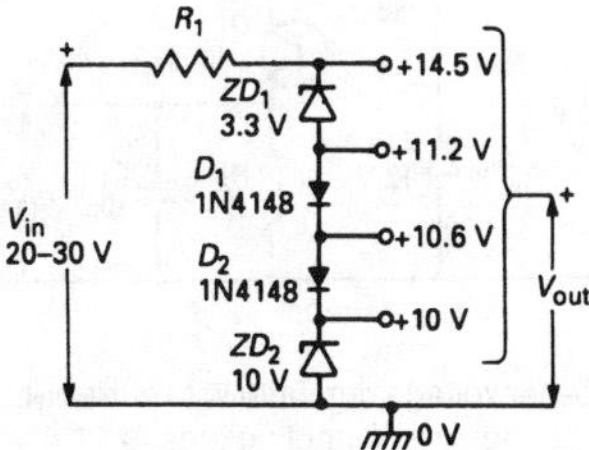

Figure 8.6. Thermally-compensated 10 volt zener reference, with pre-regulator stage.

Finally, *Figure 8.7* shows how zeners and ordinary silicon diodes can be wired in series to give 'odd-ball' reference voltage values. Each silicon diode 'drops' about 600mV at a forward current of 5mA, and has a temperature coefficient of –2mV/°C.

Regulator current boosting

The simple voltage regulator of *Figure 8.2* can supply maximum output load currents of a few tens of mA. Higher outputs can be obtained by wiring a current-boosting emitter follower buffer stage between the

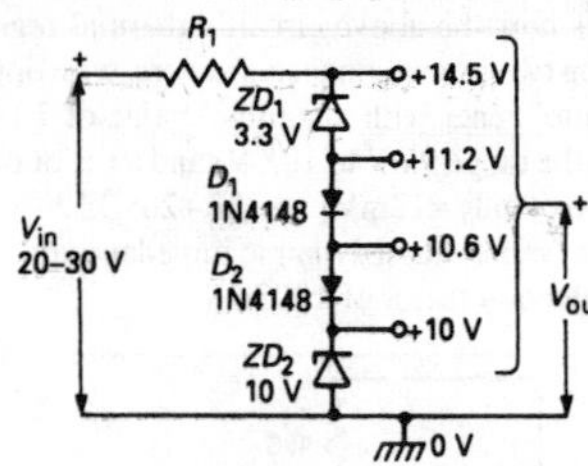

Figure 8.7. Zeners and ordinary silicon diodes can be combined to give odd-ball reference voltage values.

zener output and the load, as shown in *Figure 8.8*. This circuit reduces the zener current loading variations by a factor of about 100 (Q_1's current gain). Note, however, that the output voltage is about 600mV (equal to Q_1's base-emitter voltage drop) less than the zener voltage; this snag can be overcome either by wiring a silicon diode in series with the zener diode (to boost Q_1's input voltage by 600mV), as shown in *Figure 8.9*, or by wiring Q_1 into the feedback loop of an op-amp voltage follower stage, as in *Figure 8.10*.

Note that the output load current of each of the above three circuits is limited to about 100mA by the power rating of Q_1; higher currents can be obtained by replacing Q_1 with a power Darlington transistor.

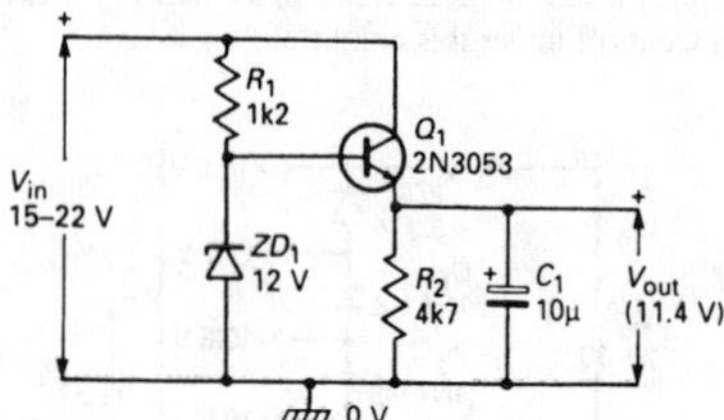

Figure 8.8. Zener voltage regulator with current-boosting series-pass output stage; nominal output voltage is 11.4 volts.

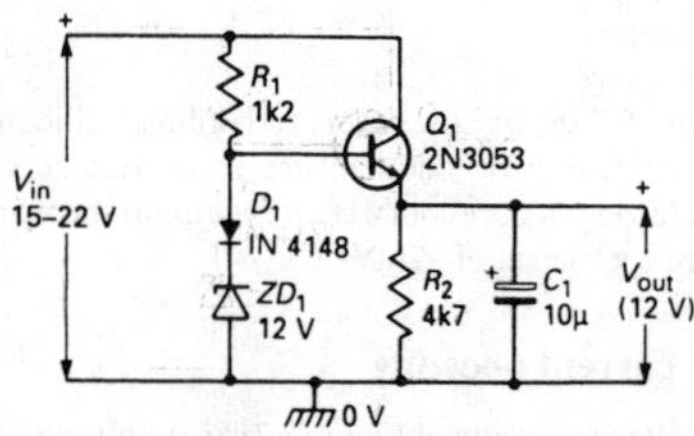

Figure 8.9. This modified series-pass circuit gives an output of 12 volts.

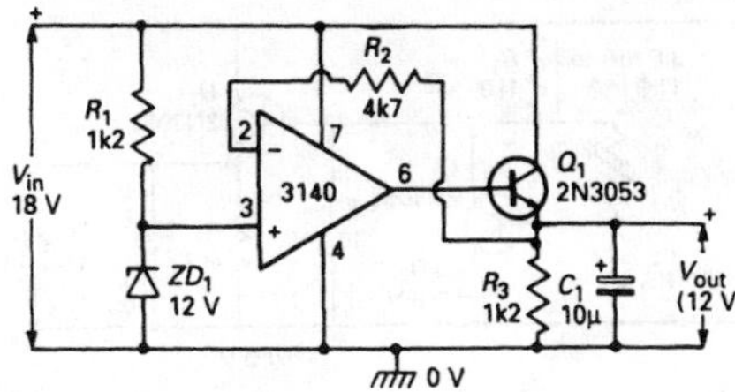

Figure 8.10. 12 volt regulator incorporating an op-amp in its series-pass network.

Variable voltage circuits

Figures 8.11 to *8.13* show various ways of generating zener-derived variable reference or regulator voltages. In *Figure 8.11,* Q_1 is wired as a modified common emitter amplifier, and gives an output of (1 + [RV_1/ R_2]) times Q_1's base-to-ground voltage, which equals the sum of Q_1's base-emitter junction voltage (600mV) and the ZD_1 voltage (6.2V), i.e., 6.8V total. The circuit's output is thus variable from 6.8V to 13.6V via RV_1. Note that Q_1's base-emitter junction has a –2mV/°C t.c. and ZD_1 has a +2mV/°C one, and these cancel each other and give a final near-zero t.c. at the circuit's output.

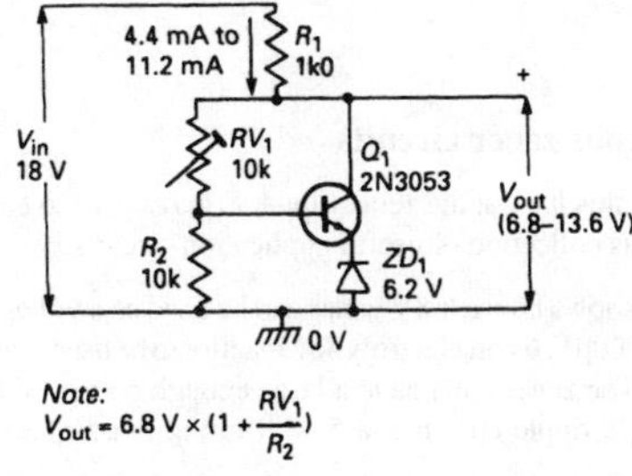

Figure 8.11. Variable zener voltage reference, with near-zero temperature coefficient.

Figure 8.12 shows the above circuit modified for use as a variable voltage regulator that gives a current-boosted output via series-pass transistor Q_2. In this case ZD_1 is a 7.5V type and has a t.c. of +3.7mV/ °C, thus giving an 8.1V to 16.2V reference with a +1.7mV/°C coefficient to the input (base) of Q_2, which gives an output that is 600mV and –2mV/°C less than this, thus giving a final output of 7.5V to 15.6V with a near-zero temperature coefficient.

Finally, *Figure 8.13* shows a simple way of generating a stable 0–12V output via RV_1 and the Q_1 current-booster stage. D_1 boosts the effective zener voltage by 600mV, to counter the 600mV loss of Q_1's base-emitter junction. The final output impedance of this circuit is fairly high (typically a few tens of ohms), being roughly equal to the output impedance of RV_1 slider divided by the current gain of Q_1.

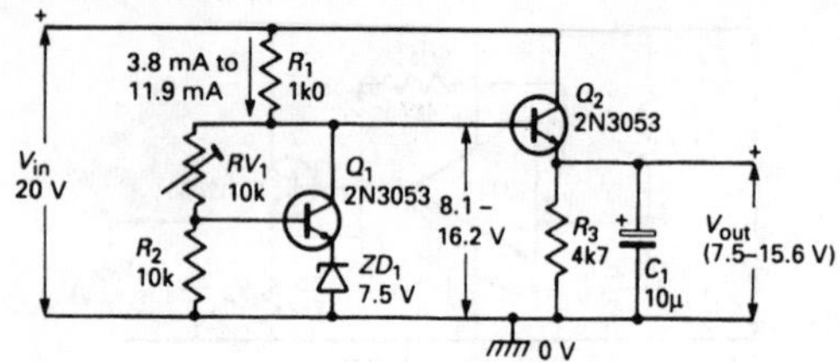

Figure 8.12. Variable voltage regulator, with near-zero temperature coefficient.

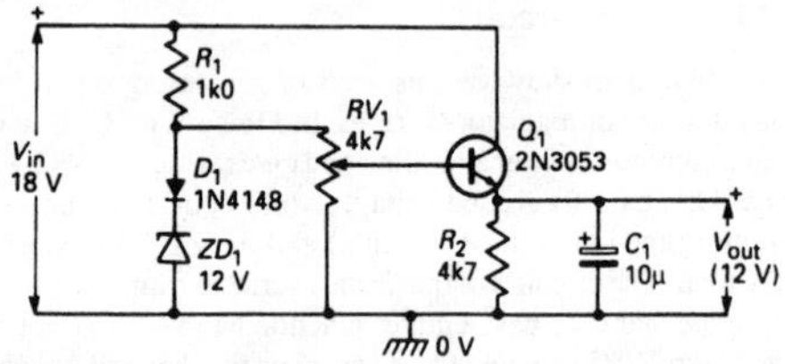

Figure 8.13. Wide-range (0–12V) voltage reference/regulator circuit.

Miscellaneous zener circuits

To complete this look at the zener diode, *Figures 8.14* to *8.18* show a miscellaneous collection of useful application circuits.

Figure 8.14 shows how a 6.8V zener can be used as a voltage dropper to enable a 1000µF, 6 volt electrolytic capacitor to be used with a 12 volt d.c. supply. The zener must have a large enough power (V x I) rating to handle C_1's ripple currents; a 5 watt rating is adequate for most purposes.

Figure 8.15 shows a 5.6V zener used as a voltage dropper to enable a 6 volt relay to be used with a 12V d.c. supply. This circuit also helps improve the relay's effective on/off ratio. Suppose the basic relay

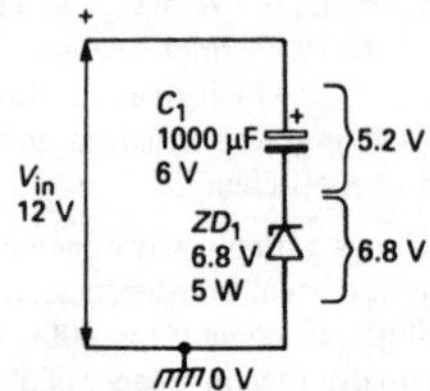

Figure 8.14. Zener voltage dropper used with electrolytic capacitor.

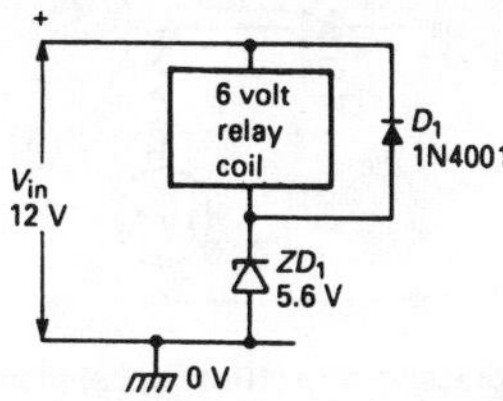

Figure 8.15. Zener voltage dropper used with relay coil.

normally turns on at 5V and off at 2.5V, thus giving a 2:1 on/off ratio; in this circuit it will turn on at 10.6V and off at 8.1V, thus giving a 1.3:1 on/off ratio.

Figure 8.16 shows a resistor and zener diode used as a half-wave limiter in an a.c. circuit; the waveform's positive halves limit at ZD_1's rated voltage value, and the negative ones its –600mV 'junction diode' value.

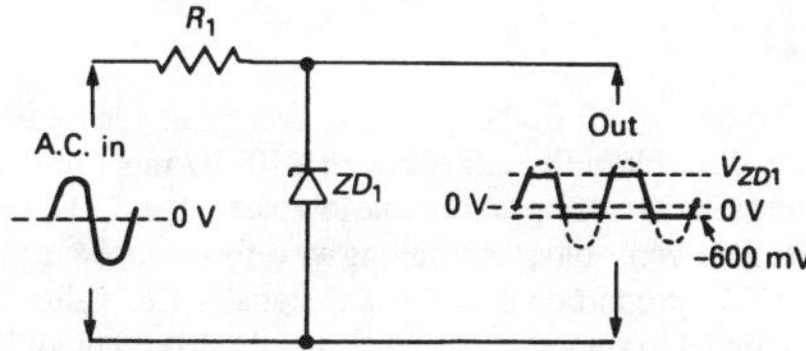

Figure 8.16. Half-wave zener limiter.

Figure 8.17 shows a resistor and two inversely series-connected zener diodes used as a full-wave limiter in an a.c. circuit. In this case the positive halves of the waveform are limited by the sum of ZD_1's voltage value and ZD_2's 600mV 'diode' value, and the negative halves by the sum of ZD_2's voltage value and ZD_1's 600mV 'diode' value.

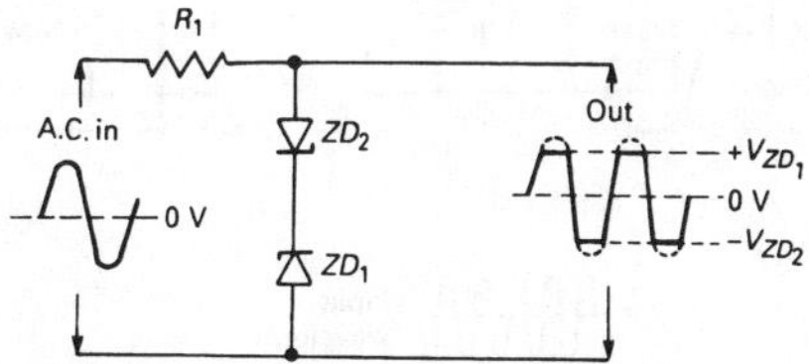

Figure 8.17. Full-wave zener limiter.

Finally, *Figure 8.18* shows a zener, a multiplier resistor (R_1), and a 1mA f.s.d. moving coil meter used to make a suppressed-zero meter, that in this case spans the range 10V to 15V. The zener sets the minimum voltage reading of the meter (10V), and R_1 is given a value of 1000Ω per volt to set its span (5V) and thus its f.s.d. value (15V).

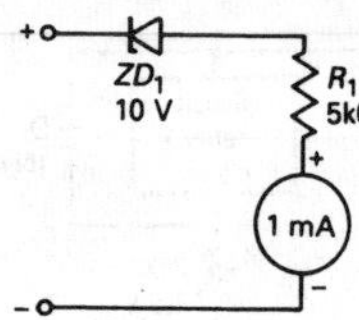

Figure 8.18. Suppressed-zero (10–15 volts) meter.

Signal diode circuits

Any diode that has a low knee voltage value and is able to efficiently 'detect' or rectify low-level RF signals can be called a 'signal' diode. Most germanium and Schottky types can be used as signal diodes; silicon types cannot. The most obvious applications of signal diodes is in RF probes, sniffers, and tell-tales in instrumentation applications, and as detectors in radio circuits. The next three sections describe simple circuits of these types.

RF probes

Figure 8.19 shows an RF-to-d.c. converter probe suitable for use with an electronic d.c. volt/millivoltmeter with a 10MΩ input impedance. C_1 and D_1 form a diode clamp that ties the low part of the C_1–D_1 junction waveform to near-zero volts; the resulting waveform thus has a positive mean d.c. value proportional to the RF signal's a.c. value. R_1–C_2 convert this signal to smooth d.c., and R_1 and the 10MΩ input impedance of the electronic voltmeter form a potential divider that gives form-factor correction and (ideally) makes this voltage equal to the

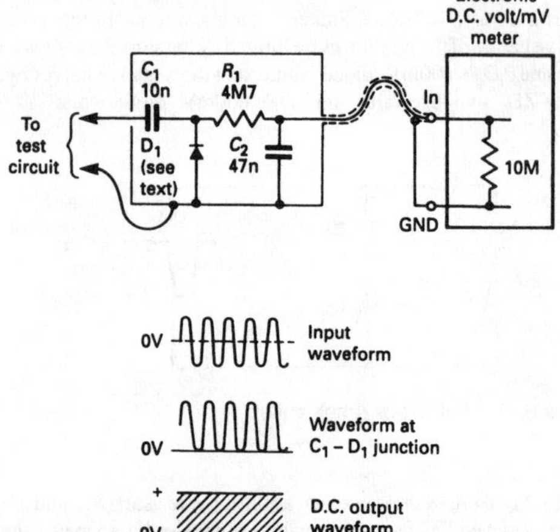

Figure 8.19. RF probe; gives a signal strength reading on a high-impedance (10MΩ) d.c. voltmeter.

r.m.s. value of a sinewave input signal, assuming that the signal amplitude is greater than a couple of volts. D_1 must be a sensitive germanium signal diode, such as an IN34A or OA91, etc., in which case the probe may have a useful bandwidth that extends well above 100MHz, and gives a useful performance as a 'relative value' indicator at signal levels down to about 200mV. The circuit can be made to give a negative (rather than positive) d.c. output by reversing D_1's polarity.

The *Figure 8.20* circuit is a modification of the above type of probe, with R_1 replaced with a sensitive (IN34A or OA91, etc.) germanium diode, so that C_2 charges to the peak value of the C_1–D_1 junction signal and gives a positive d.c. output voltage proportional to the peak-to-peak value of the RF input signal. Note that the electronic voltmeter's input resistance acts as a discharge path for C_2, and influences the probe's ability to follow rapid variation in input signal levels.

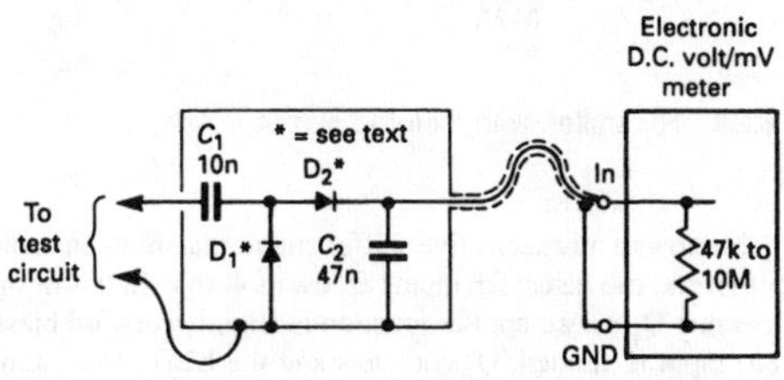

Figure 8.20. RF peak-to-peak voltage-detecting probe.

Figure 8.21 shows a simple AM 'demodulator' type of RF probe, which has an input impedance of about 10k. The probe's output consists of the demodulated AM signal superimposed on a d.c. component proportional to the amplitude of the RF input carrier wave. If desired, this d.c. component can be removed by taking the output via a blocking capacitor. This type of probe is useful in 'signal tracer' applications.

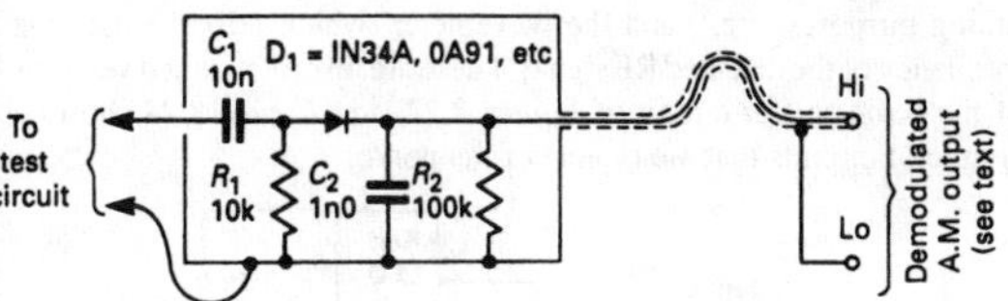

Figure 8.21. AM demodulator probe.

RF sniffers & tell-tales

RF sniffers and tell-tales are probes that enable the user to inspect an RF signal by probing into its radiated field, rather than by making direct contact with the signal source.

An RF sniffer is an untuned gadget that simply detects and indicates the presence of any reasonably powerful RF field, but conveys no special information as to its strength or frequency. *Figure 8.22* shows a simple RF sniffer with a metered output; it is not a very sensitive design and needs an input (across L_1) of a few hundred millivolts to give a reasonable reading, but operates to above 100MHz.

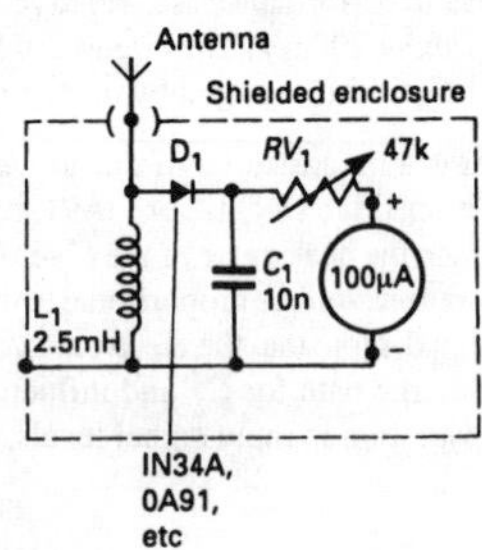

Figure 8.22. RF sniffer, with metered output.

Figure 8.23 shows a more sensitive sniffer circuit that gives an audio-visual output and can detect RF inputs as low as 40mV. In this design D_2 ensures that Q_1's base-emitter junction is slightly forward biased when zero input is applied; Q_1 conducts and the LED glows dimly under this condition. Any RF signals picked up by L_1 are detected/demodulated via D_1 and RF-filtered by C_1–R_1, and further increase both Q_1's forward bias and the LED's brightness level; if the RF input signal is amplitude modulated an amplified version of its modulation signal appears across R_2 and can be heard on a pair of earphones connected across the circuit's output terminals. This circuit gives a useful performance to above 100MHz.

An RF 'tell-tale' is an instrument that detects the presence of an RF field and presents the user with useful information about it. The two best-known versions of such an instrument are the *field strength meter*, which gives the user a reading of the *relative* strength of the signal (for tuning purposes, etc,), and the wavemeter, which tells the user the frequency of the detected RF signal. These are, in effect, tuned versions of the basic sniffer circuit of *Figure 8.22*, and *Figure 8.24* shows a couple of circuits that are typical of the genre.

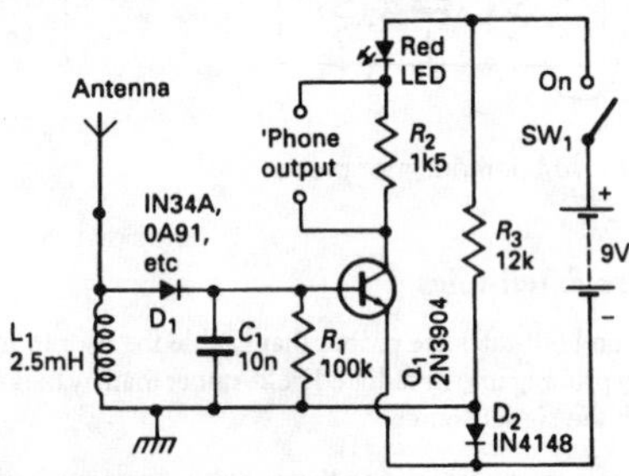

Figure 8.23. RF sniffer with audio-visual output.

The L_1 (or T_1) and C_1 values of the *Figure 8.24* circuits must be chosen to suit the frequency-band of interest; often, the tuning coil is externally mounted and used as an antenna. A field-strength meter is usually designed to operate over only a narrow spread of frequencies, and C_1 may be a trimmer capacitor. A wavemeter is usually designed to operate over a very wide frequency band; it may use a set of plug-in coils, and its tuning capacitor has a calibrated tuning scale that gives a direct reading of the tuned frequency.

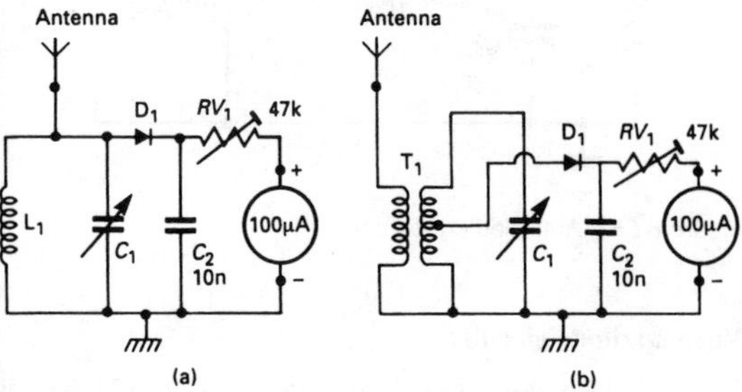

Figure 8.24. Typical wavemeter/field-strength-meter circuits.

RF detectors

A 'detector' circuit is one that extracts the superimposed audio signal from a modulated RF carrier wave and presents it in a useful form. The simplest circuit of this type is the 'crystal' set of *Figure 8.25*, in which D_1 rectifies the L_1–VC_1 tuned signal and feeds it to a set of earphones, which ignore the signal's residual RF components and reproduce its AF contents. This circuit is not very sensitive or selective, and needs a good antenna and ground connection to give reasonable results. Its performance can be enhanced by feeding its output to an audio amplifier, as in the circuit of *Figure 8.26*, which can be turned on and off by connecting or removing the earphones.

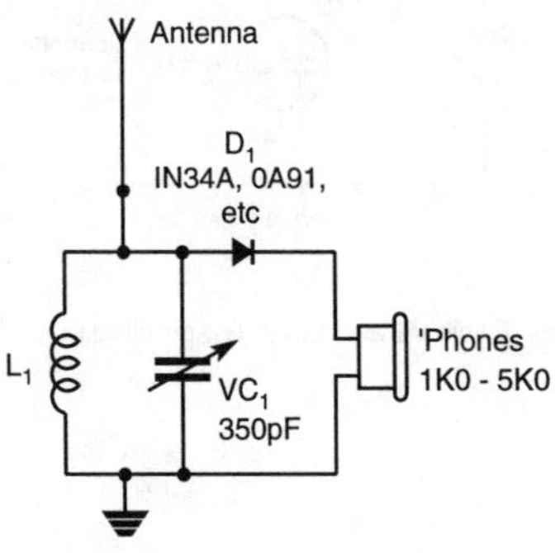

Figure 8.25. Simple crystal set.

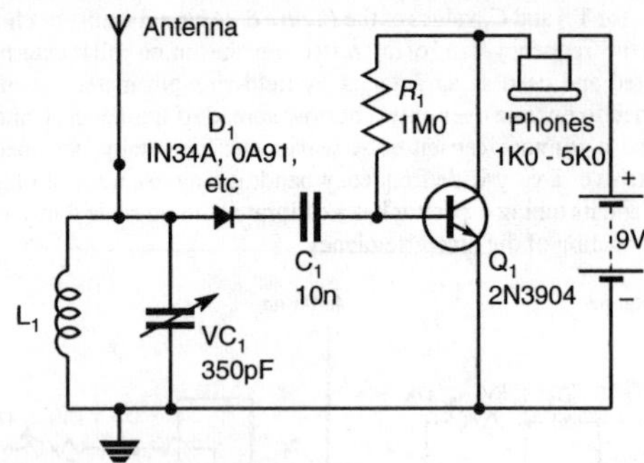

Figure 8.26. Amplified crystal set.

Varicap diode circuits

To complete this look at special diode circuits, *Figure 8.27* shows a basic varicap diode usage circuit. The diode is reverse biased via R_1 and a stable external control voltage (typically variable from zero to 10V), and the varicap is coupled to an external circuit via blocking capacitor C_1. D_1's capacitance is maximum at zero bias, and decreases as bias is increased.

Ordinary silicon diodes have maximum (zero bias) capacitances of a few pF and have typical maximum-to-minimum *'C'* ratios of about 2:1, but true varicap diodes (which are often available as matched pairs) are available with maximum values of about 500pF and *'C'* ratios of 20:1 (i.e., *'C'* can be voltage-controlled from 25pF to 500pF). They are widely used in voltage-controlled tuning applications, etc.

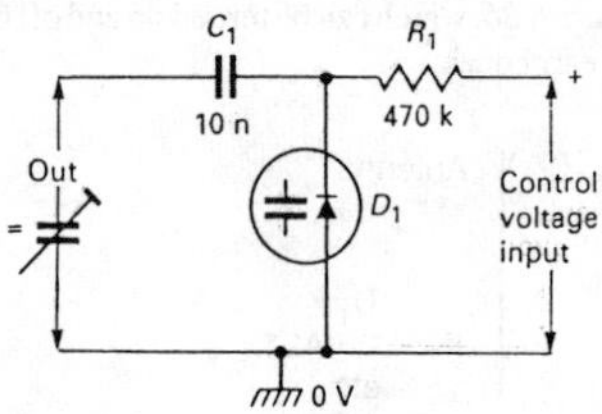

Figure 8.27. Basic varicap diode usage circuit.

9 Transistor principles

The bipolar transistor is the most important element used in modern electronics, and forms the basis of most linear and digital ICs and op-amps, etc. In its discrete form it can function as either a digital switch or as a linear amplifier, and is available in many low-, medium- and high-power forms. This chapter looks at basic transistor characteristics and circuit configurations; Chapters 10 to 12 show many practical application circuits.

Bipolar transistor basics

A bipolar transistor is a three-terminal (base, emitter, and collector) current-amplifying device in which a small input current can control the magnitude of a much larger output current. The term 'bipolar' means that the device is made from semiconductor materials in which conduction relies on both positive and negative (majority and minority) charge carriers.

A transistor is made from a three-layer sandwich of n-type and p-type semiconductor materials, with the base terminal connected to the central layer, and the collector and emitter terminals connected to the outer layers. If it uses an n–p–n construction sandwich, as in *Figure 9.1(a)*, it is known as an npn transistor and uses the standard symbol of *Figure 9.1(b)*; if it uses a p–n–p structure, as in *Figure 9.2(a)*, it is known as a pnp transistor and uses the symbol of *Figure 9.2(b)*.

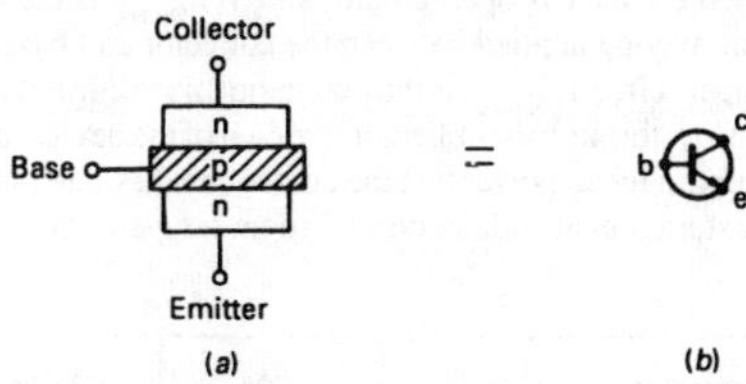

Figure 9.1. Basic construction *(a)* and symbol *(b)* of npn transistor.

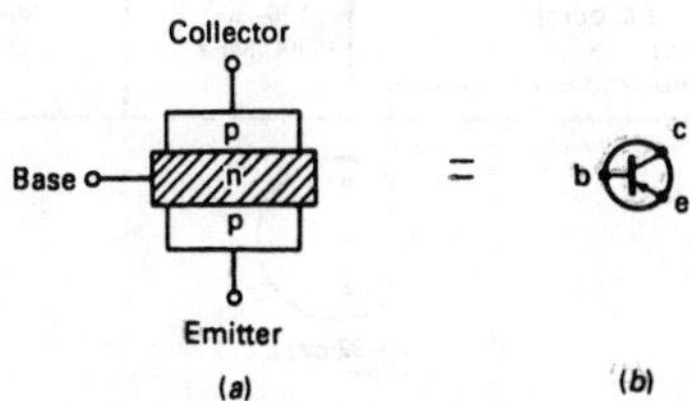

Figure 9.2. Basic construction *(a)* and symbol *(b)* of pnp transistor.

In use, npn and pnp transistors each need a power supply of the appropriate polarity, as shown in *Figure 9.3*. An npn device needs a supply that makes the collector positive to the emitter; its output or main-terminal signal current flows from collector to emitter and its amplitude is controlled by an input current that flows from base to emitter via an external current-limiting resistor (R_b) and a positive bias voltage. A pnp transistor needs a negative supply; its main-terminal current flows from emitter to collector, and is controlled by an emitter-to-base input current that flows to a negative bias voltage.

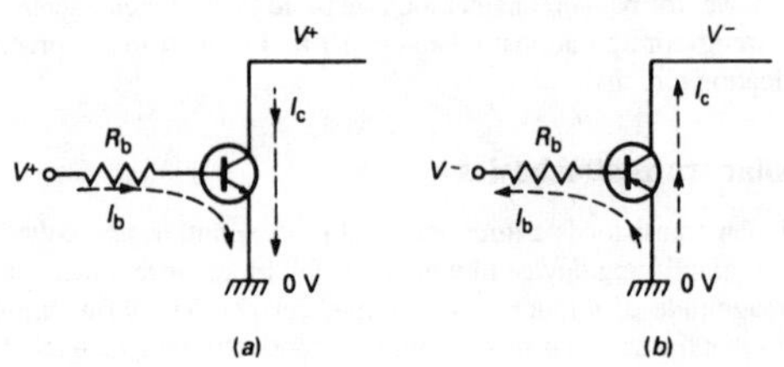

Figure 9.3. Polarity connections to *(a)* npn and *(b)* pnp transistors.

A wide variety of bipolar transistor types are available. *Figure 9.4* lists the basic characteristics of two typical general-purpose low-power types, the 2N3904 (npn) and the 2N3906 (pnp), which are each housed in a TO-92 plastic case. Note from this list that $V_{CEO(max)}$ is the maximum voltage that may be applied between the collector and emitter when the base is open-circuit, and $V_{CBO(max)}$ is the maximum voltage that may be applied between the collector and base when the emitter is open-circuit. $I_{C(max)}$ is the maximum mean current that can be allowed to flow through the collector terminal of the device, and $P_{T(max)}$ is the maximum mean power that the device can dissipate, without the use of an external heat sink, at normal room temperature.

Parameter	2N3904	2N3906
Transistor type	npn	pnp
I_C (max)	200 mA	−200 mA
V_{CEO} (max)	40 V	−40 V
V_{CBO} (max)	60 V	−40 V
P_T (max)	310 mW	310 mW
h_{fe} (= a.c. beta)	100–300	100–300
f_T (typ) = gain/bandwidth product	300 MHz	250 MHz

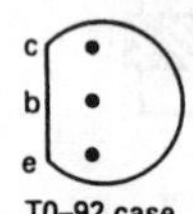

Figure 9.4. General characteristics and outlines of the 2N3904 and 2N3906 low-power transistors.

One of the most important parameters of the transistor is its forward current transfer ratio, or h_{fe}; this is the current-gain or output/input current ratio of the device (typically 100 to 300 in the two devices listed). Finally, the f_T figure indicates the available gain/bandwidth product frequency of the device, i.e., if the transistor is used in a voltage feedback configuration that provides a voltage gain of x100, the bandwidth is 100th of the f_T figure, but if the voltage gain is reduced to x10 the bandwidth increases to $f_T/10$, etc.

Transistor characteristics

To get the maximum value from a transistor, the user must understand both its static (d.c.) and dynamic (a.c.) characteristics. *Figure 9.5* shows the static equivalent circuits of npn and pnp transistors; a zener diode is inevitably formed by each of a transistor's n–p or p–n junctions, and the transistor is thus (in static terms) equal to a pair of reverse-connected zener diodes wired between the collector and emitter terminals, with the base terminal wired to their 'common' point. In most low-power transistors the base-to-emitter junction has a typical zener value in the range 5V to 10V; the base-to-collector junction's typical zener value is in the range 20V to 100V.

Thus, the transistor's base-emitter junction acts like an ordinary diode when forward biased and as a zener when reverse biased. If the transistor is a silicon type its forward-biased junction passes little current until the bias voltage rises to about 600mV, but beyond this value the current increases rapidly. When forward biased by a fixed current, the junction's forward voltage has a thermal coefficient of about –2mV/°C. When the transistor is used with the emitter open-circuit, the base-to-collector junction acts like that just described, but has a greater zener value. If the transistor is used with its base open-circuit, the collector-to-emitter path acts like a zener diode wired in series with an ordinary diode.

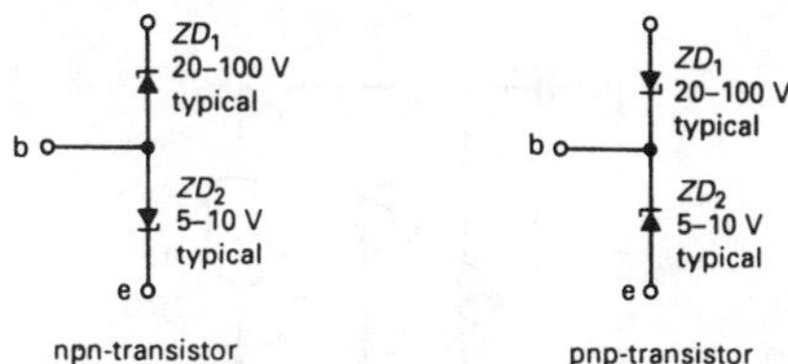

Figure 9.5. Static equivalent circuits of npn and pnp transistors.

The transistor's dynamic characteristics can be understood with the aid of *Figure 9.6*, which shows the typical forward transfer characteristics of a low-power npn silicon transistor with a nominal h_{fe} value of 100. Thus, when the base current (I_b) is zero, the transistor passes only a slight collector leakage current. When the collector voltage is greater than a few hundred millivolts the collector current is almost directly proportional to base current, and is little influenced by the collector voltage value. The device can thus be used as a constant-current generator by feeding a fixed bias current into the base, or can be used as a linear amplifier by superimposing the input signal on a nominal input current.

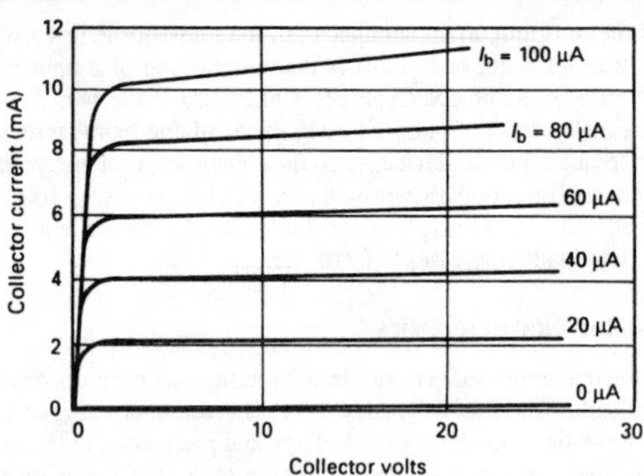

Figure 9.6. Typical transfer characteristics of low-power npn transistor with h_{fe} value of 100 nominal.

Practical applications

A transistor can be used in a variety of different basic circuit configurations, and the remainder of this chapter presents a brief summary of the most important of these. Note that although all circuits are shown using npn transistor types, they can be used with pnp types by simply changing circuit polarities, etc.

Diodes and switches

The base-emitter or base-collector junction of a silicon transistor can be used as a diode/rectifier or as a zener diode by using it in the appropriate polarity. *Figure 9.7* shows two alternative ways of using an

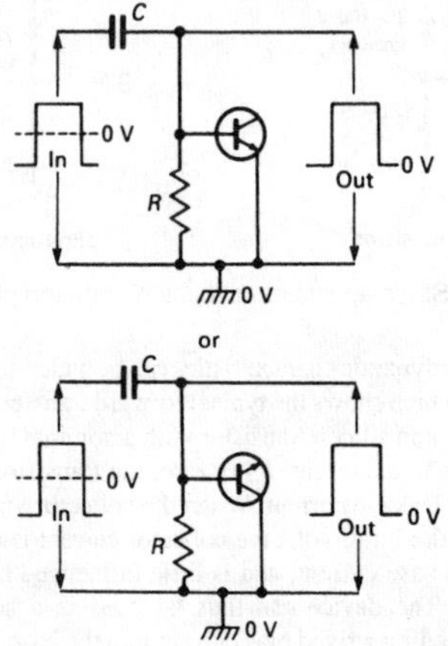

Figure 9.7. Clamping diode circuit, using npn transistor as diode.

npn transistor as a diode clamp that converts an a.c.-coupled rectangular input waveform into a rectangular output that swings between zero and a positive voltage value, i.e., which 'clamps' the output signal to the zero-volts reference point.

Figure 9.8 shows an npn transistor used as a zener diode that converts an unregulated supply voltage into a fixed-value regulated output with a typical value in the range 5V to 10V, depending on the individual transistor. Only the base-emitter junction is suitable for use in this application.

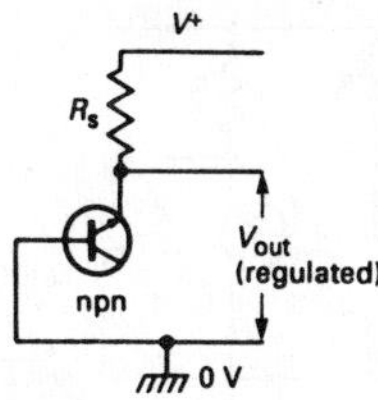

Figure 9.8. A transistor used as a zener diode.

Figure 9.9 shows a transistor used as a simple electronic switch or digital inverter. Its base is driven (via R_b) by a digital input that is at either zero volts or at a positive value, and load R_L is connected between the collector and the positive supply rail. When the input voltage is zero the transistor is cut off and zero current flows through the load, so the full supply voltage appears between the collector and emitter. When the input is high the transistor switch is driven fully on (saturated) and maximum current flows in the load, and only a few hundred millivolts are developed between collector and emitter. The output voltage is an inverted form of the input signal.

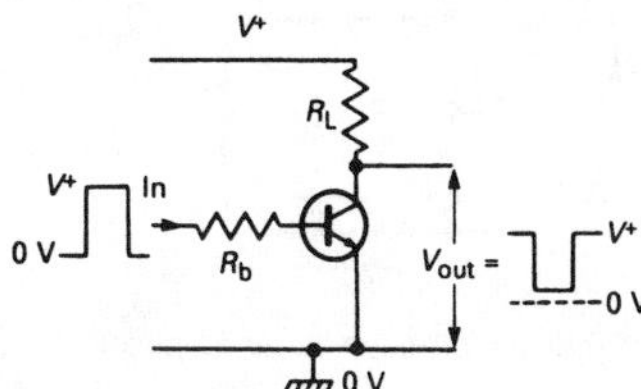

Figure 9.9. Transistor switch or digital inverter.

Linear amplifiers

A transistor can be used as a linear current or voltage amplifier by feeding a suitable bias current into its base and then applying the input signal between an appropriate pair of terminals. The transistor can in this case be used in any one of three basic operating modes, each of which provides a unique set of characteristics. These three modes are known as 'common-emitter' *(Figure 9.10)*, 'common-base' *(Figure 9.11)*, and 'common-collector' *(Figures 9.12* and *9.13)*.

In the common-emitter circuit of *Figure 9.10*, load R_L is wired between the collector and the positive supply line, and a bias current is fed into the base via R_b, whose value is chosen to set the collector at a quiescent half-supply voltage value (to provide maximum undistorted output signal voltage swings). The input signal is applied between base and emitter via C_1, and the output signal (which is phase-inverted relative to the input) is taken between the collector and emitter. This circuit gives a medium-value input impedance and a fairly high overall voltage gain.

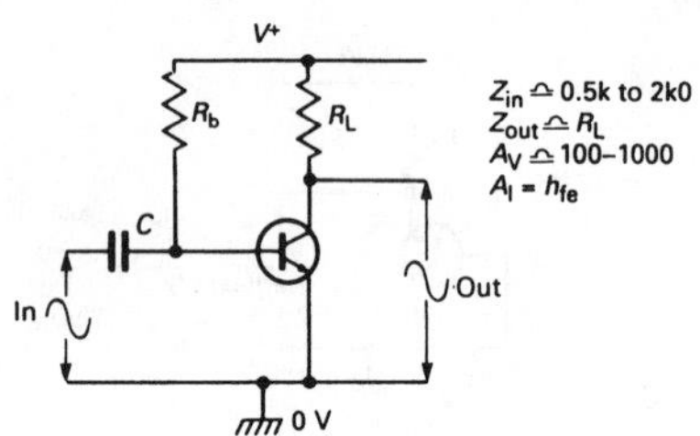

Figure 9.10. Common-emitter linear amplifier.

In the common-base circuit of *Figure 9.11* the base is biased via R_b and is a.c.-decoupled (or a.c.-grounded) via C_b. The input signal is effectively applied between the emitter and base via C_1, and the amplified but non-inverted output signal is effectively taken from between the collector and base. This circuit features good voltage gain, near-unity current gain, and a very low input impedance.

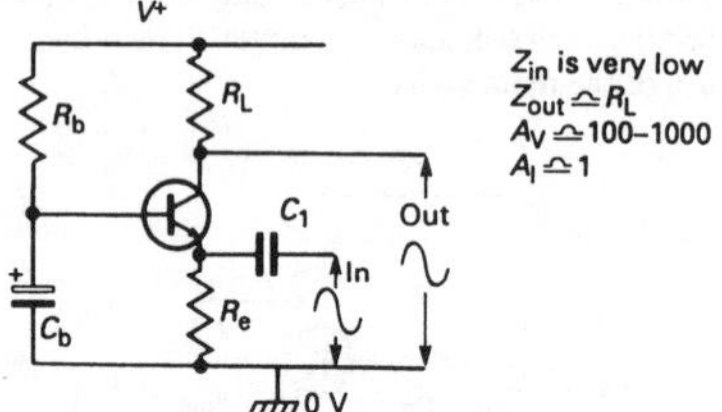

Figure 9.11. Common-base linear amplifier.

In the common-collector circuit of *Figure 9.12* the collector is shorted to the low-impedance positive supply rail and is thus at 'ground' impedance level. The input signal is applied between base and ground (collector), and the non-inverted output is taken from between emitter and ground (collector). This circuit gives near-unity overall voltage gain, and its output 'follows' the input signal; it is thus known as a d.c.-voltage follower (or emitter follower); it has a very high input impedance (equal to the product of the R_L and h_{fe} values). Note that this circuit can be modified for a.c. use by simply biasing the transistor to half-supply volts and a.c.-coupling the input signal to the base, as shown in *Figure 9.13*.

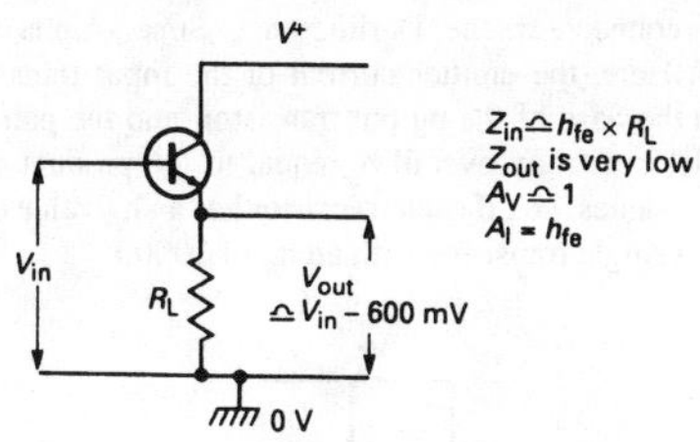

Figure 9.12. D.c. common-collector linear amplifier or voltage follower.

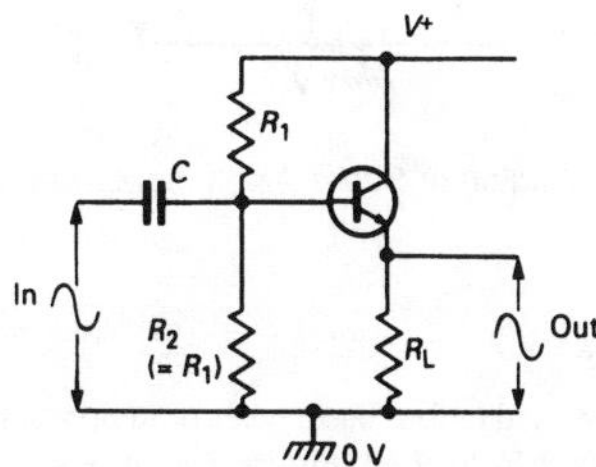

Figure 9.13. A.c. common-collector amplifier or voltage follower.

The chart of *Figure 9.14* summarizes the performances of the basic amplifier configurations. Thus, the common-collector amplifier gives near-unity overall voltage gain and a high input impedance, while the common-emitter and common-base amplifiers both give high values of voltage gain but have medium to low values of input impedance.

The Darlington connection

The input impedance of the *Figure 9.12* emitter follower circuit equals the product of its R_L and h_{fe} values; if an ultra-high input impedance is wanted it can be obtained by replacing the single transistor with a pair

	Common collector	Common emitter	Common base
Z_{in}	High($\simeq h_{fe} \times R_L$)	Medium ($\simeq$1k0)	Low ($\simeq$40R)
Z_{out}	Very low	$\simeq R_L$	$\simeq R_L$
A_V	$\simeq 1$	High	High
A_I	$\simeq h_{fe}$	$\simeq h_{fe}$	$\simeq 1$
Cut-off frequency	Medium	Low	High
Voltage phase shift	Zero	180°	Zero

Figure 9.14. Comparative performances of the three basic configurations.

of transistors connected in the 'Darlington' or Super-Alpha mode, as in *Figure 9.15*. Here, the emitter current of the input transistor feeds directly into the base of the output transistor, and the pair act like a single transistor with an overall h_{fe} equal to the product of the two individual h_{fe} values, i.e., if each transistor has an h_{fe} value of 100, the pair act like a single transistor with an h_{fe} of 10000.

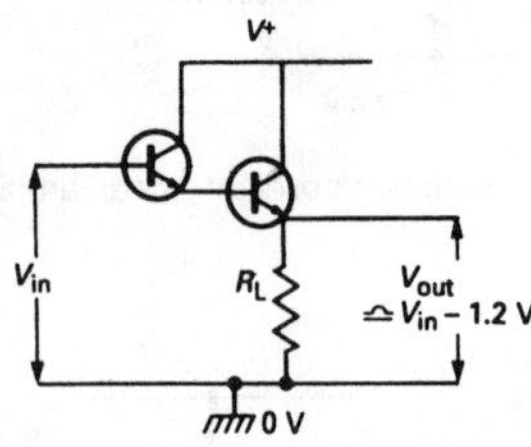

Figure 9.15. Darlington or Super-Alpha d.c. emitter follower.

Multivibrators

Transistors can be used in four basic types of multivibrator circuit, as shown in *Figures 9.16* to *9.19*. *Figure 9.16* is a simple manually-triggered cross-coupled bistable multivibrator, in which the base bias of each transistor is derived from the collector of the other, so that one transistor automatically turns off when the other turns on, and *vice versa*. Thus, the output can be driven low by briefly turning Q_2 off via S_2; the circuit automatically locks into this state until Q_1 is turned off via S_1, at which point the output locks into the high state, and so on.

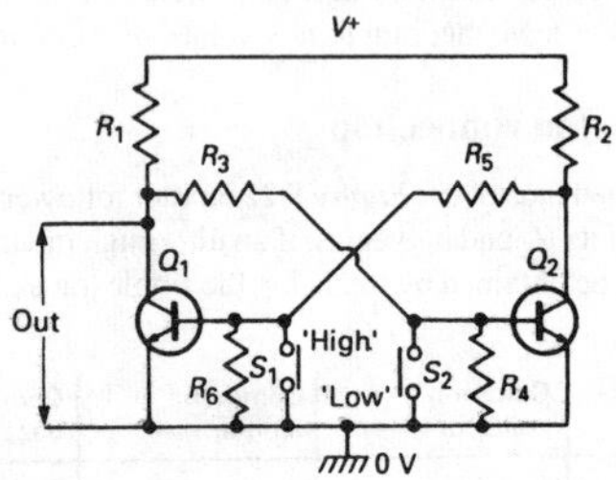

Figure 9.16. Manually-triggered bistable multivibrator.

Figure 9.17 shows a monostable multivibrator or one-shot pulse generator circuit; its output is normally low, but switches high for a pre-set period (determined by C_1–R_5) if Q_1 is briefly turned off via S_1.

Figure 9.18 shows an astable multivibrator or free-running square-wave generator; the square wave's on and off periods are determined by C_1–R_4 and C_2–R_3.

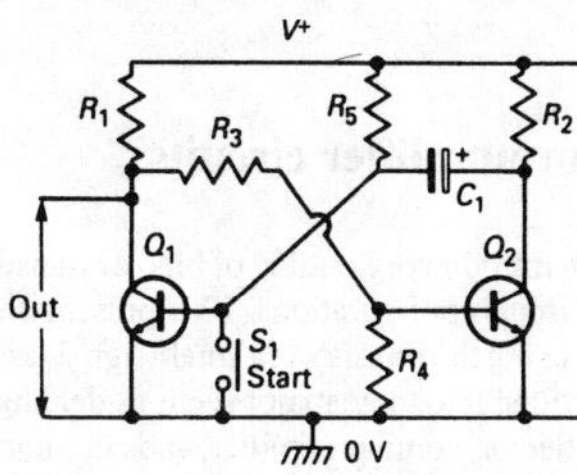

Figure 9.17. Manually-triggered monostable multivibrator.

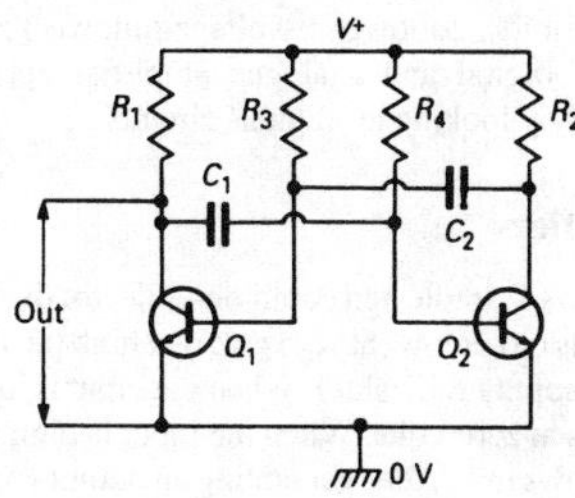

Figure 9.18. Astable multivibrator or free-running square wave generator.

Finally, *Figure 9.19* shows a Schmitt trigger or sine-to-square waveform converter. The circuit action is such that Q_2 switches abruptly from the on state to the off state, or *vice versa*, as Q_1 base goes above or below pre-determined 'trigger' voltage levels.

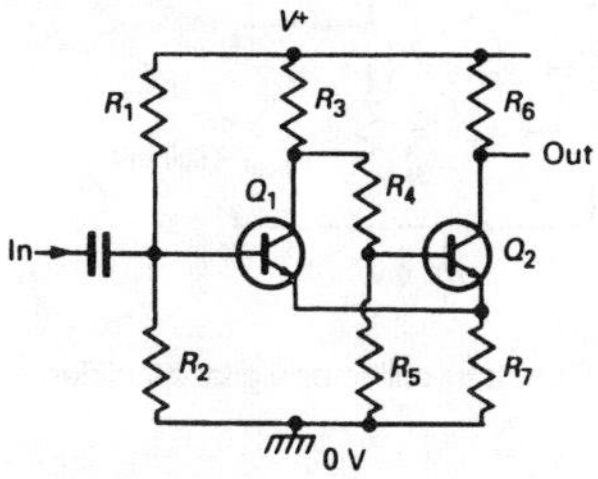

Figure 9.19. Schmitt trigger or sine-to-square waveform converter.

10 Transistor amplifier circuits

Chapter 9 gave an introductory outline of bipolar transistor characteristics and basic circuit configurations. This present chapter looks at practical ways of using the transistor in small-signal 'amplifier' applications, and is divided into three major sections dealing, in sequence, with common-collector, common-emitter, and common-base circuits.

Common-collector amplifier circuits

The common-collector amplifier (also known as the grounded-collector amplifier or emitter follower or voltage follower) can be used in a wide variety of digital and analogue amplifier applications. This section starts off by looking at 'digital' circuits.

Digital amplifiers

Figure 10.1 shows a simple npn common-collector digital amplifier in which the input is either low (at zero volts) or high (at a V_{peak} value not greater than the supply rail value). When the input is low Q_1 is cut off, and the output is at zero volts. When the input is high Q_1 is driven on and current I_L flows in R_L, thus generating an output voltage across R_L; intrinsic negative feedback makes this output voltage take up a value one base-emitter junction volt-drop (about 600mV) below the input V_{peak} value. Thus, the output voltage 'follows' (but is 600mV less than) the input voltage.

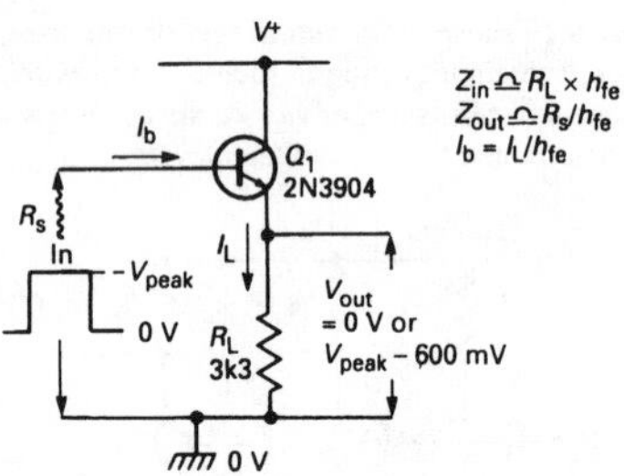

Figure 10.1. Common-collector digital amplifier.

This circuit's input (base) current equals the I_L value divided by Q_1's h_{fe} value (nominally 200 in the 2N3904), and its input impedance equals $R_L \times h_{fe}$, i.e., nominally 660k in the example shown. The circuit's output impedance equals the input signal source impedance (R_S) value divided by h_{fe}. Thus, the circuit has a high input and low output impedance, and acts as a unity-voltage-gain 'buffer' circuit.

If this buffer circuit is fed with a fast input pulse its output may have a deteriorated trailing edge, as shown in *Figure 10.2*. This is caused by the presence of stray capacitance (C_S) across R_L; when the input pulse switches high Q_1 turns on and rapidly 'sources' (feeds) a charge current into C_S, thus giving an output pulse with a sharp leading edge, but when the input signal switches low again Q_1 switches off and is thus unable to 'sink' (absorb) the charge current of C_S, which thus discharges via R_L and makes the output pulse's trailing edge decay exponentially, with a time constant equal to the C_S-R_L product.

Note from the above that an npn emitter follower can efficiently source, but not sink, high currents; a pnp emitter follower gives the opposite action, and can efficiently sink, but not source, high currents.

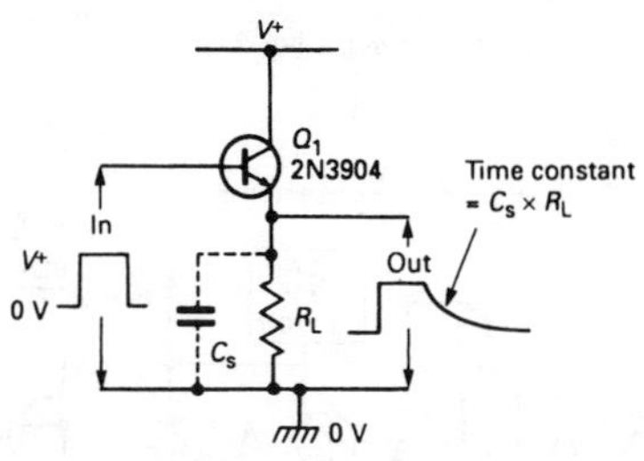

Figure 10.2. Effect of C_S on the output pulses.

Relay drivers

If the basic *Figure 10.1* switching circuit is used to drive inductive loads such as coils or loudspeakers, etc., it must be fitted with a diode protection network to limit inductive switch-off back-emfs to safe values. One very useful inductor-driving circuit is the relay driver, and a number of examples of this are shown in *Figures 10.3* to *10.7*.

The relay in the npn driver circuit of *Figure 10.3* can be activated via a digital input or via switch SW_1; it turns on when the input signal is high or SW_1 is closed, and turns off when the input signal is low or SW_1 is open. Relay contacts RLA/1 are available for external use, and the circuit can be made self-latching by wiring a spare set of normally-open relay contacts (RLA/2) between Q_1's collector and emitter, as shown dotted. *Figure 10.4* is a pnp version of the same circuit; in this case the relay can be turned on by closing SW_1 or by applying a 'zero' input signal. Note in *Figure 10.3* that D_1 damps relay switch-off back-emfs by preventing this voltage from swinging below the zero-volts rail value; optional diode D_2 can be used to stop this voltage swinging above the positive rail.

The *Figures10.3* and *10.4* circuits effectively increase the relay current sensitivity by a factor of about 200 (the h_{fe} value of Q_1), e.g., if the relay has a coil resistance of 120R and needs an activating current of 100mA, the circuit's input impedance is 24k and the input operating current requirement is 0.5mA.

Sensitivity can be further increased by using a Darlington pair of transistors in place of Q_1, as shown in *Figure 10.5*, but the emitter 'following' voltage of Q_2 will be 1.2V (two base-emitter volt drops)

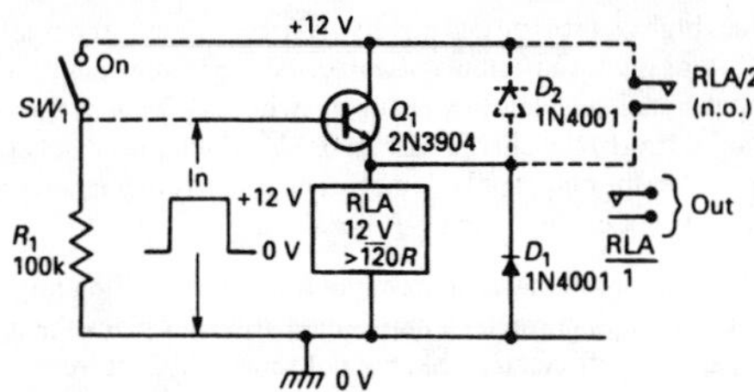

Figure 10.3 Emitter-follower relay driver.

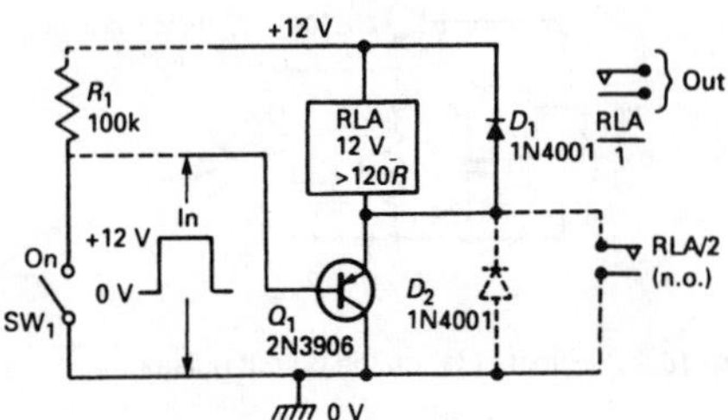

Figure 10.4. Pnp version of the relay driver.

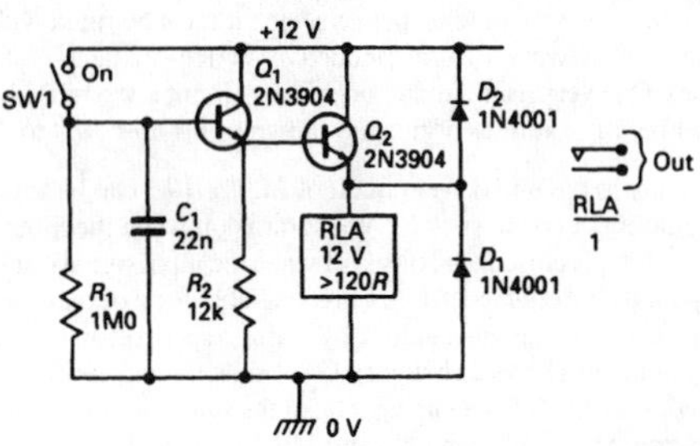

Figure 10.5. Darlington version of the npn relay driver.

below the base input voltage of Q_1. This circuit has an input impedanc of about 500k and needs an input operating current of 24μA; C_1 protect the circuit against activation via high-impedance transient voltages such as those induced by lightening flashes, RFI, etc.

The Darlington buffer is useful in relay-driving *C–R* time-delay de signs such as those shown in *Figures 10.6* and *10.7*, in which C_1–R generate an exponential waveform that is fed to the relay via Q_1–Q thus making the relay change state some delayed time after the suppl is initially connected. With an R_1 value of 120k the circuits giv

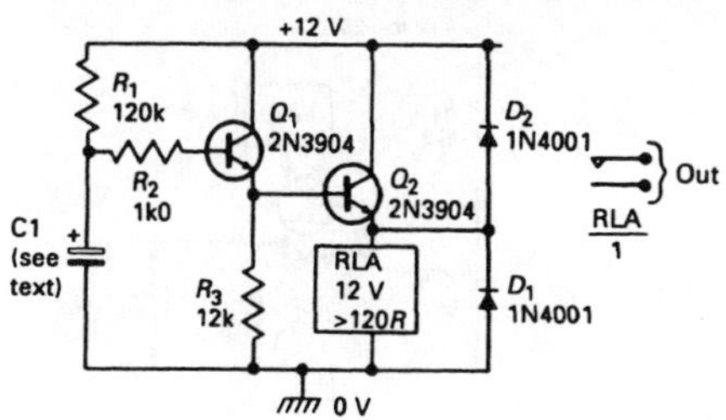

Figure 10.6. Delayed-switch-on relay driver.

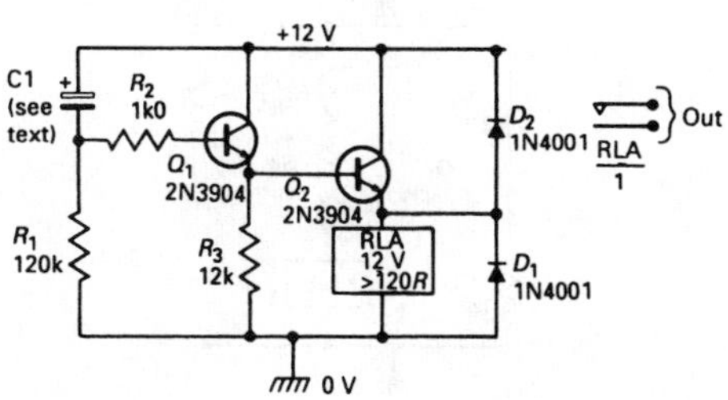

Figure 10.7 Auto-turn-off time-delay circuit.

operating delays of roughly 0.1 seconds per μF of C_1 value, i.e., a 10 second delay if $C_1 = 100\mu F$, etc. The *Figure 10.6* circuit makes the relay turn on some delayed time after its power-supply is connected. The *Figure 10.7* circuit makes the relay turn on as soon as the supply is connected, but turn off again after a fixed delay.

Constant-current generators

A constant-current generator (CCG) is a circuit that generates a constant load current irrespective of wide variations in load resistance. A bipolar transistor can be used as a CCG by using it in the common-collector mode shown in *Figure 10.8*. Here, R_1-ZD_1 apply a fixed 5.6V 'reference' to Q_1 base, making 5V appear across R_2, which thus passes 5mA via Q_1's emitter; a transistor's emitter and collector currents are inherently almost identical, so a 5mA current also flows in any load connected between Q_1's collector and the positive supply rail, provided that its resistance is not so high that Q_1 is driven into saturation; these two points thus act as 5mA 'constant-current' terminals.

This circuit's constant-current value is set by Q_1's base voltage and the R_2 value, and can be altered by varying either of these values. *Figure 10.9* shows how the basic circuit can be 'inverted' to give a ground-referenced constant-current output that can be varied from about 1mA to 10mA via RV_1.

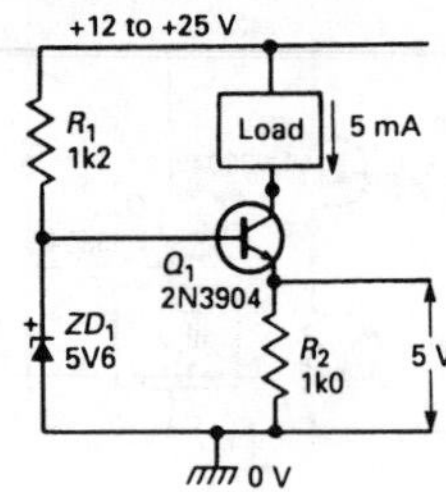

Figure 10.8. Simple 5mA constant-current generator.

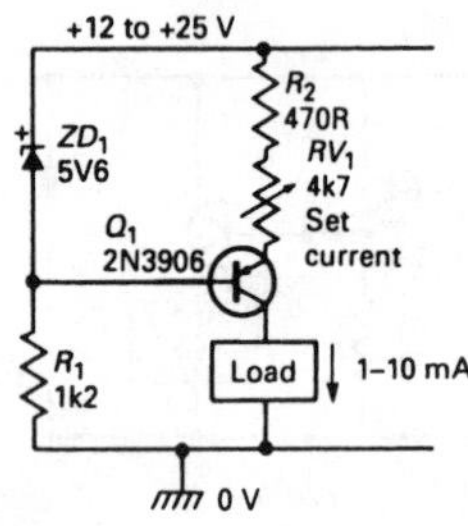

Figure 10.9. Ground-referenced variable (1mA–10mA) constant-current generator.

In many practical CCG applications the circuit's most important feature is its high dynamic output impedance or 'current constancy', the precise current magnitude being of minor importance; in such cases the basic *Figures10.8* and *10.9* circuits can be used. If greater precision is needed, the 'reference' voltage accuracy must be improved; one way of doing this is to replace R_1 with a 5mA constant-current generator, as indicated in *Figure 10.10* by the 'double circle' symbol, so that the zener current (and thus voltage) is independent of supply voltage variations. A red LED acts as an excellent reference voltage generator, and has a very low temperature coefficient, and can be used in place of a zener, as shown in *Figure 10.11*. In this case the LED generates roughly 2.0V, so only 1.4V appears across R_1, which has its value reduced to 270R to give a constant-current output of 5mA.

The CCG circuits of *Figures 10.8* to *10.11* are all '3-terminal' designs that need both supply and output connections. *Figure 10.12* shows a 2-terminal CCG that consumes a fixed 2mA when wired in series with an external load. Here, ZD_1 applies 5.6V to the base of Q_1, which (via R_1) thus generates a constant collector current of 1mA; this current drives ZD_2, which thus develops a very stable 5.6V on the base of Q_2, which in turn generates a constant collector current of about 1mA, which drives ZD_1. The circuit thus acts as a closed loop current regulator that consumes a total of 2mA. R_3 acts as a start-up resistor that provides the transistors with initial base current.

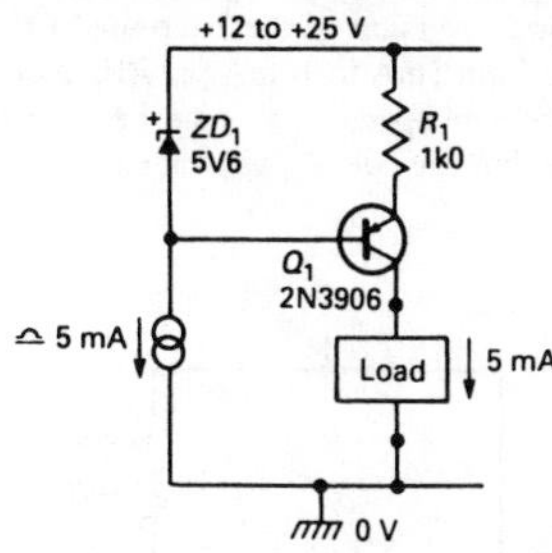

Figure 10.10. Precision constant-current generator.

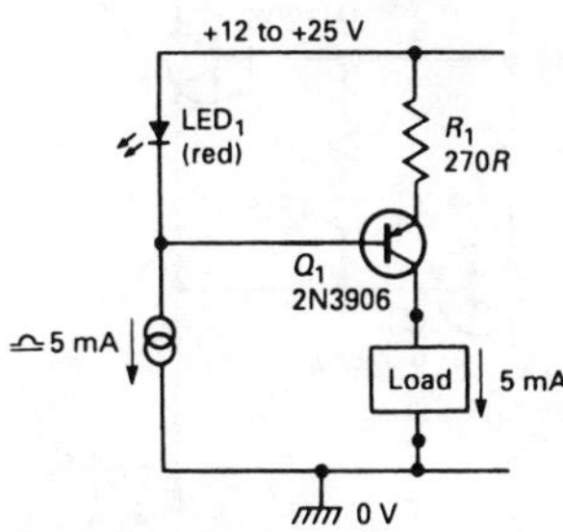

Figure 10.11. Thermally-stabilised constant-current generator, using a LED as a voltage reference.

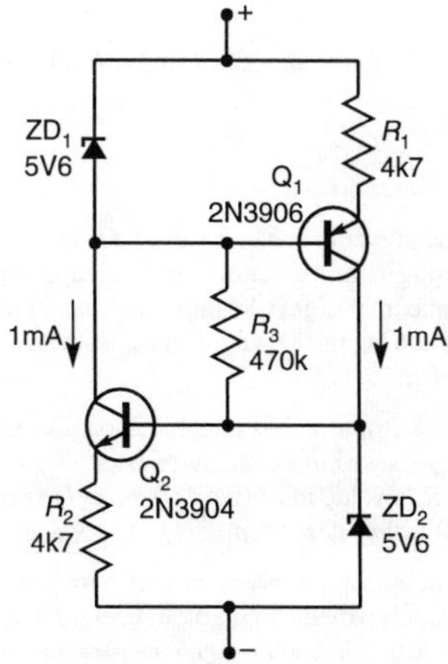

Figure 10.12. 2-terminal 2mA constant-current generator.

Figure 10.13 shows a version of the 2-terminal CCG in which the current is variable from 1mA to 10mA via RV_1. Note that these two circuits need a minimum operating voltage, between their two terminals, of about 12V, but can operate with maximum ones of 40V.

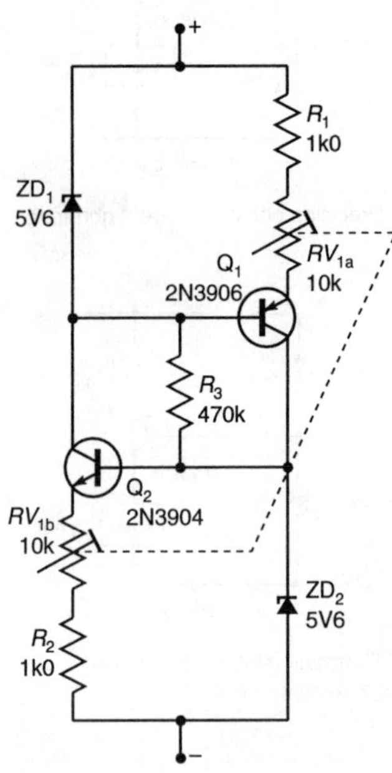

Figure 10.13. 2-terminal variable (1mA–10mA) constant-current generator.

Linear amplifiers

A common-collector circuit can be used as an a.c.-coupled linear amplifier by biasing its base to a quiescent half-supply voltage value (to accommodate maximal signal swings) and a.c.-coupling the input signal to its base and taking the output signal from its emitter, as shown in *Figures 10.14* and *10.15*.

Figure 10.14 shows the simplest possible version of the linear emitter follower, with Q_1 biased via a single resistor (R_1). To achieve half-supply biasing, R_1's value must (ideally) equal Q_1's input resistance; the biasing level is thus dependent on Q_1's h_{fe} value.

Figure 10.15 shows an improved circuit in which R_1–R_2 apply a quiescent half-supply voltage to Q_1 base, irrespective of variations in Q_1's h_{fe} values. Ideally, R_1 should equal the paralleled values of R_2 and R_{IN}, but in practice it is adequate to simply make R_1 low relative to R_{IN} and to make R_2 slightly larger than R_1.

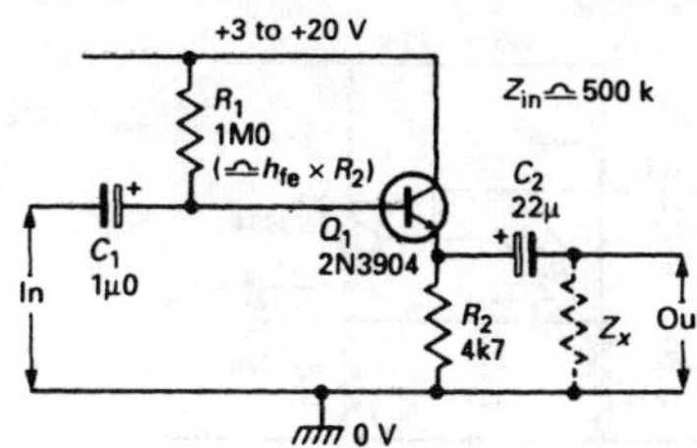

Figure 10.14. Simple emitter follower.

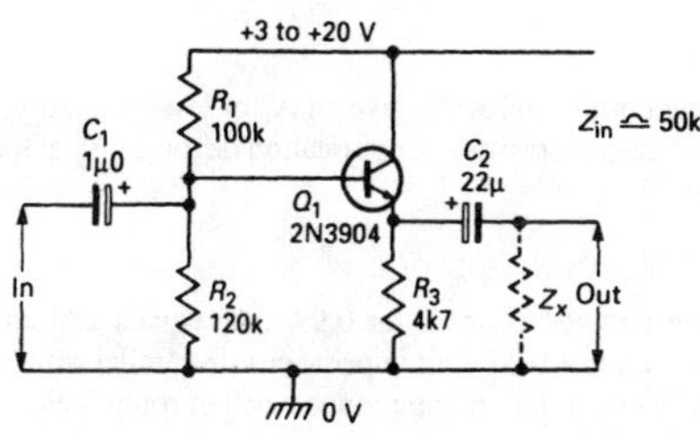

Figure 10.15. High-stability emitter follower.

In these two circuits the input impedance *looking directly into Q_1 base* equals h_{fe} x Z_{load}, where Z_{load} is the parallel impedance of R_2 and external output load Z_X. Thus, the base impedance value is roughly 1M0 when Z_X is infinite. The input impedance of the *complete* circuit equals the parallel impedances of the base impedance and the bias network; thus, the *Figure 10.14* circuit gives an input impedance of about 500k, and that of *Figure 10.15* is about 50k. Both circuits give a voltage gain (A_V) that is slightly below unity, the actual gain being given by:

$$A_V = Z_{load}/(Z_b + Z_{load}),$$

where $Z_b = 25/I_e\Omega$, where I_e is the emitter current in mA. Thus, at an operating current of 1mA these circuits give a gain of 0.995 when Z_{load} = 4k7, or 0.975 when Z_{load} = 1k0.

Bootstrapping

The *Figure 10.15* circuit's input impedance can easily be boosted by using the 'bootstrapping' technique of *Figure 10.16*. Here, 47k resistor R_3 is wired between the R_1–R_2 biasing network junction and Q_1 base, and the input signal is fed to Q_1 base via C_1. Note, however, that Q_1's output is fed back to the R_1–R_2 junction via C_2, and near-identical signal voltages thus appear at both ends of R_3; very little signal current thus flows in R_3, which appears (to the input signal) to have a far greater impedance than its true resistance value.

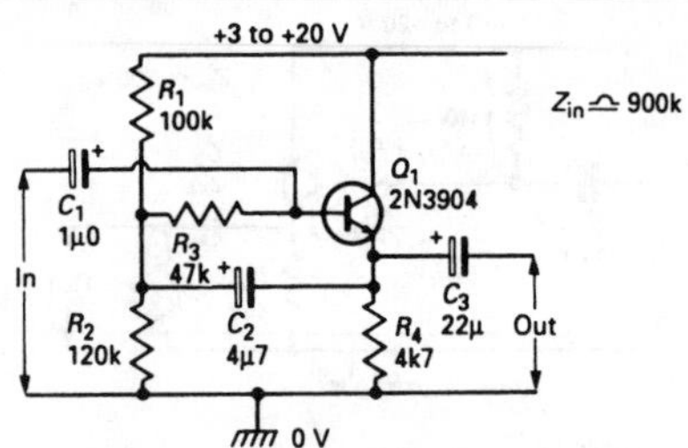

Figure 10.16. Bootstrapped emitter follower.

All practical emitter followers give an A_V of less than unity, and this value determines the resistor 'amplification factor' or A_R of the circuit, as follows:

$$A_R = 1/(1 - A_V).$$

Thus, if the circuit has an A_V of 0.995, A_R equals 200 and the R_3 impedance is almost 10M. This impedance is in parallel with R_{IN}, so the *Figure 10.16* circuit has an input impedance of roughly 900k.

The input impedance of the *Figure 10.16* circuit can be increased even more by using a pair of Darlington-connected transistors in place of Q_1 and increasing the value of R_3, as shown in *Figure 10.17*, which gives a measured input impedance of about 3M3.

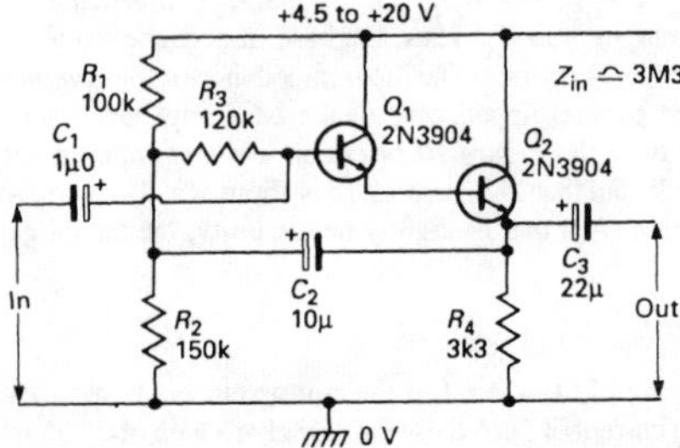

Figure 10.17. Bootstrapped Darlington emitter follower.

An even greater input impedance can be obtained by using the bootstrapped 'complementary feedback pair' circuit of *Figure 10.18*, which gives an input impedance of about 10M. In this case Q_1 and Q_2 are in fact both wired as common emitter amplifiers, but they operate with virtually 100% negative feedback and thus give an overall voltage gain of almost exactly unity: this 'pair' of transistors thus acts like a near-perfect Darlington emitter follower.

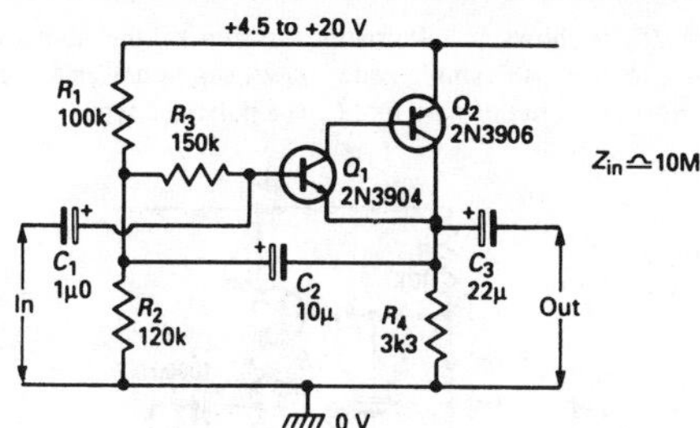

Figure 10.18. Bootstrapped complementary feedback pair.

Complementary emitter followers

It was pointed out earlier that an npn emitter follower can source current but cannot sink it, and that a pnp emitter follower can sink current but cannot source it; i.e., these circuits can handle unidirectional output currents only. In many applications, a 'bidirectional' emitter follower circuit (that can source and sink currents with equal ease) is required, and this action can be obtained by using a complementary emitter follower configuration in which npn and pnp emitter followers are effectively wired in series. *Figures 10.19* and *10.21* show basic circuits of this type.

The *Figure 10.19* circuit uses a dual ('split') power supply and has its output direct-coupled to a grounded load. The series-connected npn and pnp transistors are biased at a quiescent 'zero volts' value via the R_1–D_1–D_2–R_2 potential divider, with each transistor slightly forward biased via silicon diodes D_1 and D_2, which have characteristics inherently similar to those of the transistor base-emitter junctions; C_2 ensures that identical input signals are applied to the transistor bases, and R_3 and R_4 protect the transistors against excessive output currents. The circuit's action is such that Q_1 sources currents into the load when the input goes positive, and Q_2 sinks load current when the input goes negative. Note that input capacitor C_1 is a non-polarised type.

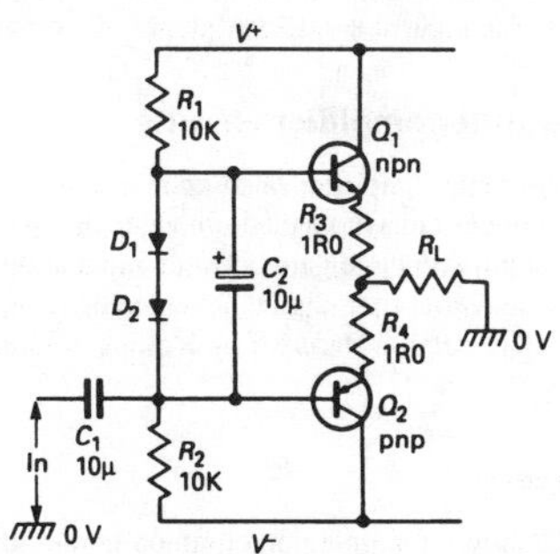

Figure 10.19. Complementary emitter follower using split supply and direct-coupled output load.

Figure 10.20 shows an alternative version of the above circuit, designed for use with a single-ended power supply and an a.c.-coupled output load; note in this case that C_1 is a polarized type.

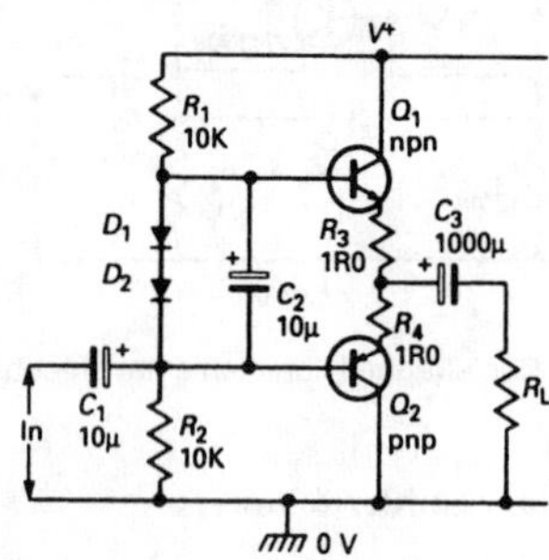

Figure 10.20. Complementary emitter follower, using single-ended supply and a.c.-coupled load.

The amplified diode

In the *Figures 10.19* and *10.20* circuits Q_1 and Q_2 are slightly forward biased (to minimise cross-over distortion problems) via silicon diodes D_1 and D_2; in practice, the diode currents (and thus the transistor forward bias voltages) are usually adjustable over a limited range. If these basic circuits are modified for use with Darlington transistor stages a total of four biasing diodes are needed; in such cases the diodes are usually replaced by a transistor 'amplified diode' stage, as shown in *Figure 10.21*. Here, Q_5's collector-to-emitter voltage equals the Q_5 base-emitter volt drop (about 600mV) multiplied by $(RV_1+R_3)/R_3$; thus, if RV_1 is set to zero ohms, 600mV are developed across Q_5, which thus acts like a single silicon diode, but if RV_1 is set to 47k about 3.6V is developed across Q_5, which thus acts like six series-connected silicon diodes. RV_1 can thus be used to precisely set the Q_5 volt drop and thus adjust the quiescent current values of the Q_2–Q_3 output stages.

Common-emitter amplifier circuits

The common-emitter amplifier (also known as the common-earth or grounded-emitter circuit) has a medium value of input impedance and provides substantial voltage gain between input and output. It can be used in a wide variety of digital and analogue voltage amplifier applications. This section starts off by looking at 'digital' application circuits.

Digital circuits

Figure 10.22 shows a simple npn common-emitter digital amplifier, inverter, or switch, in which the input signal is at either zero volts or a substantial positive value. When the input is zero the transistor is cut off and the output is at full positive supply rail value. When the input is high

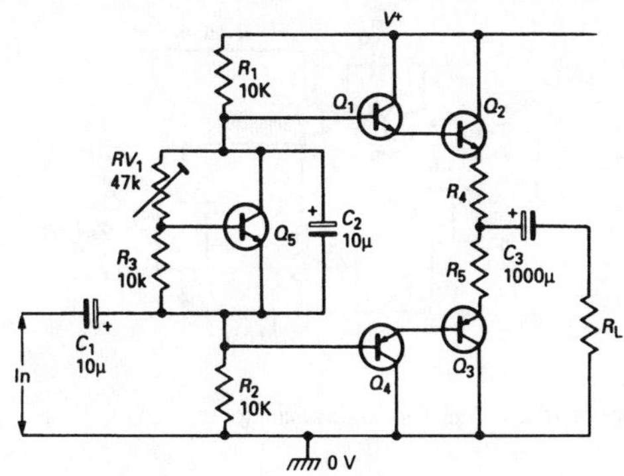

Figure 10.21. Darlington complementary emitter follower, with biasing via an amplified diode (Q_5).

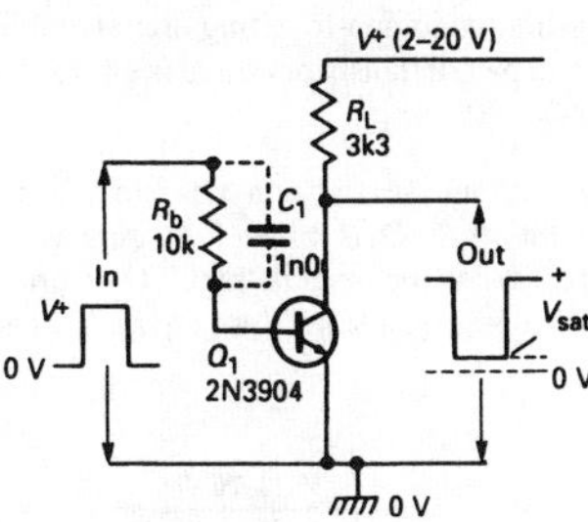

Figure 10.22. Digital inverter/switch (npn).

the transistor is biased on and collector current flows via R_L, thus pulling the output low; if the input voltage is large enough, Q_1 is driven fully on and the output drops to a 'saturation' value of a few hundred mV. Thus, the output signal is an amplified and inverted version of the input signal.

In *Figure 10.22*, R_b limits the input base-drive current to a safe value; the circuit's input impedance is slightly greater than the R_b value, which also influences the rise and fall times of the output signal; the greater the R_b value, the worse these become. This snag can be overcome by shunting R_b with a 'speed-up' capacitor (typical value about 1n0), as shown dotted in the diagram. In practice, R_b should be as small as possible, consistent with safety and input-impedance requirements, and must not exceed R_L x h_{fe}.

Figure 10.23 shows a pnp version of the digital inverter/switch circuit. Q_1 switches fully on, with its output a few hundred mV below the positive supply value, when the input is at zero, and turns off (with its output at zero volts) when the input rises to within less than 600mV of the positive supply rail.

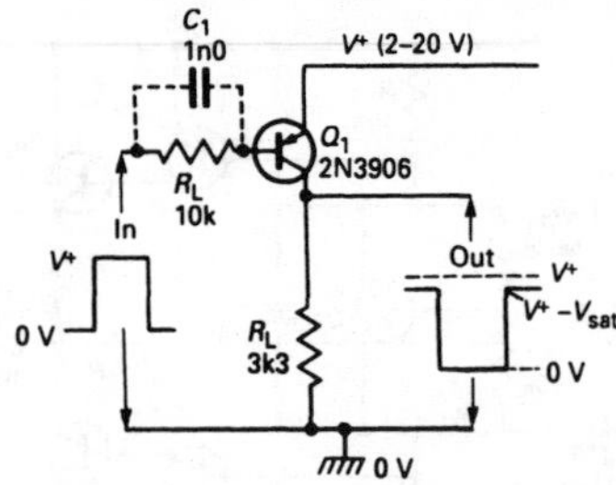

Figure 10.23. Digital inverter/switch (pnp).

The sensitivity of the *Figures 10.21* or *10.22* circuit can be increased by replacing Q_1 with a pair of Darlington-connected transistors. Alternatively, a very-high-gain non-inverting digital amplifier/switch can be made by using a pair of transistors wired in either of the ways shown in *Figures 10.24* or *10.25*.

The *Figure 10.24* circuit uses two npn transistors. When the input is at zero volts Q_1 is cut off, so Q_2 is driven fully on via R_2, and the output is low (saturated). When the input is 'high', Q_1 is driven to saturation and pulls Q_2 base to less than 600mV, so Q_2 is cut off and the output is high (at V+).

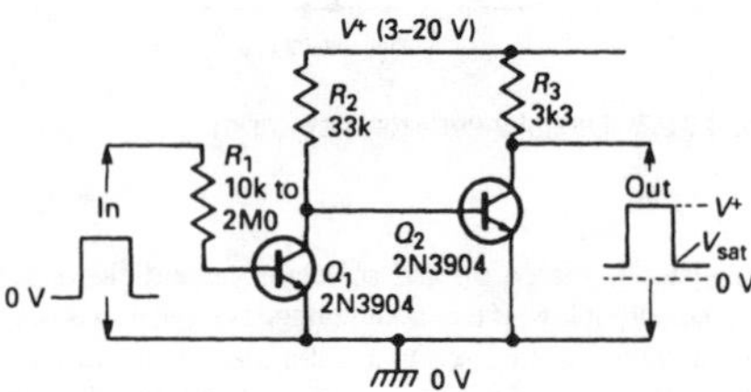

Figure 10.24. Very-high-gain non-inverting digital amplifier/switch using npn transistors.

The *Figure 10.25* circuit uses one npn and one pnp transistor. When the input is at zero volts Q_1 is cut off, so Q_2 is also cut off (via R_2–R_3) and the output is at zero volts. When the input is 'high', Q_1 is driven on and pulls Q_2 into saturation via R_3; under this condition the output takes up a value a few hundred mV below the positive supply rail value.

Figure 10.26 shows (in basic form) how a complementary pair of the *Figure 10.25* circuits can be used to make a D.C.-motor direction-control network, using a dual power supply. The circuit operates as follows.

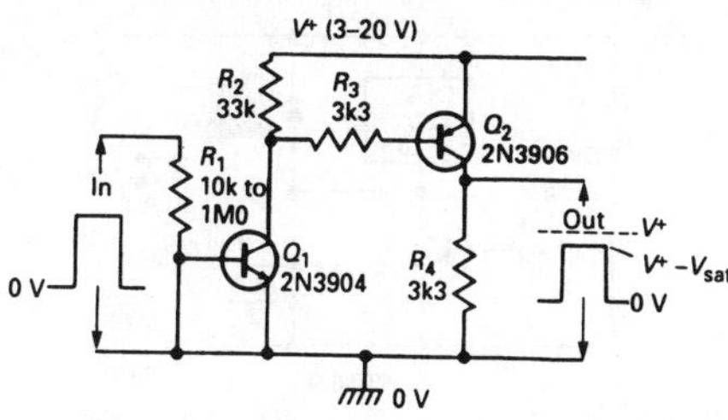

Figure 10.25. Alternative non-inverting digital amplifier/switch using an npn-pnp pair of transistors.

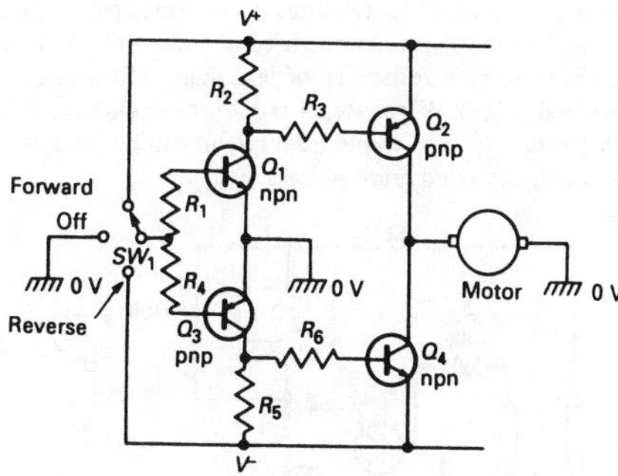

Figure 10.26. D.C.-motor direction-control circuit.

When SW_1 is set to forward, Q_1 is driven on via R_1, and pulls Q_2 on via R_3, but Q_3 and Q_4 are cut off; the 'live' side of the motor is thus connected (via Q_2) to the positive supply rail under this condition, and the motor runs in a forward direction.

When SW_1 is set to off, all four transistors are cut off, and the motor is inoperative.

When SW_1 is set to reverse, Q_3 is biased on via R_4, and pulls Q_4 on via R_6, but Q_1 and Q_2 are cut off; the 'live' side of the motor is thus connected (via Q_4) to the negative supply rail under this condition, and the motor runs in the reverse direction.

Relay drivers

The basic digital circuits of *Figures 10.22* to *10.25* can be used as efficient relay drivers if fitted with suitable diode protection networks. *Figures 10.27* to *10.29* show examples of such circuits.

The *Figure 10.27* circuit raises a relay's current sensitivity by a factor of about 200, and greatly increases its voltage sensitivity. R_1 gives base drive protection, and can be larger than 1k0 if desired. The relay is turned on by a positive input voltage.

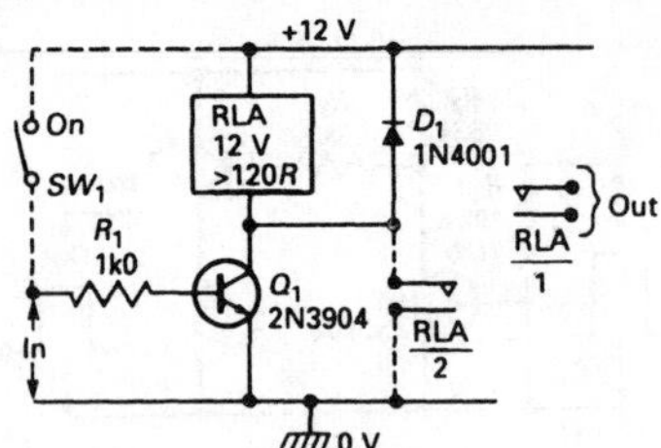

Figure 10.27. Simple relay-driving circuit.

The current sensitivity of the relay can be raised by a factor of about 20000 by replacing Q_1 with a Darlington-connected pair of transistors. *Figure 10.28* shows this technique used to make a circuit that can be activated by placing a resistance of less than 2M0 across a pair of stainless metal probes. Water, steam and skin contacts have resistances below this value, so this simple little circuit can be used as a water, steam or touch-activated relay switch.

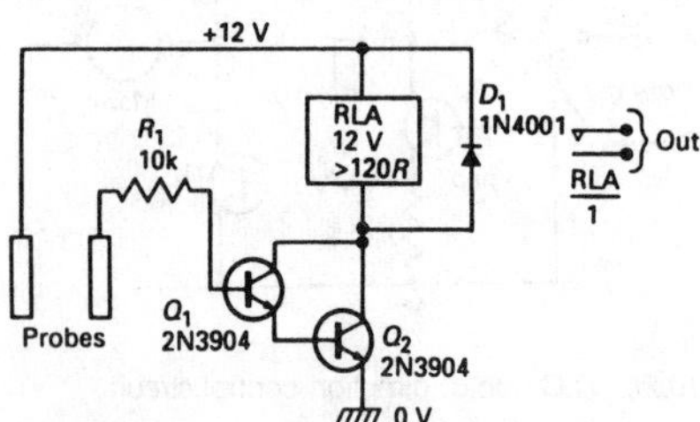

Figure 10.28. Touch, water or steam-activated relay switch.

Figure 10.29 shows another ultra-sensitive relay driver, based on the *Figure 10.25* circuit, that needs an input of only 700mV at 40µA to activate the relay; R_2 ensures that Q_1 and Q_2 turn fully off when the input terminals are open circuit.

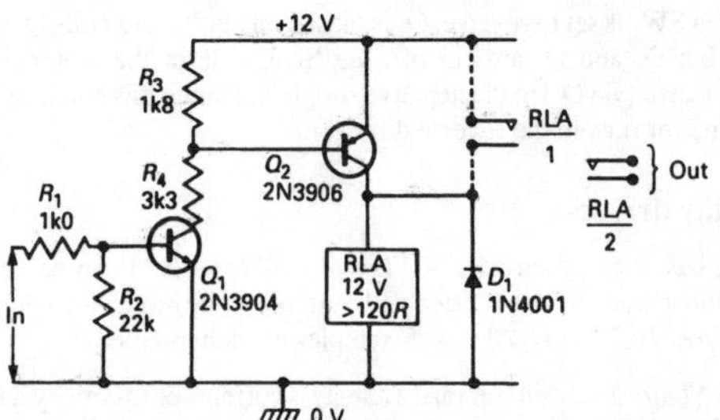

Figure 10.29. Ultra-sensitive relay driver (needs an input of 700mV at 40µA).

Linear biasing circuits

A common-emitter circuit can be used as a linear a.c. amplifier by applying a d.c. bias current to its base so that its collector takes up a quiescent half-supply voltage value (to accommodate maximal undistorted output signal swings), and by then feeding the a.c. input signal to its base and taking the a.c. output from its collector; *Figure 10.30* shows such a circuit.

The first step in designing a circuit of the *Figure 10.30* type is to select the value of load resistor R_2. The lower this is, the higher will be the amplifier's upper cut-off signal frequency (due to the smaller shunting effects of stray capacitance on the effective impedance of the load), but the higher will be Q_1's quiescent operating current. In the diagram, R_2 has a compromise value of 5k6, which gives an upper '3dB down' frequency of about 120kHz and a quiescent current consumption of 1mA from a 12V supply.

To bias the output to half-supply volts, R_1 needs a value of $R_2 \times 2h_{fe}$, and (assuming a nominal h_{fe} of 200) this works out at about 2M2 in the example shown.

The formulae for input impedance (looking into Q_1 base) and voltage gain are both given in the diagram. In the example shown, the input impedance is roughly 5k0, and is shunted by R_1; the voltage gain works out at about x200, or 46dB.

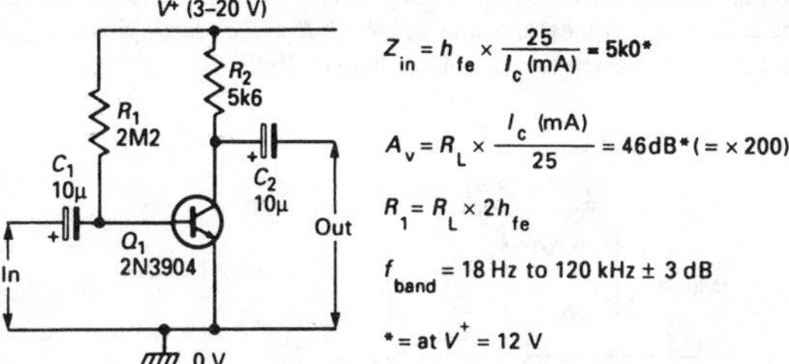

Figure 10.30 Simple npn common-emitter amplifier.

The quiescent biasing point of the *Figure 10.30* circuit depends on Q_1's h_{fe} value. This weakness can be overcome by modifying the circuit as in *Figure 10.31*, where biasing resistor R_1 is wired in a d.c. feedback mode between collector and base and has a value of $R_2 \times h_{fe}$. The feedback action is such that any shift in the output level (due to variations in h_{fe}, temperature, or component values) causes a counter-change in the base-current biasing level, thus tending to cancel the original shift.

The *Figure 10.31* circuit has the same values of bandwidth and voltage gain as the *Figure 10.30* design, but has a lower total value of input impedance. This is because the a.c. feedback action reduces the apparent impedance of R_1 (which shunts the 5k0 base impedance of Q_1)

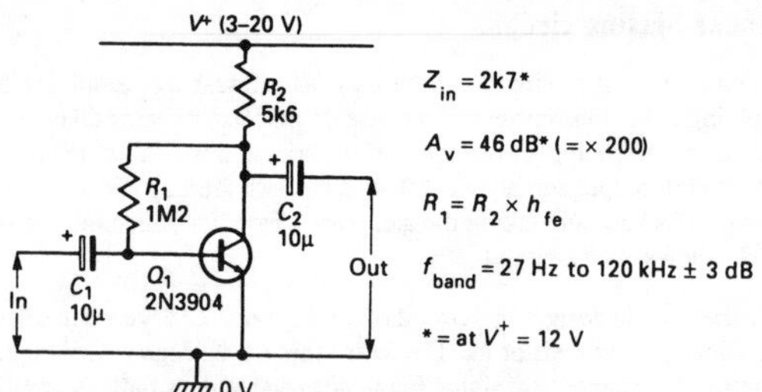

Figure 10.31. Common-emitter amplifier with feedback biasing.

by a factor of 200 (= A_V), thus giving a total input impedance of 2k7. If desired, the shunting effects of the biasing network can be eliminated by using two feedback resistors and a.c.-decoupling them as shown in *Figure 10.32.*

Finally, the ultimate in biasing stability is given by the 'potential-divider biasing' circuit of *Figure 10.33*. Here, potential divider R_1–R_2 sets a quiescent voltage slightly greater than V+/3 on Q_1 base, and voltage follower action causes 600mV less than this to appear on Q_1 emitter. V+/3 is thus developed across 5k6 emitter resistor R_3, and (since Q_1's emitter and collector currents are almost identical) a similar voltage is dropped across R_4, which also has a value of 5k6, thus setting the collector at a quiescent value of 2V+/3. R_3 is a.c.-decoupled via C_2, and the circuit gives an a.c. voltage gain of 46dB.

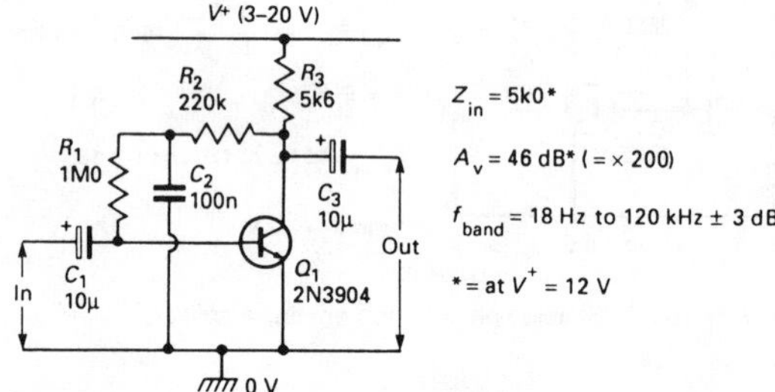

Figure 10.32. Amplifier with a.c.-decoupled feedback biasing.

Circuit variations

Figures 10.34 to *10.37* show some useful common-emitter amplifier variations. *Figure 10.34* shows the basic *Figure 10.33* design modified to give an a.c. voltage gain of x10; the gain actually equals the R_4 collector load value divided by the effective 'emitter' impedance value, which in this case (since R_3 is decoupled by series-connected C_2–R_5) equals the value of the base-emitter junction impedance in series with the paralleled values of R_3 and R_5, and works out at roughly 560Ω, thus giving a voltage gain of x10. Alternative gain values can be obtained by altering the R_5 value.

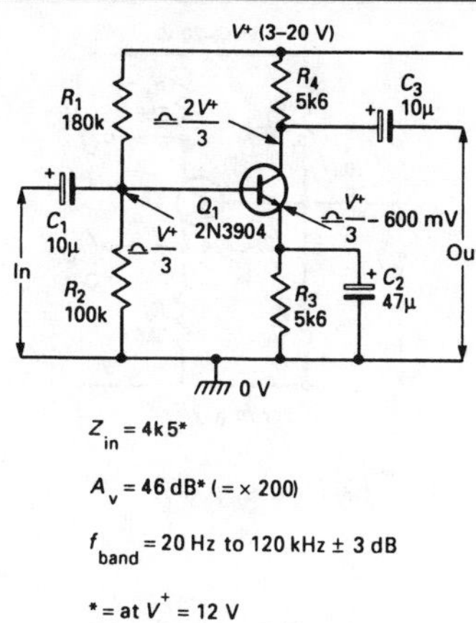

Figure 10.33. Amplifier with voltage-divider biasing.

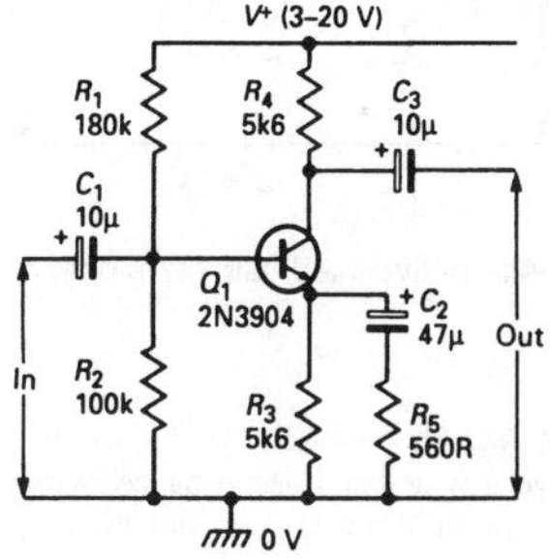

Figure 10.34. Fixed-gain (x10) common-emitter amplifier.

Figure 10.35 shows a useful variation of the above design. In this case R_3 equals R_4, and is not decoupled, so the circuit gives unity voltage gain. Note, however, that this circuit gives two unity-gain output signals, with the emitter output in phase with the input and the collector signal in anti-phase; this circuit thus acts as a unity-gain phase splitter.

Figure 10.36 shows another way of varying circuit gain. This design gives high voltage gain between Q_1 collector and base, but R_2 gives a.c. feedback to the base, and R_1 is wired in series between the input signal and Q_1 base; the net effect is that the circuit's voltage gain (between input and output) equals R_2/R_1, and works out at x10 in this particular case.

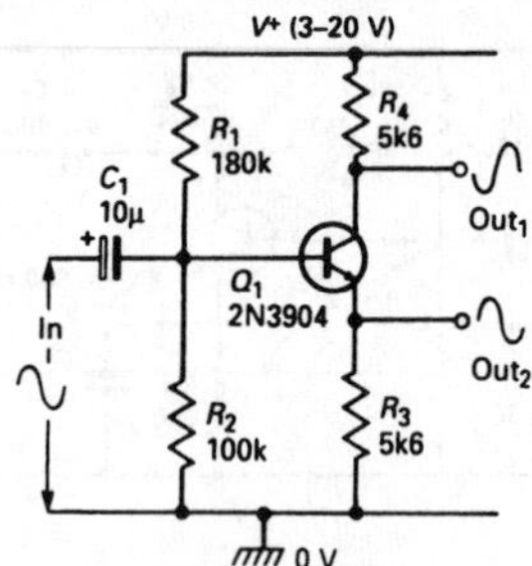

Figure 10.35. Unity-gain phase splitter.

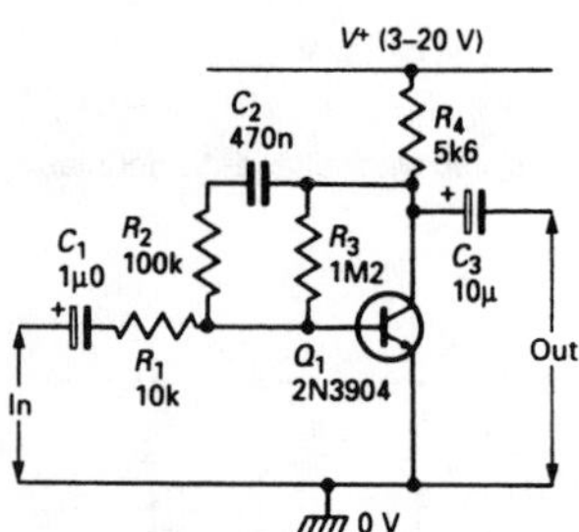

Figure 10.36. Alternative fixed-gain (x10) amplifier.

Finally, *Figure 10.37* shows how the *Figure 10.31* design can be modified to give a wide-band performance by wiring d.c.-coupled emitter follower buffer between Q_1 collector and the output terminal, to minimise the shunting effects of stray capacitance on R_2 and thus extending the upper bandwidth to several hundred kHz.

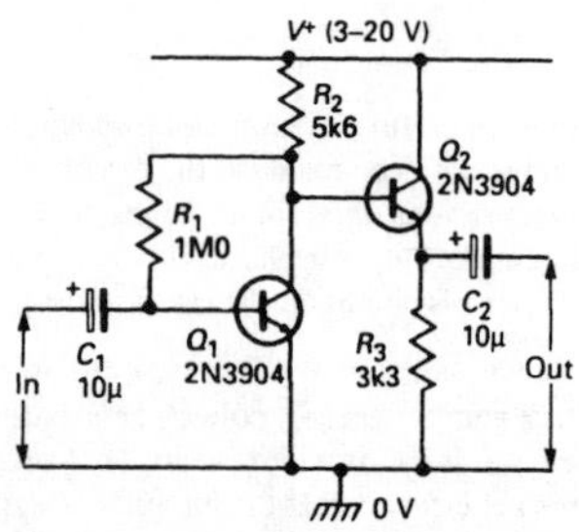

Figure 10.37. Wide-band amplifier.

High-gain circuits

A single-stage common-emitter amplifier circuit can not give a voltage gain much greater than 46dB when using a resistive collector load; a multi-stage circuit must be used if higher gain is needed. *Figures 10.38* to *10.40* show three useful high-gain two-transistor voltage amplifier designs.

The *Figure 10.38* circuit acts like a direct-coupled pair of common-emitter amplifiers, with Q_1's output feeding directly into Q_2 base, and gives an overall voltage gain of 76dB (about x6150) and an upper –3dB frequency of 35kHz. Note that feedback biasing resistor R_4 is fed from Q_2's a.c.-decoupled emitter (which 'follows' the quiescent collector voltage of Q_1), rather than directly from Q_1 collector, and that the bias circuit is thus effectively a.c.-decoupled. *Figure 10.39* shows an alternative version of the above design, using a pnp output stage; its performance is the same as that of *Figure 10.38*.

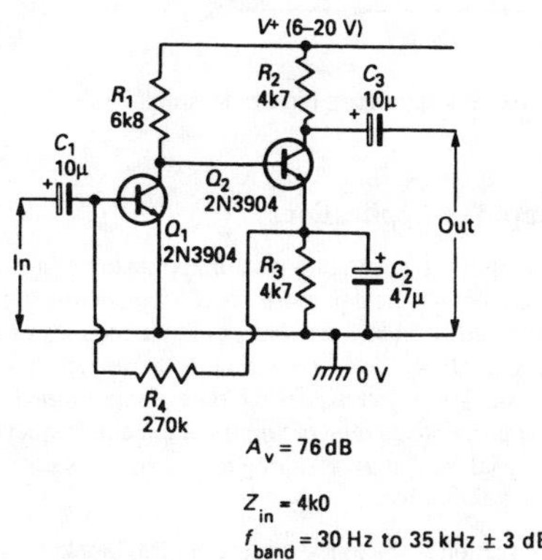

Figure 10.38. High-gain two-stage amplifier.

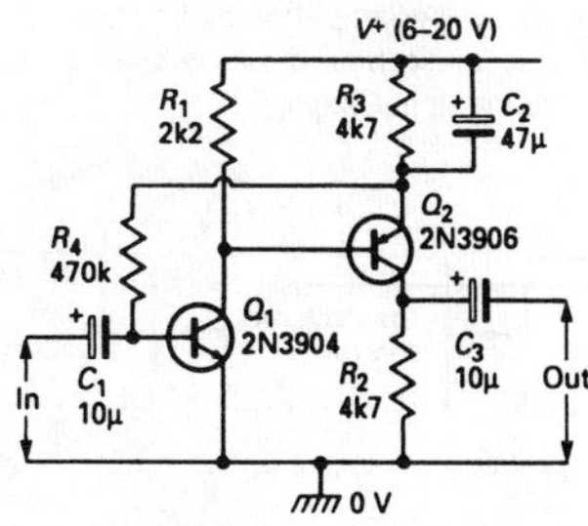

Figure 10.39. Alternative high-gain two-stage amplifier.

The *Figure 10.40* circuit gives a voltage gain of about 66dB. Q_1 is a common-emitter amplifier with a split collector load (R_2–R_3), and Q_2 is an emitter follower and feeds its a.c. output signal back to the R_2–R_3 junction via C_3, thus 'bootstrapping' the R_3 value (as described earlier) so that it acts as a high a.c. impedance; Q_1 thus gives a very high voltage gain. This circuit's bandwidth extends up to about 32kHz, but its input impedance is only 330Ω.

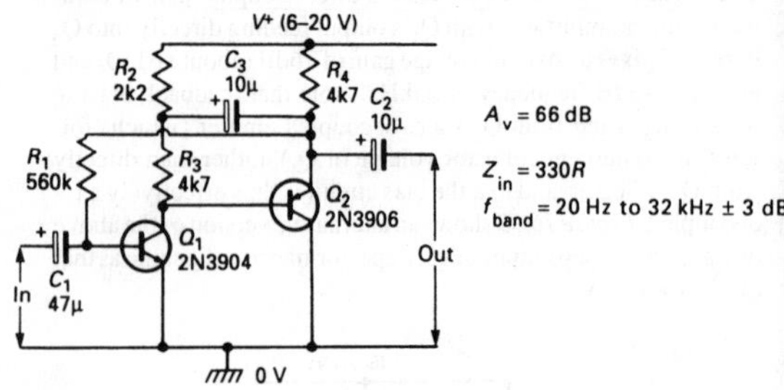

Figure 10.40. Bootstrapped high-gain amplifier.

Audio amplifier applications

Transistor amplifiers have many useful applications in stereo audio systems. For most practical purposes, each channel of a stereo system can be broken down into three distinct circuit sections or blocks, as shown in *Figure 10.41*. The first section is the selector/pre-amplifier block. It lets the user select the desired type of input signal source and applies an appropriate amount of amplification and frequency correction to the signal, so that the resulting output signal is suitable for use by the second circuit block.

The second section is the tone-/volume-control block, which lets the user adjust the system's frequency characteristics and output signal amplitude to suit personal tastes; this section may contain additional filter circuits and gadgets, such as scratch and rumble filters and audio mixer circuitry, etc. Its output is fed to the system's final section, the audio power amplifier, which drives the loudspeakers.

A variety of practical pre-amplifier and tone-control circuits are described in the next few sections of this chapter; audio power amplifier circuits are dealt with in Chapter 12.

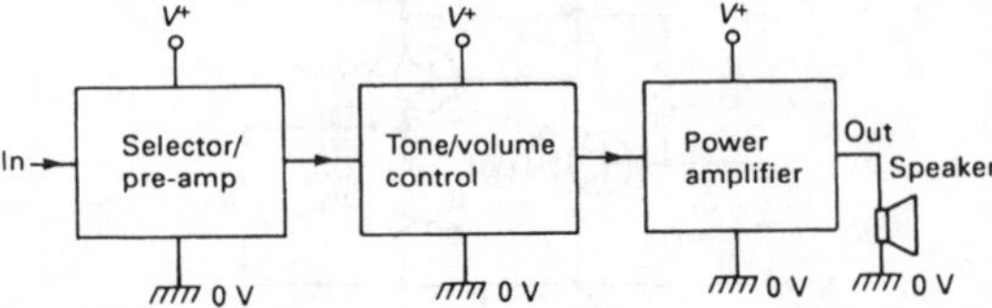

Figure 10.41. Basic elements of one channel of an audio amplifier system.

Simple pre-amps

The basic function of an audio pre-amplifier is that of modifying the input signal characteristics so that they give the level frequency response and nominal 100mV mean output amplitude needed to drive the amplifier's tone-control system. If the input comes from a radio tuner or a tape player, etc., the signal characteristics are usually such that they can be fed directly to the tone-control sections, by-passing the pre-amplifier circuit, but if derived from a microphone or pick-up they usually need modification via a pre-amp stage.

Microphones and pick-ups are usually either magnetic or ceramic/crystal devices. Magnetic types usually have a low output impedance and a low signal sensitivity (about 2mV nominal); their outputs thus need to be fed to high-gain pre-amplifier stages. Ceramic/crystal types usually have a high output impedance and a high sensitivity (about 100mV nominal); their outputs thus need to be fed to a high-impedance pre-amp stage with near-unity voltage gain.

Most microphones have a flat frequency response and can be used with simple pre-amp stages. *Figure 10.42* shows a unity-gain pre-amp that can be used with most high-impedance ceramic/crystal microphones. It is an emitter follower circuit with a bootstrapped (via C_2–R_3) input network, and has an input impedance of about 2M0; its supply is decoupled via C_5–R_5.

Figures 10.43 and *10.44* show pre-amp circuits that can be used with magnetic microphones. The single-stage circuit of *Figure 10.43* gives 46dB (x200) of voltage gain, and can be used with most magnetic microphones. The two-stage circuit of *Figure 10.44* gives 76dB of voltage gain, and is meant for use with magnetic microphones with very low sensitivity.

RIAA pre-amp circuits

If a constant-amplitude 20Hz to 20kHz variable-frequency signal is recorded on a phonograph disc (record) using conventional stereo recording equipment, and the record is then replayed, it generates the highly non-linear frequency response curve shown in *Figure 10.45*; the dotted line shows the 'idealised' shape of this curve, and the solid line

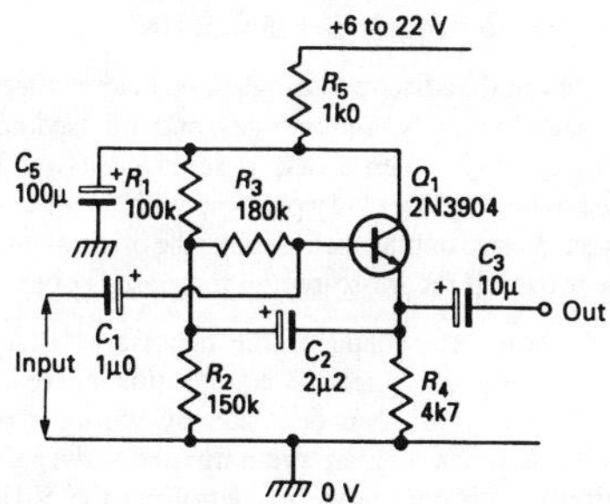

Figure 10.42. High-impedance pre-amp for use with ceramic/crystal microphones.

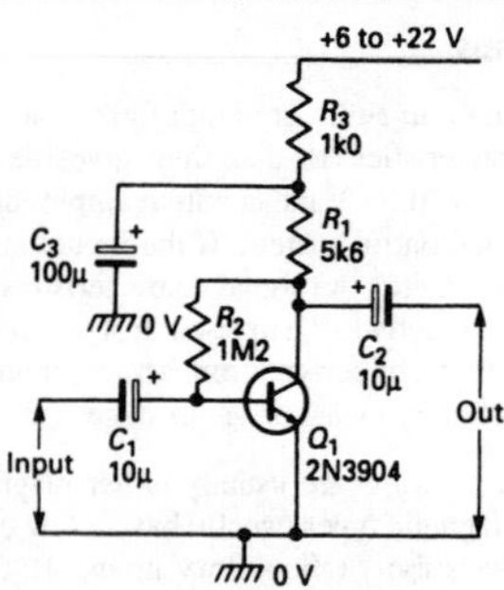

Figure 10.43. Magnetic microphone pre-amp, giving 46dB of gain.

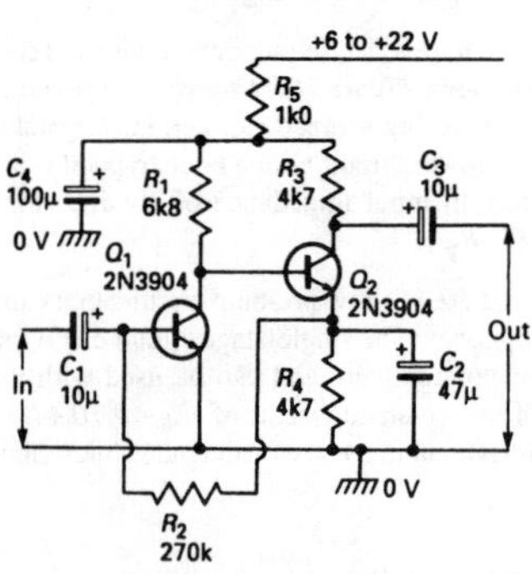

Figure 10.44. Magnetic microphone pre-amp, giving 76dB of gain.

shows its practical form. The 'idealised' response is flat between 500Hz and 2120Hz, but rises at a rate of 6dB/octave (20dB/decade) above 2120Hz, and falls at a 6dB/octave rate between 500Hz and 50Hz; the response is flat to frequencies below 50Hz.

These responses enable disc recordings to be made with good signal-to-noise ratios and wide dynamic ranges, and are used on all normal records. Consequently, when a disc is replayed its output must be passed to the power amplifier via a pre-amp with an equalisation curve that is the exact inverse of that used to make the original disc recording, so that a linear overall record-to-replay response is obtained.

Figure 10.46 shows the shape of the necessary 'RIAA' (Record Industry Association of America) equalisation curve. A practical RIAA equalisation circuit can be made by wiring a pair of *C–R* feedback networks into a standard pre-amp (so that the gain falls as the frequency rises), with one network controlling the 50Hz to 500Hz response, and the other the 2120Hz to 20kHz response. *Figure 10.47* shows such an amplifier.

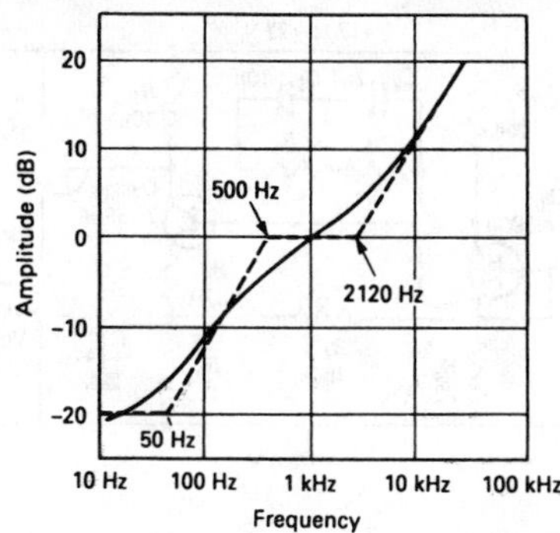

Figure 10.45. Typical phonograph disc frequency response playback curve.

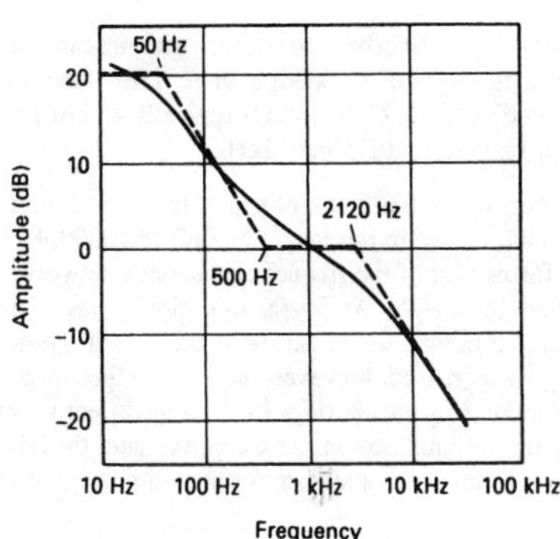

Figure 10.46. RIAA playback equalisation curve.

The *Figure 10.47* circuit can be used with any magnetic pick-up cartridge. It gives a 1V output from a 6mV input at lkHz, and provides equalisation that is within 1dB of the RIAA standard between 40Hz and 12kHz. The actual pre-amp is designed around Q_1 and Q_2, with C_2–R_5 and C_3–R_6 forming the feedback equalisation network. Q_3 as an emitter-follower buffer stage, and drives optional volume control RV_1.

Ceramic/crystal pick-ups usually give a poorer reproduction quality than magnetic types, but give output signals of far greater amplitude: they can thus be used with a very simple type of equalisation pre-amp, and are consequently found in many ‘popular’ record player systems. *Figures 10.48* and *10.49* show alternative phonograph pre-amplifier circuits that can be used with ceramic or crystal pick-up cartridges: in each case, the pre-amp/equaliser circuit is designed around Q_1, and Q_2 is an emitter follower output stage that drives optional volume control RV_1.

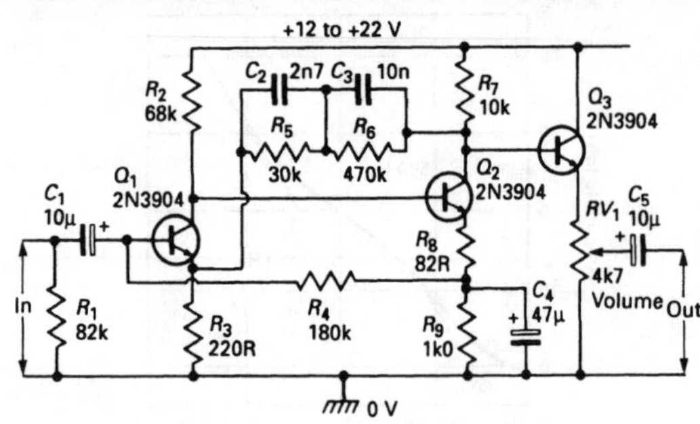

Figure 10.47. RIAA equalisation pre-amp, for use with magnetic pick-up cartridges.

The *Figure 10.48* circuit can be used with any pick-up cartridge that has a capacitance in the 1000pF to 10000pF range. Two-stage equalisation is provided via C_1–R_2 and C_2–R_3, and is typically within 1.6dB of the RIAA standard between 40Hz and 12kHz.

The alternative *Figure 10.49* circuit can only be used with pick-ups with capacitance values in the range 5000pF to 10000pF, since this capacitance forms part of the frequency response network: the other part is formed by C_1–R_3. At 50Hz, this circuit has a high input impedance (about 600k), and causes only slight cartridge loading: as the frequency is increased, however, the input impedance decreases sharply, thus increasing the cartridge loading and effectively reducing the circuit gain. The equalisation curve approximates the RIAA standard, and the performance is adequate for many practical applications.

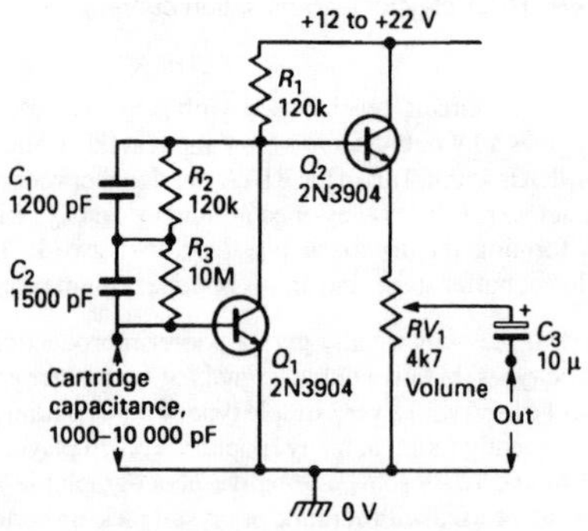

Figure 10.48. RIAA phonograph equaliser for ceramic pick-up cartridges.

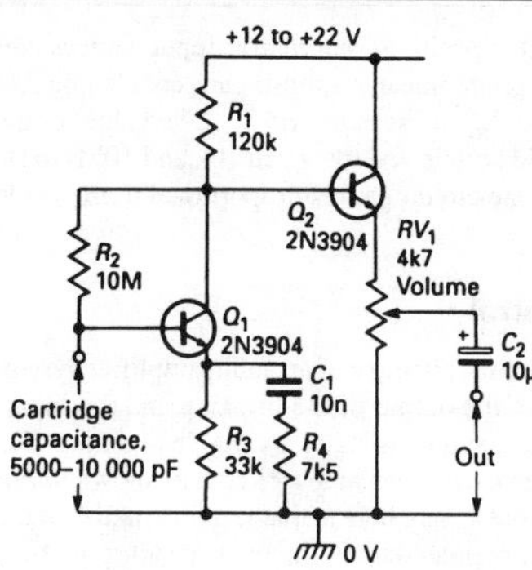

Figure 10.49. Alternative RIAA phonograph equaliser for ceramic cartridges.

A universal pre-amp

Most audio amplifiers use pre-amps with variable characteristics, e.g., a high-gain linear response for use with magnetic microphones, low-gain linear response for use with a radio tuner, and high-gain RIAA equalisation for use with a magnetic pick-up cartridge, etc.

To meet this requirement, it is normal to fit the system with a single 'universal' pre-amp circuit of the type shown in *Figure 10.50*. This is basically a high-gain linear amplifier that can have its characteristics altered by switching alternative types of resistor/filter network into its feedback loops.

Thus, when the selector switch is set to the 'MAG P.U' position, S_{1a} connects the input to the magnetic pick-up cartridge, and S_{1b} connects the C_4–R_7–C_5 RIAA equalisation network into the feedback loop. In the

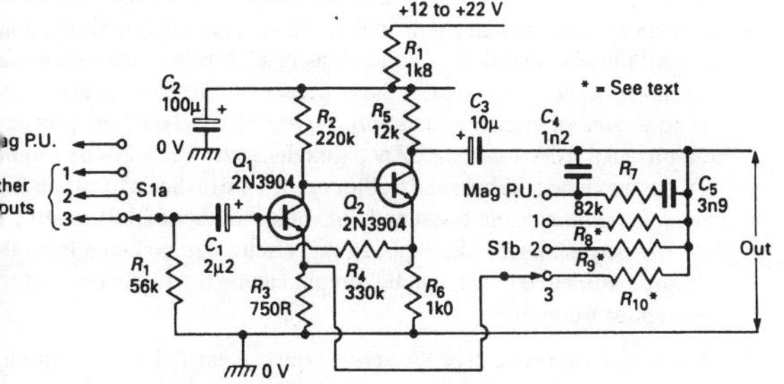

Figure 10.50. Universal pre-amplifier circuit.

remaining switch positions, alternative input sources are selected vi S_{1a}, and appropriate linear-response gain-controlling feedback resis tors (R_8, R_9 and R_{10}) are selected via S_{1b}. The values of these feedbac resistors should be selected (between 10k and 10M) to suit individua requirements; the circuit gain is proportional to the feedback resisto value.

Volume control

The volume control circuitry of an audio amplifier system is normall placed between the output of the pre-amp and the input of the tone control circuitry, and consists of a variable potential divider or 'pot' This pot can form part of an active circuit, as shown in *Figure 10.47* t *Figure 10.49*, but a snag here is that rapid variation of the control ca briefly apply d.c. potentials to the next circuit, possibly upsetting i bias and generating severe signal distortion.

Figure 10.51 shows the ideal form and location of the volume contro It is fully d.c.-isolated from the pre-amp's output via C_1, and from th input of the tone-control circuitry via C_2; variation of RV_1 slider thu has no effect on the d.c. bias levels of either circuit. RV_1 should be 'log' type of pot.

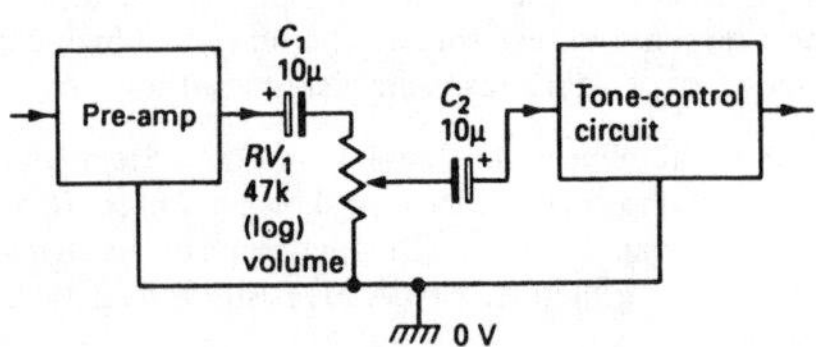

Figure 10.51. Ideal form and location of the volume control.

Tone controls

A tone control network lets the user alter the frequency response of th amplifier system to suit a personal mood or requirement. Simple tor control networks consist of collections of *C–R* filters, through whic the audio signals are passed; these networks are passive, and caus some degree of signal attenuation. *Figure 10.52* shows the practic circuit of a passive tone control network that gives about 20dB of sign attenuation when the bass and treble controls are in the flat position, an gives maximum treble boost and cut values of about 20dB relative the flat performance. The input to this circuit can be taken from th circuit's volume control, and the output can be fed to the input of th main power amplifier.

A tone control network of the above type can easily be wired into th feedback path of a transistor amplifier so that the system gives a overall signal gain (rather than attenuation) when its controls are in t flat position. *Figure 10.53* shows an 'active' tone control circuit of th type.

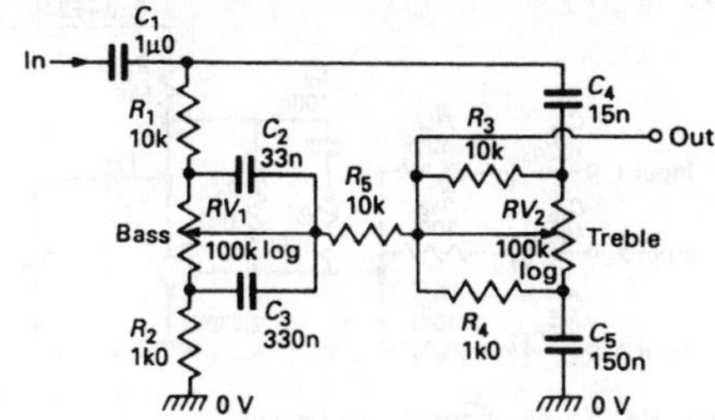

Figure 10.52. Passive bass and treble tone control network.

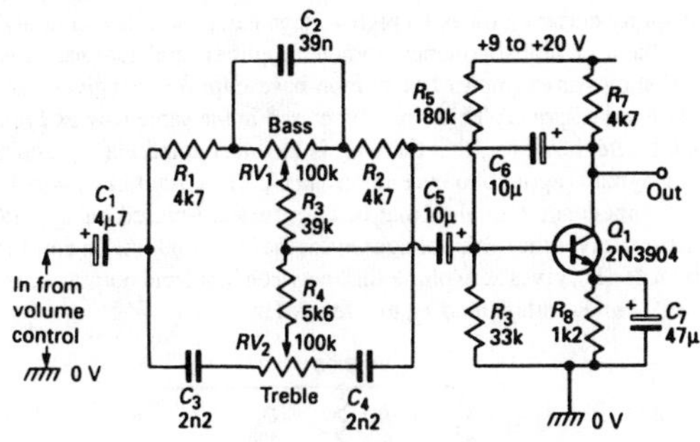

Figure 10.53. Active bass and treble tone control circuit.

An audio mixer

One useful gadget that can be fitted in the area of the volume-/tone-control section of an audio amplifier is a multi-channel audio mixer, which enables several different audio signals to be mixed together to form a single composite output signal. This can be useful in, for example, enabling the user to hear the 'emergency' sounds of a front-door or baby-room microphone, etc., while listening to normal entertainment sources.

Figure 10.54 shows a simple 3-channel audio mixer that gives unity gain between the output and each input. Each input channel comprises a single 100n capacitor (C_1) and 100k resistor (R_1), and presents an input impedance of 100k. The circuit can be given any desired number of input channels by simply adding more C_1 and R_1 components. In use, the mixer should be placed between the output of the tone-control circuitry and the input of the main power amplifier, with one input taken from the tone-control output and the others taken from the desired signal sources.

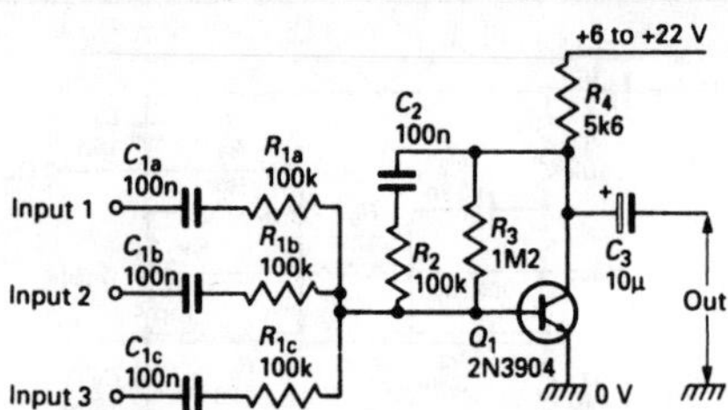

Figure 10.54. Three-channel audio mixer.

Common-base amplifier circuits

The common-base amplifier has a very low input impedance, gives near-unity current gain and a high voltage gain, and is used mainly in wide-band or high-frequency voltage amplifier applications. *Figure 10.55* shows an example of a common-base amplifier that gives a good wide-band response. This circuit is biased in the same way as *Figure 10.33*; note, however, that the base is a.c.-decoupled via C_1, and the input signal is applied to the emitter via C_3. The circuit has a very low input impedance (equal to that of Q_1's forward-biased base-emitter junction), gives the same voltage gain as the common-emitter amplifier (about 46dB), gives zero phase shift between input and output, and has a –3dB bandwidth extending to a few MHz.

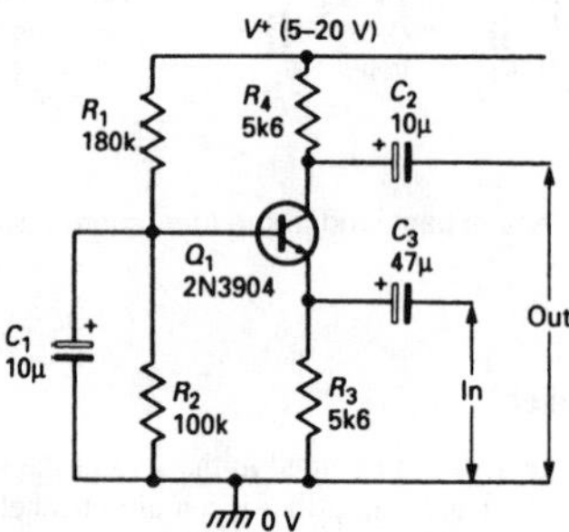

Figure 10.55. Common-base amplifier.

Figure 10.56 shows an excellent wideband amplifier, the 'cascode' circuit, which gives the wide bandwidth benefit of the common-base amplifier together with the medium input impedance of the common-emitter amplifier. This is achieved by wiring Q_1 and Q_2 in series, with Q_1 connected in the common-base mode and Q_2 in the common-emitter mode. The input signal is applied to the base of Q_2, which uses Q_1 emitter as its collector load and thus gives unity voltage gain and a very wide bandwidth, and Q_1 gives a voltage gain of about 46dB. Thus, the complete circuit has in input impedance of about 1.8kΩ, a voltage gain of 46dB, and a –3dB bandwidth that extends to a few MHz.

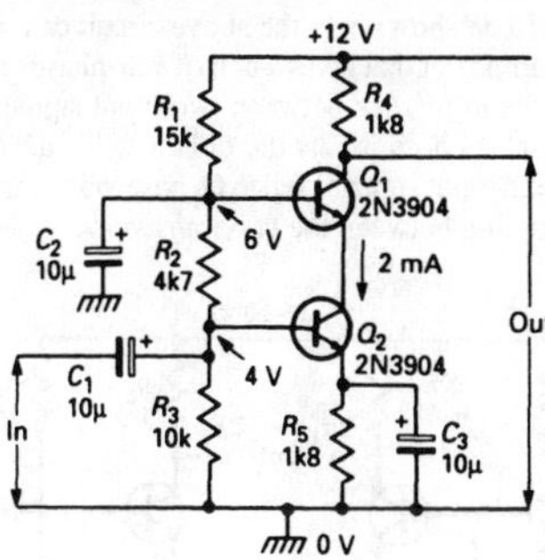

Figure 10.56. Wide-band cascode amplifier.

Figure 10.57 shows a close relative of the common-base amplifier, the 'long-tailed pair' phase splitter, which gives a pair of anti-phase outputs when driven from a single-ended input signal. Q_1 and Q_2 share a common emitter resistor (the 'tail'), and the circuit bias point is set via RV_1 so that the two transistors pass identical collector currents (giving zero difference between the two collector voltages) under quiescent conditions. Q_1 base is a.c.-grounded via C_1, and a.c.-input signals are applied to Q_2 base. The circuit acts as follows.

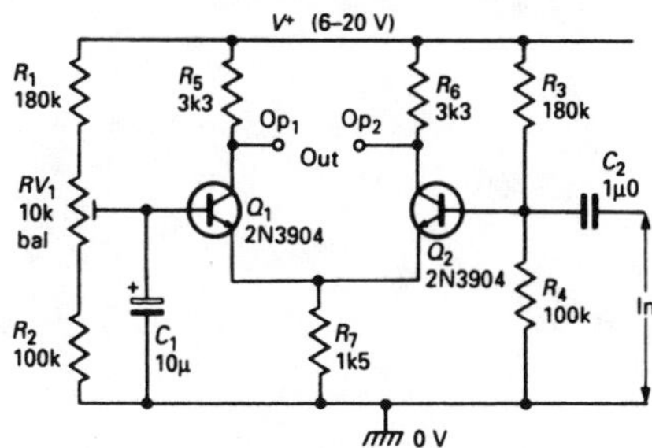

Figure 10.57. Phase splitter.

Suppose that a sinewave input signal is fed to Q_2 base. Q_2 acts as an inverting common-emitter amplifier, and when the signal drives its base upwards its collector inevitably swings downwards, and *vice versa*; simultaneously, Q_2's emitter 'follows' the input signal, and as its emitter voltage rises it inevitably reduces the base-emitter bias of Q_1, thus making Q_1's collector voltage rise, etc. Q_1 thus operates in the common-base mode and gives the same voltage gain as Q_2, but gives a non-inverting amplifier action; this 'phase-splitter' circuit thus generates a pair of balanced anti-phase output signals.

Finally, *Figure 10.58* shows how the above circuit can be made to act as a differential amplifier that gives a pair of anti-phase outputs that are proportional to the *difference* between two input signals; if identical signals are applied to both inputs the circuit will (ideally) give zero output. The second input signal is fed to Q_1 base via C_1, and the R_7 'tail' provides the coupling between the two transistors.

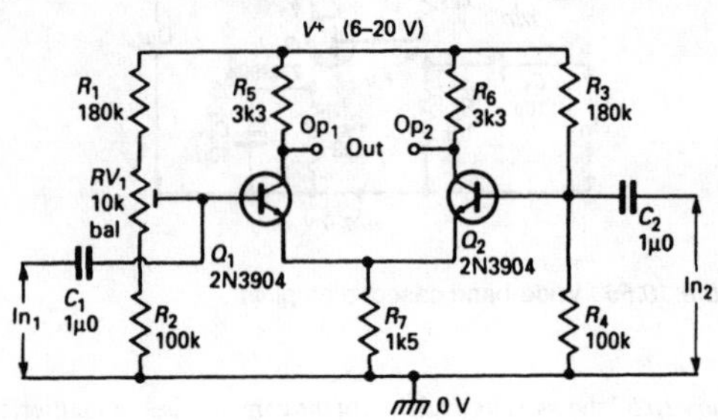

Figure 10.58. Simple differential amplifier or long-tailed pair.

11 Transistor waveform generator circuits

The two most widely used basic types of transistor waveform generator are the 'oscillator' types that produce sinewaves and use transistors as linear amplifying elements, and the 'multivibrator' types that generate square or rectangular waveforms and use transistors as digital switching elements. Both types of generator are described in this chapter.

Oscillator basics

To generate reasonably pure sinewaves, an oscillator has to satisfy two basic design requirements, as shown in *Figure 11.1*. First, the output of its amplifier (A1) must be fed back to its input via a frequency-selective network (A2) in such a way that the sum of the amplifier and feedback-network phase-shifts equals zero degrees (or 360°) at the desired oscillation frequency, i.e., so that $x° + y° = 0°$ (or 360°). Thus, if the amplifier generates 180° of phase shift between input and output, an additional 180° of phase shift must be introduced by the frequency-selective network.

The second requirement is that the amplifier's gain must exactly counter the losses of the frequency-selective feedback network at the desired oscillation frequency, to give an overall system gain of unity, e.g., A1 x A2 = 1. If the gain is below unity the circuit will not oscillate, and if greater than unity it will be over-driven and will generate distorted waveforms. The frequency-selective feedback network usually consists of either a *C–R* or L–C or crystal filter.

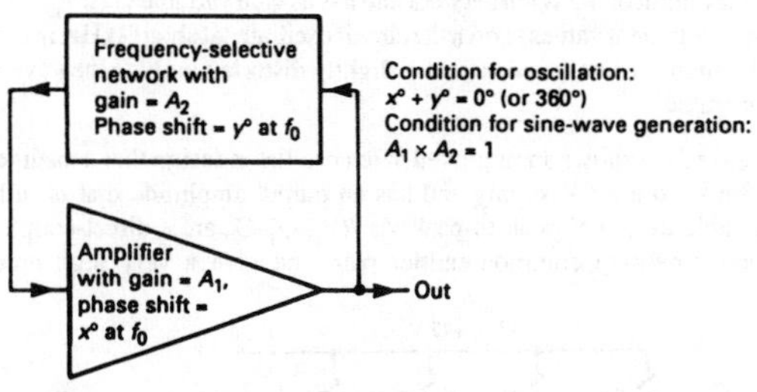

Figure 11.1. Essential circuit and conditions needed for sine-wave generation.

C–R oscillators

The simplest *C–R* sinewave oscillator is the phase-shift type, which is fully described in Chapter 5 together with (in *Figure 5.4*) a practical design example. In reality, the simple phase-shift oscillator has poor gain stability, and its operating frequency can not easily be made variable. A far more versatile *C–R* oscillator can be built using the Wien bridge network, which is also described in detail in Chapter 5.

Figure 11.2 shows the basic elements of the Wien bridge oscillator. The Wien network consists of R_1–C_1 and R_2–C_2, which have their values balanced so that C_1=C_2=C, and R_1=R_2=R. This network's phase shifts are negative at low frequencies, positive at high ones, and zero at a 'centre' frequency of 1/(6.28*CR*), at which the network has an attenuation factor of three; the network can thus be made to oscillate by connecting a non-inverting x3 high-input-impedance amplifier between its output and input terminals, as shown in the diagram.

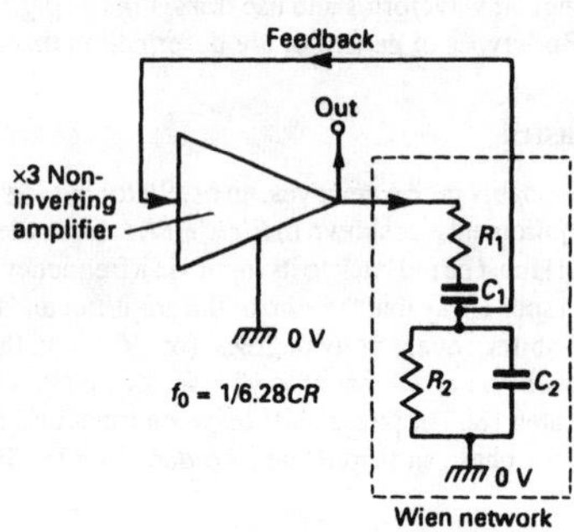

Figure 11.2. Basic Wien oscillator circuit.

Figure 11.3 shows a simple fixed-frequency Wien oscillator in which Q_1 and Q_2 are both wired as low-gain common-emitter amplifiers. Q_2 gives a voltage gain slightly greater than unity and uses Wien network resistor R_1 as its collector load, and Q_1 presents a high input impedance to the output of the Wien network and has its gain variable via RV_1. With the component values shown the circuit oscillates at about 1kHz; in use, RV_1 should be adjusted so that a slightly distorted output sinewave is generated.

Figure 11.4 shows an improved Wien oscillator design that consumes 1.8mA from a 9V supply and has an output amplitude that is fully variable up to 6V peak-to-peak via RV_2. Q_1–Q_2 are a direct-coupled complementary common-emitter pair, and give a very high input

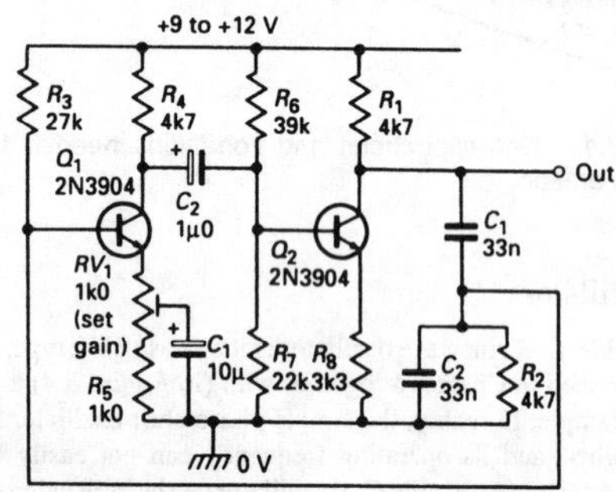

Figure 11.3. Practical 1kHz Wien oscillator.

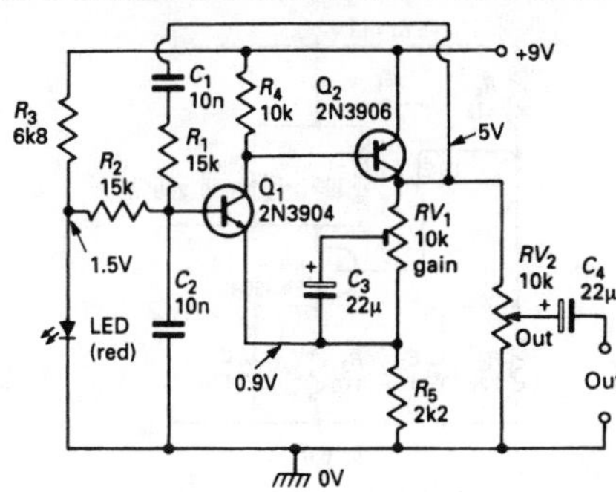

Figure 11.4. 1kHz Wien bridge sine-wave generator with variable-amplitude output.

impedance to Q_1 base, a low output impedance from Q_2 collector, and non-inverted voltage gains of x5.5 d.c. and x1 to x5.5 a.c. (variable via RV_1). The red LED generates a low-impedance 1.5V that is fed to Q_1 base via R_2 and thence biases Q_2's output to a quiescent value of +5V. The R_1–C_1 and R_2–C_2 Wien network is connected between Q_2's output and Q_1's input, and in use RV_1 is simply adjusted so that, when the circuit's output is viewed on an oscilloscope, a stable and visually 'clean' waveform is generated. Under this condition the oscillation amplitude is limited at about 6V peak-to-peak by the onset of positive-peak clipping as the amplifier starts to run into saturation. If RV_1 is carefully adjusted this clipping can be reduced to an almost imperceptible level, enabling good-quality sinewaves, with less than 0.5% THD, to be generated.

The *Figure 11.4* circuit can be modified to give limited-range variable-frequency operation by reducing the R_1 and R_2 values to 4k7 and wiring them is series with ganged 10k variable resistors. Note, however, that variable-frequency Wien oscillators are best built using op-amps or other linear ICs, in conjunction with automatic-gain-control feedback systems, and several circuits of this type are in fact shown in Volume 1 of the "Newnes Electronic Circuits Pocket Book" series.

L–C oscillators

C–R sinewave oscillators usually generate signals in the 5Hz to 500kHz range. L–C oscillators usually generate them in the 5kHz to 500MHz range, and consist of a frequency-selective L–C network that is connected into an amplifier's feedback loop.

The simplest L–C transistor oscillator is the tuned collector feedback type shown in *Figure 11.5*. Q_1 is wired as a common-emitter amplifier, with base bias provided via R_1–R_2 and with emitter resistor R_3 a.c.-decoupled via C_2. L_1–C_1 forms the tuned collector circuit, and collector-to-base feedback is provided via L_2, which is inductively coupled to L_1 and provides a transformer action; by selecting the phase of this

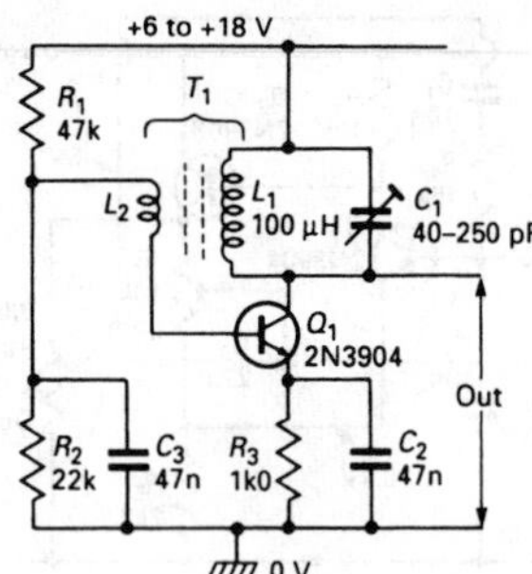

Figure 11.5. Tuned collector feedback oscillator.

feedback signal the circuit can be made to give zero loop phase shift at the tuned frequency, so that it oscillates if the loop gain (determined by the turns ratio of T_1) is greater than unity.

A feature of any L–C tuned circuit is that the phase relationship between its energising current and induced voltage varies from –90° to +90°, and is zero at a 'centre' frequency given by $f = 1/(2\pi\sqrt{LC})$. Thus, the *Figure 11.5* circuit gives zero overall phase shift, and oscillates at, this centre frequency. With the component values shown, the frequency can be varied from 1MHz to 2MHz via C_1. This basic circuit can be designed to operate at frequencies ranging from a few tens of Hz by using a laminated iron-cored transformer, up to tens or hundreds of MHz using RF techniques.

Circuit variations

Figure 11.6 shows a simple variation of the *Figure 11.5* design, the Hartley oscillator. Its L_1 collector load is tapped about 20% down from its top, and the positive supply rail is connected to this point; L_1 thus gives an auto-transformer action, in which the signal voltage at the top of L_1 is 180° out of phase with that at its low (Q_1 collector) end. The signal from the top of the coil is fed to Q_1's base via C_2, and the circuit thus oscillates at a frequency set by the L–C values.

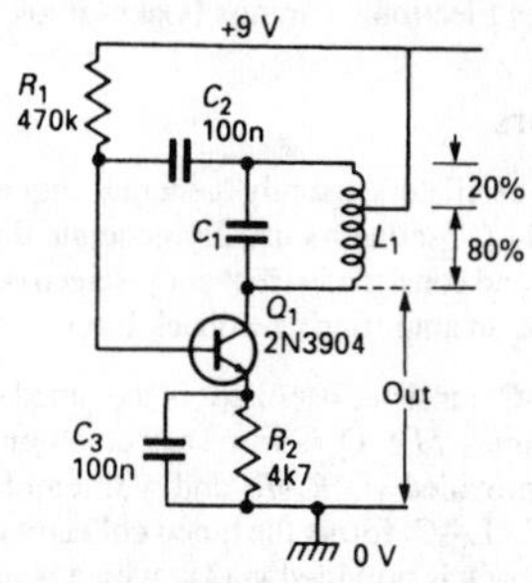

Figure 11.6. Basic Hartley oscillator.

Note from the above description that oscillator action depends on some kind of 'common signal' tapping point being made into the tuned circuit, so that a phase-splitting autotransformer action is obtained. This tapping point does not have to be made into the actual tuning coil, but can be made into the tuning capacitor, as in the Colpitts oscillator circuit shown in *Figure 11.7*. With the component values shown, this particular circuit oscillates at about 37kHz.

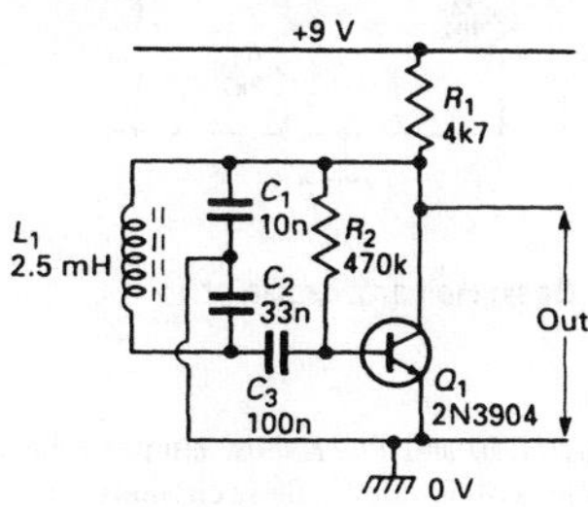

Figure 11.7. 37kHz Colpitts oscillator.

A modification of the Colpitts design, known as the Clapp or Gouriet oscillator, is shown in *Figure 11.8*. C_3 is wired in series with L_1 and has a value that is small relative to C_1 and C_2; consequently, the circuit's resonant frequency is set mainly by L_1 and C_3 and is almost independent of variations in transistor capacitances, etc. The circuit thus gives excellent frequency stability. With the component values shown it oscillates at about 80kHz.

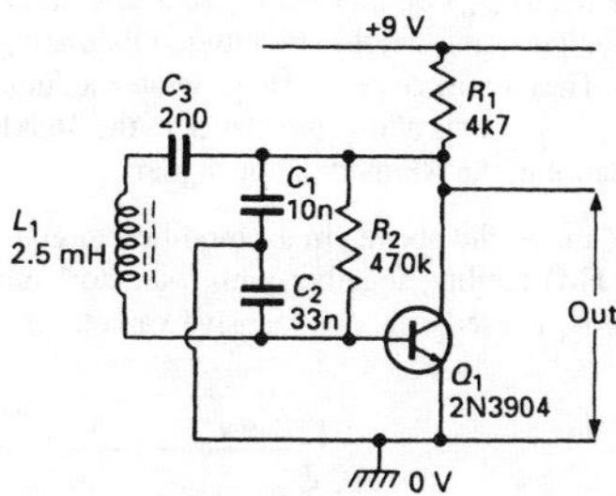

Figure 11.8. 80kHz Gouriet or Clapp oscillator.

Figure 11.9 shows a Reinartz oscillator, in which the tuning coil has three inductively-coupled windings. Positive feedback is obtained by coupling the transistor's collector and emitter signals via windings L_1 and L_2. Both of these inductors are coupled to L_3, and the circuit oscillates at a frequency determined by L_3–C_1. The diagram shows typical coil-turns ratios for a circuit that oscillates at a few hundred kHz.

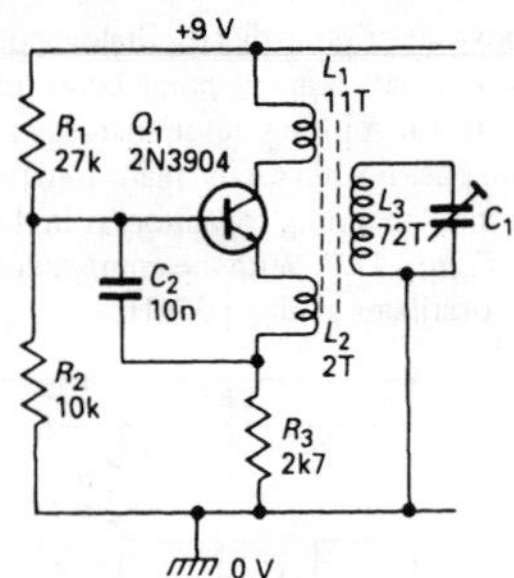

Figure 11.9. Basic Reinartz oscillator.

Finally, *Figures 11.10* and *11.11* show emitter follower versions of Hartley and Colpitts oscillators. In these circuits the transistors and L_1–C_1 tuned circuits each give zero phase shift at the oscillation frequency, and the tuned circuit gives the voltage gain necessary to ensure oscillation.

Modulation

The L–C oscillator circuits of *Figures 11.5* to *11.11* can easily be modified to give modulated (AM or FM) rather than continuous-wave (CW) outputs. *Figure 11.12*, for example, shows the *Figure 11.5* circuit modified to act as a 465kHz beat-frequency oscillator (B.F.O.) with an amplitude-modulation (AM) facility. A standard 465kHz transistor I.F. transformer (T_1) is used as the L–C tuned circuit, and an external AF signal can be fed to Q_1's emitter via C_2, thus effectively modulating Q_1's supply voltage and thereby amplitude-modulating the 465kHz carrier signal. The circuit can be used to generate modulation depths up to about 40%. C_1 presents a low impedance to the 465kHz carrier but a high impedance to the AF modulation signal.

Figure 11.13 shows the above circuit modified to give a frequency-modulation (FM) facility, together with 'varactor' tuning via RV_1. Silicon diode D_1 is used as an inexpensive varactor diode which (as

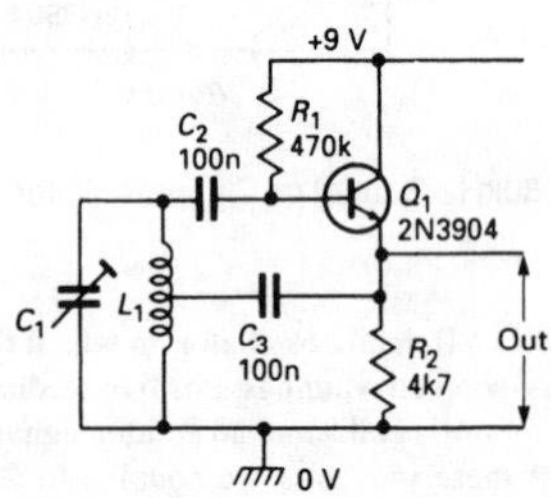

Figure 11.10. Emitter follower version of the Hartley oscillator.

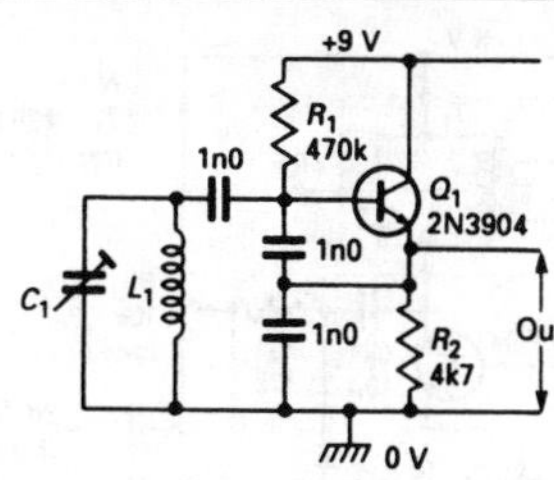

Figure 11.11. Emitter follower version of the Colpitts oscillator.

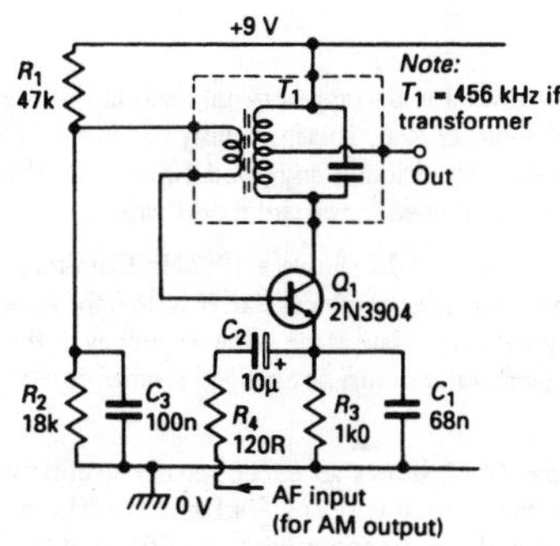

Figure 11.12. 465kHz B.F.O. with AM facility.

already pointed out in Chapters 7 and 8), when reverse biased, exhibits a capacitance (of a few tens of pF) that decreases with applied reverse voltage. D_1 and blocking capacitor C_2 are wired in series and effectively connected across the T_1 tuned circuit (since the circuit's supply rails are shorted together as far as a.c. signals are concerned). Consequently, the oscillator's centre frequency can be varied by altering D_1's capacitance via RV_1, and FM signals can be obtained by feeding an AF modulation signal to D_1 via C_3 and R_4.

Crystal oscillators

Crystal-controlled oscillators give excellent frequency accuracy and stability. Quartz crystals have typical Qs of about 100000 and provide about 1000 times greater frequency stability than a conventional L–C tuned circuit; their operating frequency (which may vary from a few kHz to 100MHz) is determined by the mechanical dimensions of the crystal, which may be cut to give either series or parallel resonant operation; series-mode devices show a low impedance at resonance; parallel-mode types show high ones at resonance.

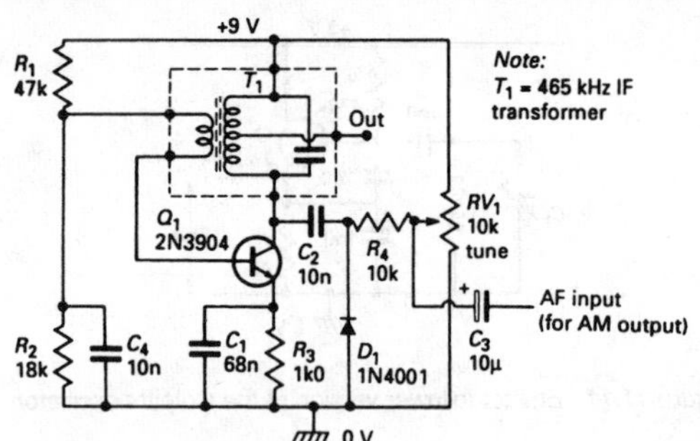

Figure 11.13. 465kHz B.F.O. with varactor tuning and FM facility.

Figure 11.14 shows a wide-range crystal oscillator designed for use with a parallel-mode crystal. This is actually a Pierce oscillator circuit, and can be used with virtually any good 100kHz to 5MHz parallel-mode crystal without need for circuit modification.

Alternatively, *Figure 11.15* shows a 100kHz Colpitts oscillator designed for use with a series-mode crystal. Note that the L_1–$C1$–C_2 tuned circuit is designed to resonate at the same frequency as the crystal, and that its component values must be changed if other crystal frequencies are used.

Finally, *Figure 11.16* shows an exceptionally useful two-transistor oscillator that can be used with any 50kHz to 10MHz series-resonant crystal. Q_1 is wired as a common-base amplifier and Q_2 is an emitter follower, and the output signal (from Q_2 emitter) is fed back to the input (Q_1 emitter) via C_2 and the series-resonant crystal. This excellent circuit will oscillate with any crystal that shows the slightest sign of life.

White noise generator

One useful 'linear' but non-sinusoidal waveform is that known as 'white noise', which contains a full spectrum of randomly generated frequencies, each with equal mean power when averaged over a unit of

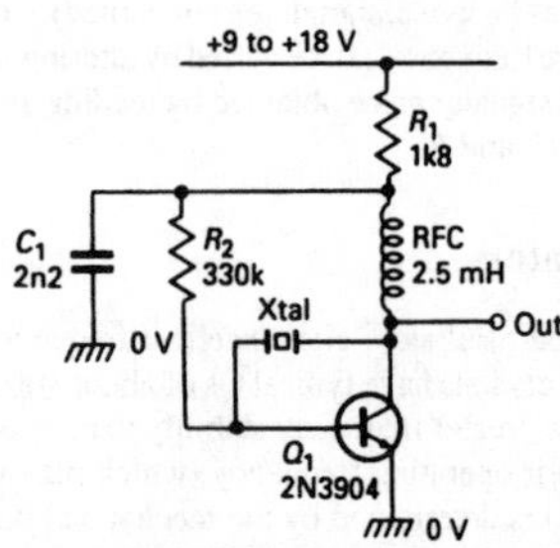

Figure 11.14. Wide-range Pierce oscillator uses parallel-mode crystal.

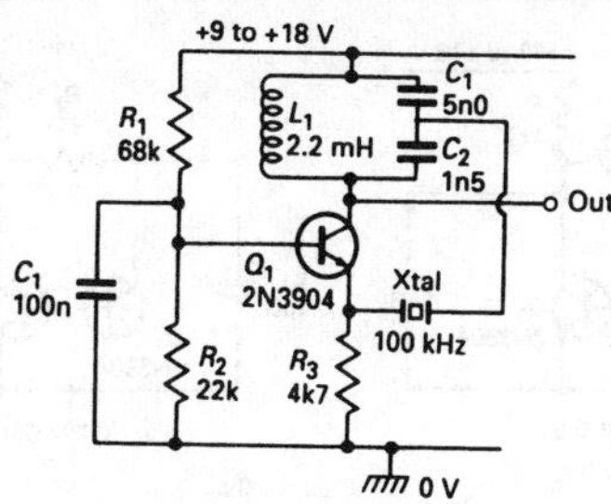

Figure 11.15. 100kHz Colpitts oscillator uses series-mode crystal.

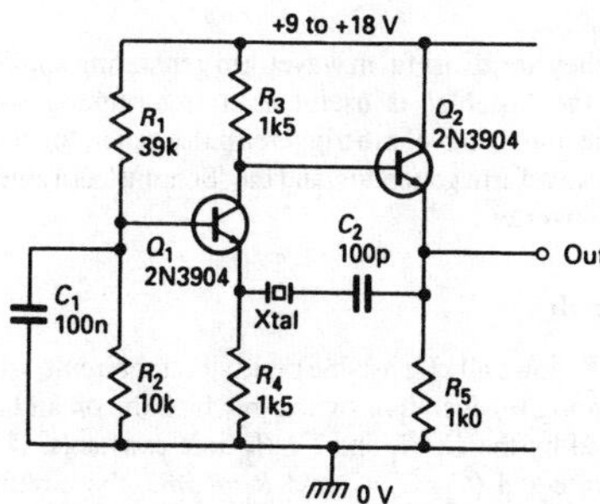

Figure 11.16. Wide-range (50kHz–10MHz oscillator can be used with almost any series-mode crystal.

time. White noise is of value in testing AF and RF amplifiers, and is widely used in special-effects sound generator systems.

Figure 11.17(a) shows a simple white noise generator that relies on the fact that all Zener diodes generate substantial white noise when operated at a low current. R_2 and ZD_1 are wired in a negative-feedback loop between the collector and base of common-emitter amplifier Q_1, thus stabilising the circuits d.c. working levels, and the loop is a.c.-decoupled via C_1. ZD_1 thus acts as a white noise source that is wired in series with the base of Q_1, which amplifies the noise to a useful level of about 1V0 peak-to-peak. Any 5V6 to 12V Zener diode can be used in this circuit.

Figure 11.17(b) is a simple variation of the above design, with the reverse-biased base-emitter junction of a 2N3904 transistor (which 'zeners' at about 6V) used as the noise-generating Zener diode.

Multivibrator circuits

Multivibrators are two-state circuits that can be switched between one state and the other via a suitable trigger signal, which may be generated either internally or externally. There are four basic types of 'multi'

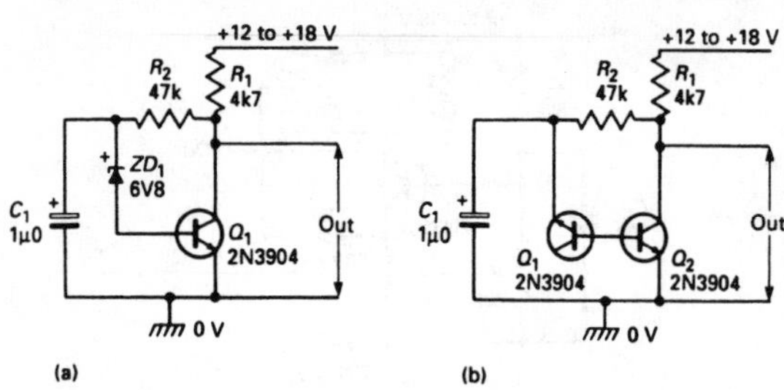

Figure 11.17. Transistor-Zener *(a)* and two-transistor *(b)* white-noise generators.

circuit, and they are all useful in waveform generating applications. Of these four, the 'astable' is useful as a free-running square-wave generator, the 'monostable' as a triggered pulse generator, the 'bistable' as a stop/go waveform generator, and the 'Schmitt' as a sine-to-square waveform converter.

Astable basics

Figure 11.18 shows a 1kHz astable multivibrator circuit, which acts as a self-oscillating regenerative switch in which the on and off periods are controlled by the C_1–R_1 and C_2–R_2 time constants. If these time constants are equal (C_1=C_2=C, and R_1=R_2=R), the circuit acts as a square-wave generator and operates at a frequency of about 1/(1.4CR); the frequency can be decreased by raising the C or R values, or *vice versa*, or can be made variable by using twin-gang variable resistors (in series with 10k limiting resistors) in place of R_1 and R_2. Outputs can be taken from either collector, and the two outputs are in anti-phase.

The *Figure 11.18* circuit's operating frequency is almost independent of supply-rail values in the range 1V5 to 9V0; the upper voltage limit is set by the fact that, as the transistors change state at the end of each half-cycle, the base-emitter junction of the 'off' one is reverse biased by an amount almost equal to the supply voltage and will 'zener' (and upset the timing action) if this voltage exceeds the junction's reverse breakdown voltage value. This problem can be overcome by wiring a silicon diode in series with the input of each transistor, to raise its effective 'zener' value to that of the diode, as shown in *Figure 11.19*. This 'protected' circuit can be used with any supply in the range 3V to 20V, and gives a frequency variation of only 2% when the supply is varied from 6V0 to 18V. This variation can be reduced to a mere 0.5% by wiring an additional 'compensation' diode in series with the collector of each transistor, as shown in the diagram.

Astable variations

The basic *Figure 11.18* astable circuit can be usefully modified in several ways, either to improve its performance or to alter the type of output waveform that it generates. Some of the most popular of these variations are shown in *Figures 11.20* to *11.25*.

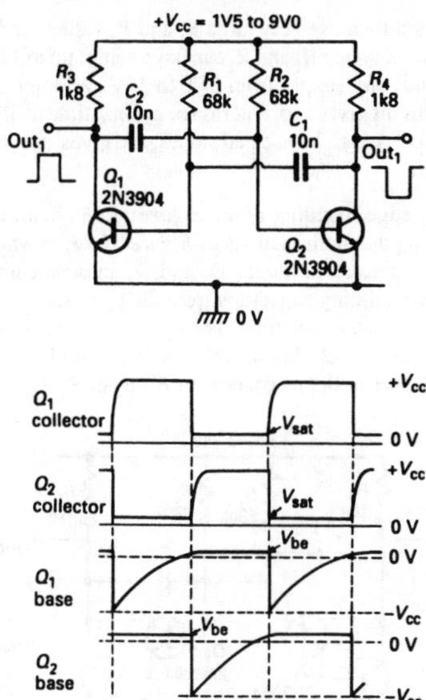

Figure 11.18. Circuit and waveforms of basic 1kHz astable multivibrator.

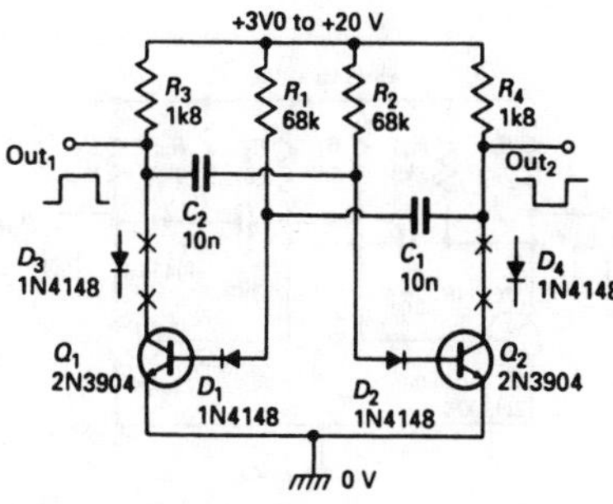

Figure 11.19. Frequency-corrected 1kHz astable multivibrator.

One weakness of the basic *Figure 11.18* circuit is that the leading edges of its output waveforms are slightly rounded; the larger the values of timing resistors R_1–R_2 relative to collector load resistors R_3–R_4, the squarer the edges becomes. The maximum usable R_1–R_2 values are in fact limited to h_{fe} x R_3 (or R_4), and one obvious way of improving the waveforms is to replace Q_1 and Q_2 with Darlington connected pairs of transistors and then use very large R_1 and R_2 values, as in the *Figure 11.20* circuit, in which R_1 and R_2 can have values up to 12MΩ, and the circuit can use any supply from 3V0 to 18V. With the R_1–R_2 values shown the circuit gives a total period or cycling time of about 1 second per µF when C_1 and C_2 have equal values, and gives an excellent square wave output.

The leading-edge rounding of the *Figure 11.18* circuit can be eliminated by using the modifications of *Figure 11.21*, in which 'steering' or waveform-correction diodes D_1 and D_2 automatically disconnect their respective timing capacitors from the transistor collectors at the moment of transistor switching. The circuit's main time constants are set by C_1–R_1 and C_2–R_2, but the effective collector loads of Q_1 and Q_2 are equal to the parallel resistances of R_3–R_4 or R_5–R_6.

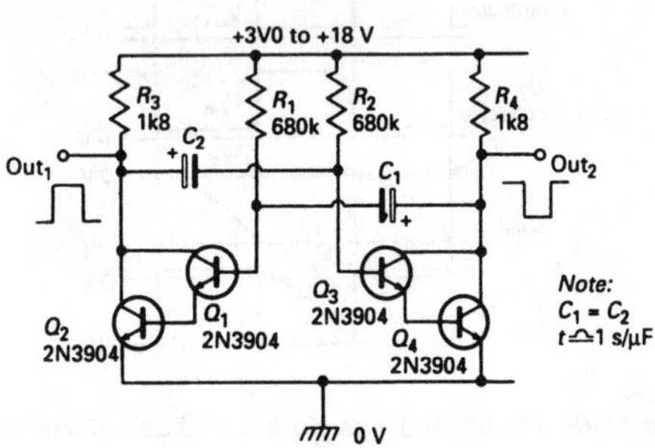

Figure 11.20. Long-period astable multivibrator.

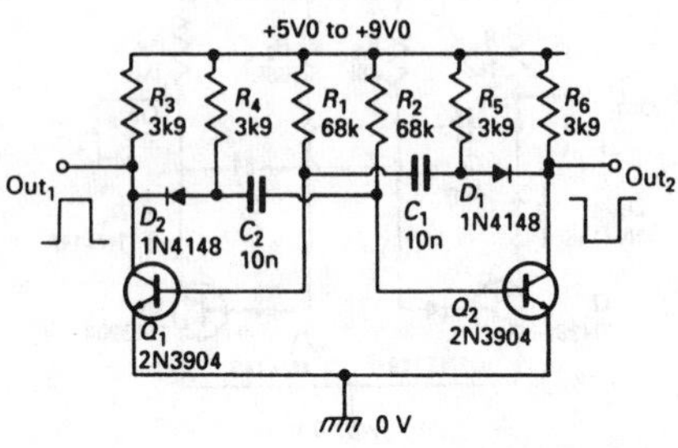

Figure 11.21. 1kHz astable with waveform correction.

A minor weakness of the basic *Figure 11.18* circuit is that if its supply is *slowly* raised from zero to its normal value, both transistors may turn on simultaneously, and the oscillator will not start. This snag can be overcome by using the 'sure-start' circuit of *Figure 11.22*, in which the timing resistors are connected to the transistor collectors in such a way that only one transistor can be on at a time.

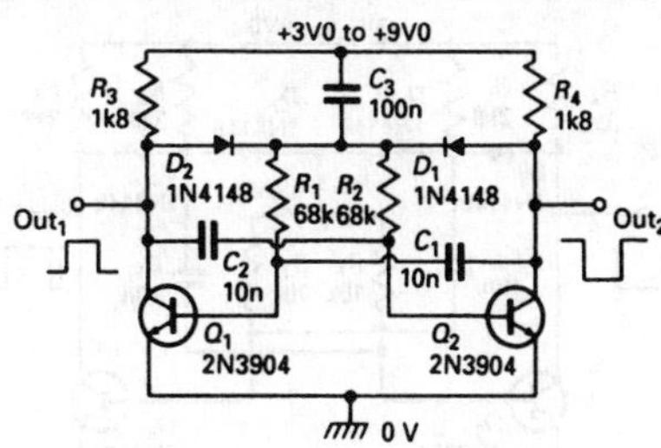

Figure 11.22. 1kHz astable with sure-start facility.

All astable circuits shown so far give symmetrical output waveforms, with a 1:1 mark/space ratio. A non-symmetrical waveform can be obtained by making one set of astable time constant components larger than the other. *Figure 11.23* shows a fixed-frequency (1100Hz) generator in which the mark/space ratio is variable from 1:10 to 10:1 via RV_1.

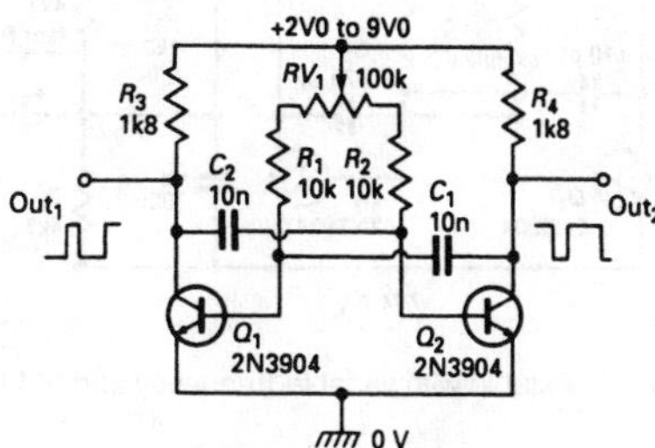

Figure 11.23. 1100Hz variable mark/space ratio generator.

The leading edges of the output waveforms of the above circuit may be objectionably rounded when the mark/space control is set to its extreme positions; also, the circuit may not start if its supply is applied too slowly. Both of these snags are overcome in the circuit of *Figure 11.24*, which is fitted with both sure-start and waveform-correction diodes.

Finally, *Figure 11.25* shows the basic astable modified so that its frequency is variable over a 2:1 range (from 20kHz down to 10kHz) via a single pot, and so that its generated waveform can be frequency modulated via an external low-frequency signal. Timing resistors R_3 and R_4 have their top ends taken to RV_1 slider, and the frequency is greatest when the slider is at the positive supply line. Frequency modulation is obtained by feeding the low-frequency signal to the tops of R_3–R_4 via C_4; C_3 presents a low impedance to the 'carrier' signal but a high impedance to the modulating one.

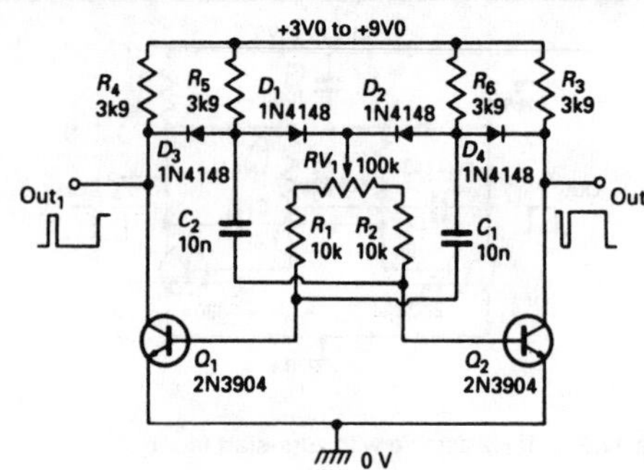

Figure 11.24. 1100Hz variable mark/space ratio generator with waveform correction and sure-start facility.

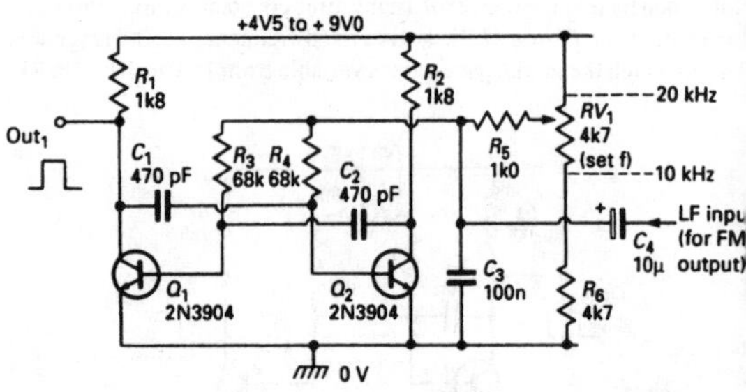

Figure 11.25. Astable with variable-frequency and FM facility.

Monostable basics

Monostable multivibrators are pulse generators, and may be triggered either electronically or manually. *Figure 11.26* shows a circuit of the latter type, which is triggered by feeding a positive pulse to Q_2 base via S_1 and R_6. This circuit operates as follows.

Normally, Q_1 is driven to saturation via R_5, so the output (Q_1 collector) is low; Q_2 (which derives its base-bias from Q_1 collector via R_3) is cut off under this condition, so C_1 is fully charged. When a start signal is applied to Q_2 base via S_1, Q_2 is driven on and its collector goes low, reverse biasing Q_1 base via C_1 and thus initiating a regenerative switching action in which Q_1 is turned off (and its output switches high) via C_1's negative charge, and Q_2 is driven on via R_1–R_3 after S_1 is released. As soon as the switching is complete C_1 starts to discharge via R_5, until its charge falls to such a low value that Q_1 starts to turn on again, thus initiating another regenerative action in which the transistors revert to their original states and the output pulse terminates, completing the action.

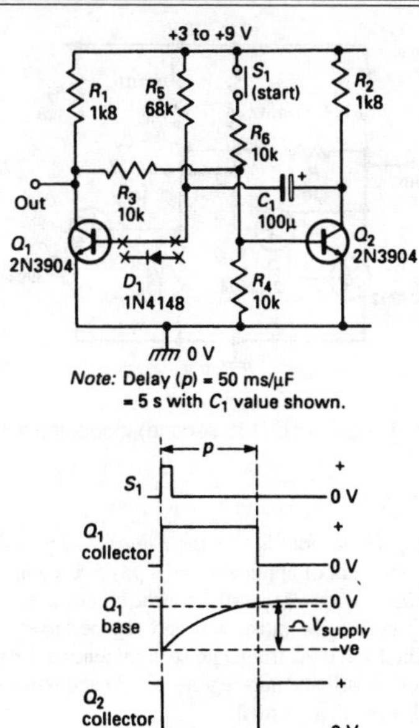

Figure 11.26. Basic manually-triggered monostable pulse generator.

Thus, a positive pulse is developed at the Q_1 output each time an input trigger signal is applied via S_1. The pulse period (P) is determined by the R_5–C_1 values, and approximates 0.7 x R_5 x C_1, where P is in µS, *C* is in µF, and *R* is in kilohms, and equals about 50mS/µF in the example shown.

In practice, the *Figure 11.26* circuit can be triggered either by applying a negative pulse to Q_1 base or a positive one to Q_2 base (as shown). Note that the base-emitter junction of Q_1 is reverse biased by a peak amount equal to V_{SUPPLY} during the operating cycle, thus limiting the maximum usable supply voltage to about 9V. Greater supply voltages can be used by wiring a silicon diode in series with Q_1 base, as shown in the diagram, to give the same 'frequency correction' action as described earlier for the astable circuit.

Long delays

The value of timing resistor R_5 must be large relative to R_2, but must be less than the product of R_1 and Q_1's h_{fe} value. Very long timing periods can be obtained by using a Darlington pair of transistors in place of Q_1, thus enabling large R_5 values to be used, as shown in the *Figure 11.27* circuit, which gives a pulse period of about 100 seconds with the component values shown.

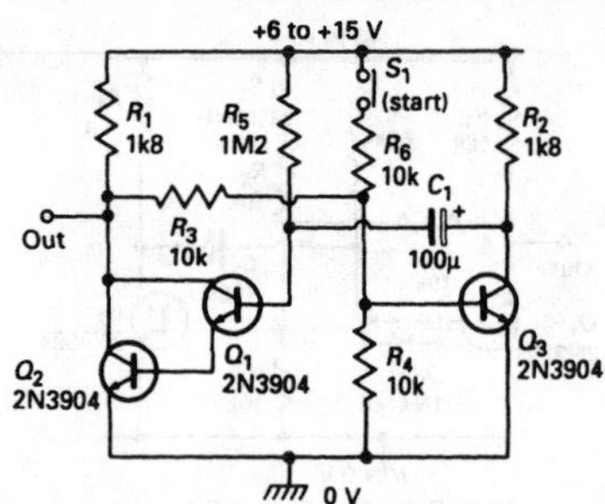

Figure 11.27. Long-period (100 second) monostable circuit.

An important point to note is that the *Figure 11.26* circuit actually triggers at the moment of application of a positive-going pulse to the base of Q_2; if this pulse is removed before the monostable completes its natural timing period the pulse will end regeneratively in the way already described, but if the trigger pulse is not removed at this time the monostable period will end non-regeneratively and will have a longer period and fall-time than normal.

Electronic triggering

Figures 11.28 and *11.29* show alternative ways of applying electronic triggering to the monostable pulse generator. In each case, the circuit is triggered by a square wave input with a short rise time; this waveform is differentiated by C_2–R_6, to produce a brief trigger pulse. In the *Figure 11.28* circuit the differentiated input signal is discriminated by D_1, to provide a positive trigger pulse on Q_2 base each time an external trigger signal is applied. In the *Figure 11.29* circuit the differentiated signal is fed to Q_3, which enables the trigger signal to be quite independent of Q_2. Note in the latter circuit that 'speed up' capacitor C_3 is wired across feedback resistor R_3, to help improve the shape of the circuit's output pulse.

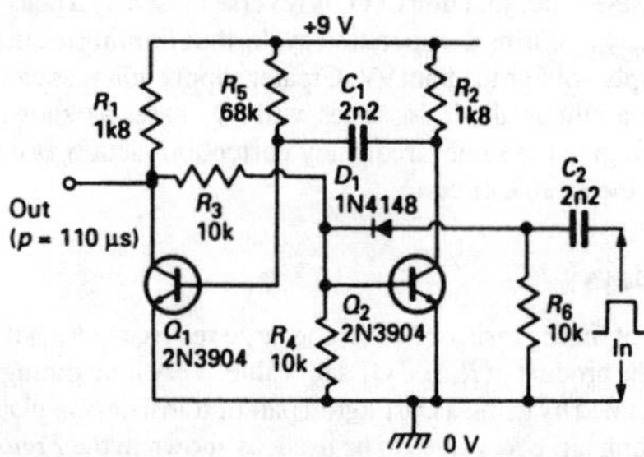

Figure 11.28. Electronically triggered monostable.

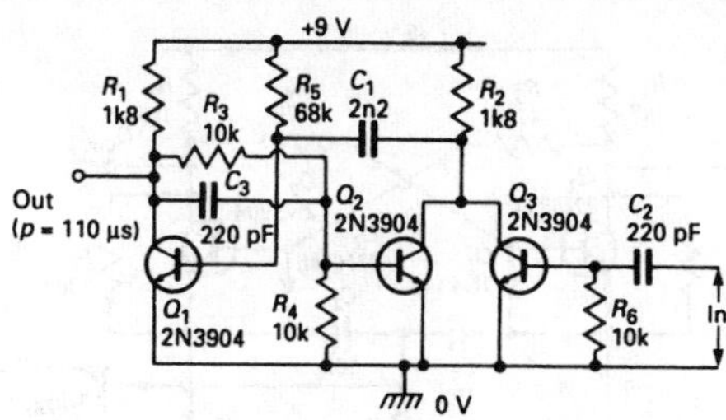

Figure 11.29. Monostable with gate-input triggering.

The *Figures 11.28* and *11.29* circuits each give an output pulse period of about 110µS with the component values shown. The period can be varied from a fraction of a µS to many seconds by choice of the C_1–R_5 values. The circuits can be triggered by sine or other non-rectangular waveforms by feeding them to the monostable input via a Schmitt trigger or similar sine/square converter circuit (see *Figure 11.32*).

Bistable circuits

Bistable multivibrators make good stop/go waveform generators, and *Figure 11.30* shows a manually-triggered version of such a circuit, which is also known as a 'R–S' (reset-set) flip-flop; its output can be 'set' to the high state by briefly closing S_1 (or by applying a negative pulse to Q_1 base), and the circuit then locks into this state until it is 'reset' to the low state by briefly closing S_2 (or by applying a negative pulse to Q_2 base); the circuit then locks into this new state until it is again set via S_1, and so on.

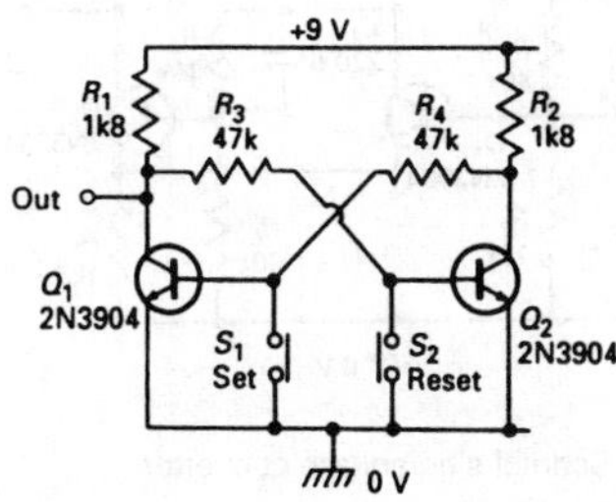

Figure 11.30. Manually-triggered R–S bistable multivibrator.

The above circuit can, by connecting two 'steering' diodes and associated components as shown in *Figure 11.31*, be modified to give a divide-by-two or 'counting' action in which it changes state each time a negative-going trigger pulse is applied; the circuit generates a pair of anti-phase outputs, known as 'Q' and 'not-Q'. This type of 'discrete' circuit is now obsolete.

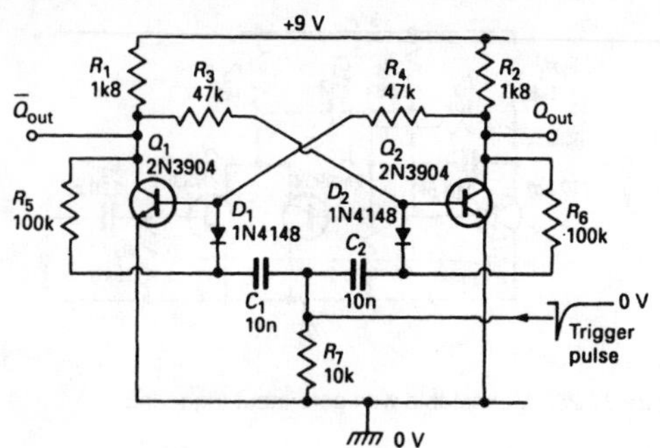

Figure 11.31. Divide-by-two bistable circuit.

The Schmitt trigger

The final member of the multivibrator family is the Schmitt trigger. This is a voltage-sensitive switching circuit that changes its output state when the input goes above or below pre-set upper and lower threshold levels; *Figure 11.32* shows it used as a sine-to-square waveform converter that gives a good performance up to a few hundred kHz and needs a sinewave input signal amplitude of at least 0.5V r.m.s. The output signal symmetry varies with input signal amplitude; RV_1 should be adjusted to give best results.

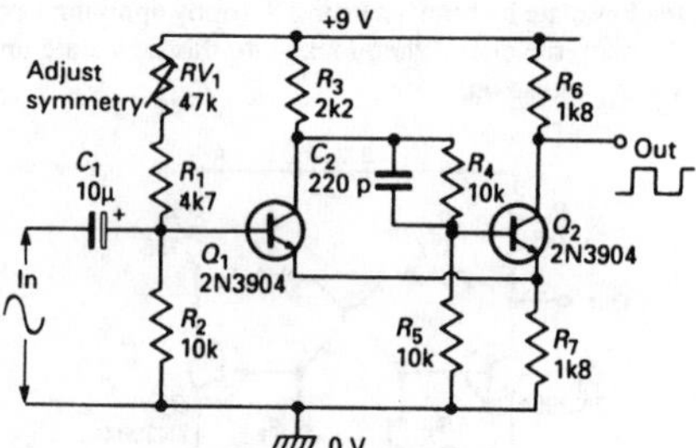

Figure 11.32. Schmitt sine/square converter.

12 Miscellaneous transistor circuits

The last three chapters looked at the operating principles and practical examples of various bipolar transistor circuits. This present chapter rounds off the 'transistor' section of the book by looking at a selection of audio power amplifiers, gadgets, and miscellaneous transistor circuits. It starts off by looking at power amplifier basics.

Power amplifier basics

A transistor power amplifier's job is that of converting a medium-level medium-impedance a.c. input signal into a high-level low-impedance state suitable for driving a low-impedance external load. This action can be achieved by operating the transistor(s) in either of two basic modes, known as 'class-A' or 'class-B'.

Figure 12.1(a) shows a basic class-A audio amplifier circuit; Q_1 is a common-emitter amplifier with a loudspeaker collector load, and is so biased that its collector current has a quiescent value half way between the desired maximum and minimum swings of output current, as shown in *Figure 12.1(b)*, so that maximal low-distortion output signal swings can be obtained. The circuit consumes a high quiescent current, and is relatively inefficient; 'efficiency' is the ratio of a.c. power feeding into the load, compared with the d.c. power consumed by the circuit, and at maximum output power is typically about 40%, falling to 4% at one tenth of maximum output, etc.

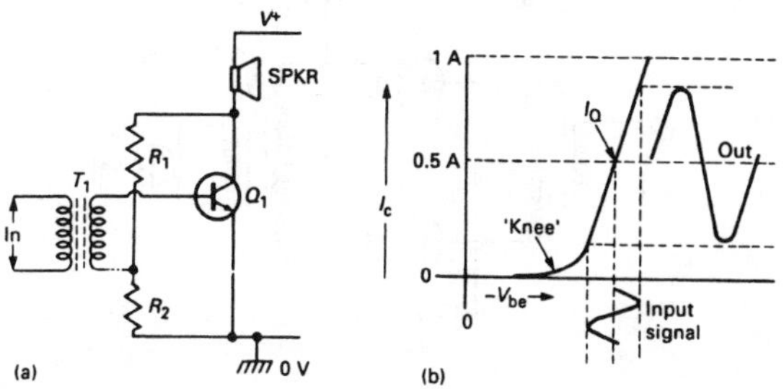

Figure 12.1. Basic circuit *(a)* and transfer characteristics *(b)* of class-A amplifier.

Figure 12.2 shows an example of a low-power high-gain general purpose class-A amplifier that draws a quiescent current of about 20mA and is suitable for driving a medium impedance (greater than 65Ω) loudspeaker or headset. Q_1 and Q_2 are wired as direct-coupled common-emitter amplifiers, and give an overall voltage gain of about 80dB. Q_1's base bias is derived (via R_2) from Q_2's emitter, which is decoupled via C_3 and thus 'follows' the *mean* collector voltage of Q_1; the bias is thus stabilised by d.c. negative feedback. Input pot RV_1 acts as the circuit's volume control.

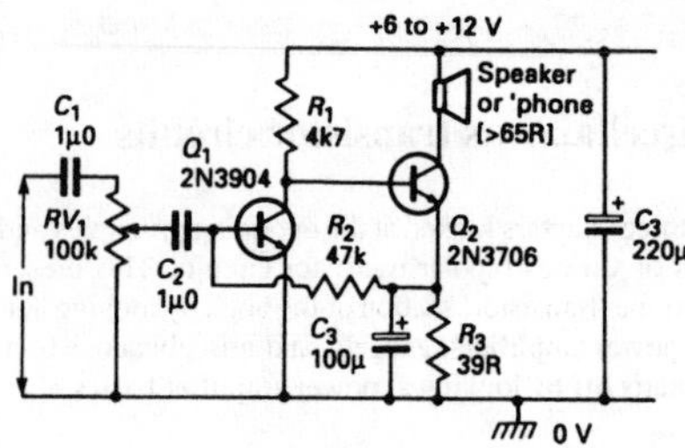

Figure 12.2. General-purpose high-gain low-power audio amplifier.

A basic class-B amplifier consists of a pair of transistors, driven in anti-phase but driving a common output load, as shown in *Figure 12.3(a)*. In this particular design Q_1 and Q_2 are wired in the common-emitter mode and drive the loudspeaker via push-pull transformer T_2, and the anti-phase input drive is obtained via phase-splitting transformer T_1. The essential features of this type of amplifier are that both transistors are cut off under quiescent conditions, that neither transistor conducts until its input drive signal exceeds its base-emitter 'knee' voltage, and that one transistor is driven on when the other is driven off, and *vice versa*. The circuit consumes near-zero quiescent current, and has high

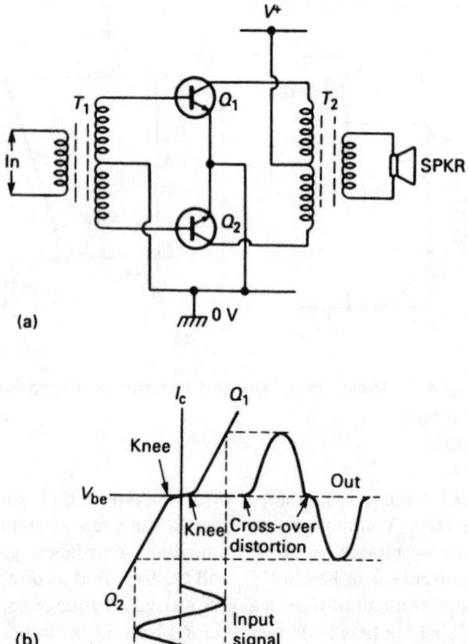

Figure 12.3. Basic circuit *(a)* and transfer characteristics *(b)* of class-B amplifier.

efficiency (up to 78.5%) under all operating conditions, but it generates severe cross-over distortion in the amplifier's output signal, as shown in *Figure 12.3(b)*; the basic class-B circuit must thus be modified if it is to be used as a practical audio power amplifier; the modified circuit is known as a 'class-AB' amplifier.

Class-AB basics

The cross-over distortion of the class-B amplifier can be eliminated by applying slight forward bias to the base of each transistor, as shown in *Figure 12.4*, so that each transistor passes a modest quiescent current. Such a circuit is known as a class-AB amplifier. Circuits of this type were widely used in early transistor power amplifier systems but are now virtually obsolete, since they require the use of transformers for input phase-splitting and output loudspeaker driving, and must have closely matched transistor characteristics if a good low-distortion performance is to be obtained.

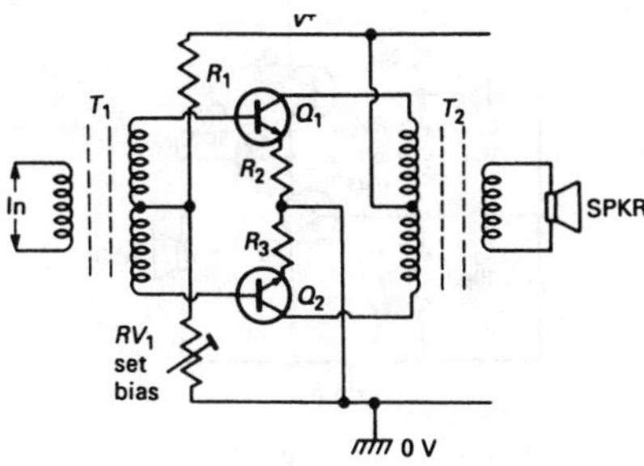

Figure 12.4. Basic circuit of class-AB amplifier.

Figure 12.5 shows the basic circuit of a class-AB amplifier that suffers from none of the snags mentioned above. It is a complementary emitter follower, and is shown using a split (dual) power supply. Q_1 and Q_2 are biased (via R_1–RV_1–R_2) so that their outputs are at zero volts and zero current flows in the speaker load under quiescent conditions, but have slight forward bias applied (via RV_1), so that they pass modest quiescent currents and thus do not suffer from cross-over distortion problems. Identical input signals are applied (via C_1 and C_2) to the bases of both emitter followers. This circuit's operation has already been fully described in Chapter 10.

The basic *Figure 12.5* circuit does not require the use of transistors with closely matched electrical characteristics, and gives direct drive to the speaker. It can be modified for use with a single-ended power supply by simply connecting one end of the speaker to either the zero or the positive supply rail, and connecting the other end to the amplifier output via a high-value blocking capacitor, as shown in *Figure 12.6*.

The basic *Figures 12.5* and *12.6* circuits form the basis of virtually all modern audio power amplifier designs, including those in IC form. Many modifications and variations can be made to the basic circuit.

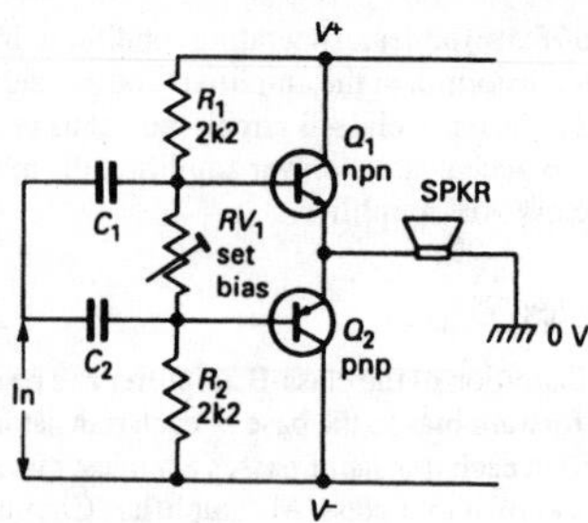

Figure 12.5. Basic class-AB amplifier with complementary emitter follower output and dual power supply.

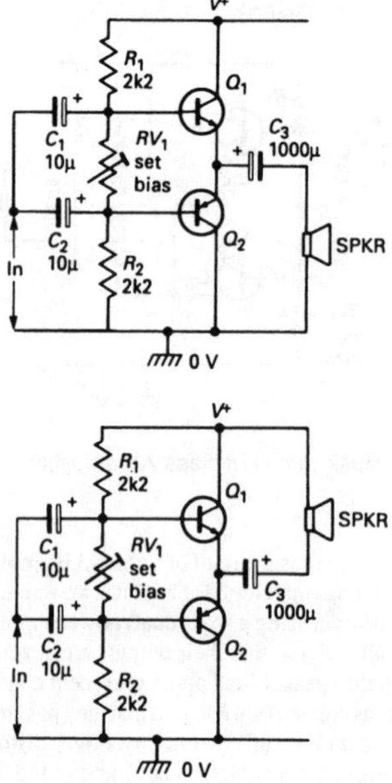

Figure 12.6. Alternative versions of the class-AB amplifier with single-ended power supply.

Circuit variations

The *Figure 12.5* circuit gives unity overall voltage gain, so an obvious circuit modification is to provide it with a voltage-amplifying driver stage, as in *Figure 12.7*. Here, common emitter amplifier Q_1 drives the Q_2–Q_3 complementary emitter followers via collector load resistor R_1 and auto-biasing silicon diodes D_1 and D_2 (the function of these diodes is fully explained in Chapter 10). Q_1's base bias is derived from the circuit's output via R_2–R_3, thus providing d.c. feedback to stabilise the circuit's operating points, and a.c. feedback to minimise signal distortion. In practice, a pre-set pot is usually wired in series with D_1–D_2, to enable the Q_2–Q_3 bias to be trimmed; low-value resistors R_4 and R_5 are wired in series with Q_2 and Q_3 emitters to prevent thermal runaway, etc.

The input impedance of the basic *Figure 12.5* circuit equals the product of the loudspeaker load impedance and the h_{fe} of Q_1 or Q_2. An obvious circuit improvement is to replace the individual Q_1 and Q_2 transistors with high-gain pairs of transistors, to increase the circuit's input impedance and enable it to be used with a driver with a high-value collector load. *Figures 12.8* to *12.10* show three alternative ways of modifying the *Figure 12.7* circuit in this way.

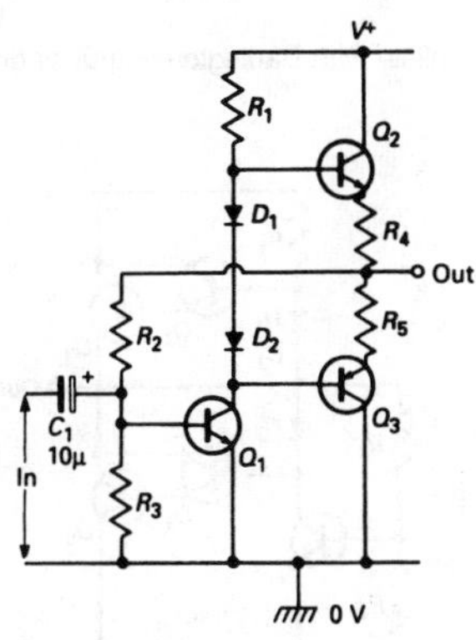

Figure 12.7. Complementary amplifier with driver and auto-bias.

In *Figure 12.8*, Q_2–Q_3 are wired as a Darlington npn pair, and Q_3–Q_4 as a Darlington pnp pair; note that four base-emitter junctions exist between Q_2 base and Q_4 base, so this output circuit must be biased via a chain of four silicon diodes.

In *Figure 12.9*, Q_2–Q_3 are wired as a Darlington npn pair, but Q_3–Q_4 are wired as a complementary pair of common-emitter amplifiers that operate with 100% negative feedback and provide unity voltage gain and a very high input impedance. This design is known as a 'quasi-complementary' output stage, and is probably the most popular of all class-AB amplifier configurations; it calls for the use of three biasing diodes.

In *Figure 12.10*, both Q_2–Q_3 and Q_4–Q_5 are wired as complementary pairs of unity-gain common-emitter amplifiers with 100% negative feedback; they are mirror images of each other, and form a complementary output stage that needs only two biasing diodes.

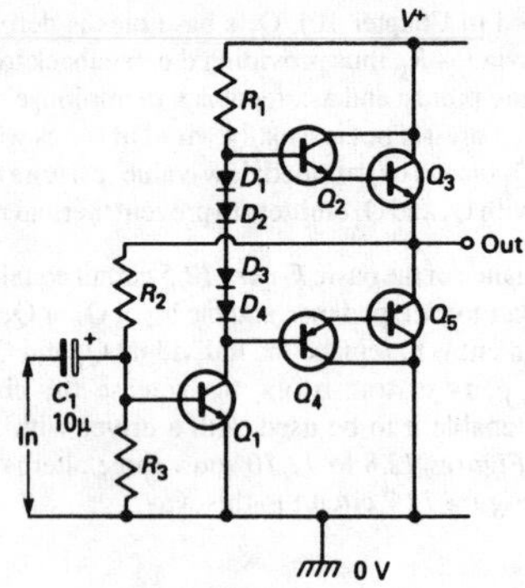

Figure 12.8. Amplifier with Darlington output stages.

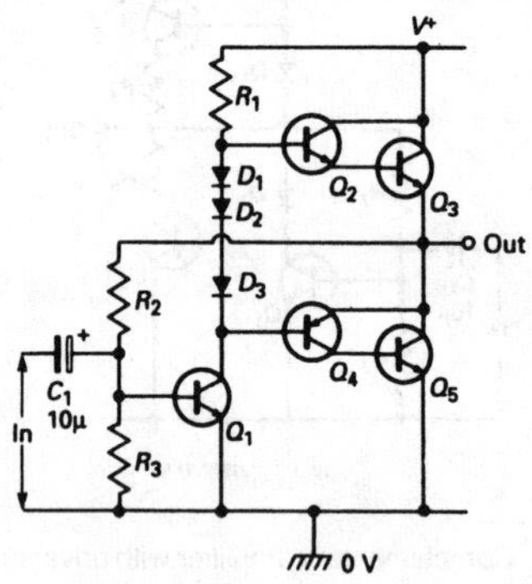

Figure 12.9. Amplifier with quasi-complementary output stages.

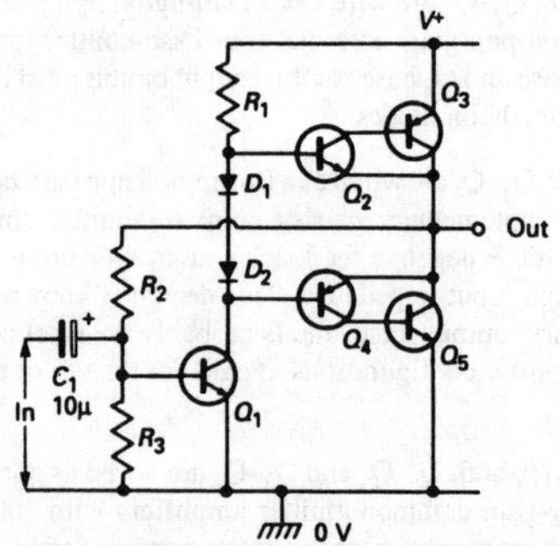

Figure 12.10. Amplifier with complementary output stages.

The circuits of *Figures 12.7* to *12.10* all call for the use of a chain of silicon biasing diodes. If desired, each of these chains can be replaced by a single transistor and two resistors, wired in the 'amplified diode' configuration described in Chapter 10 and repeated here in *Figure 12.11*. Thus, if R_1 is shorted out the circuit acts like a single base-emitter junction 'diode', and if R_1 is not shorted out it acts like $(R_1+R_2)/R_2$ series-wired diodes. *Figure 12.12* shows the circuit modified so that it acts as a fully adjustable 'amplified' silicon diode, with an output variable from 1 to 5.7 base-emitter junction voltages.

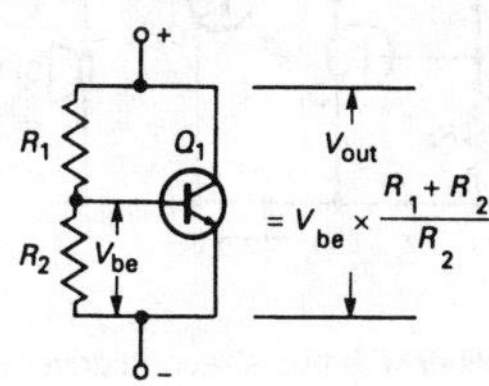

Figure 12.11. Fixed-gain amplified diode circuit.

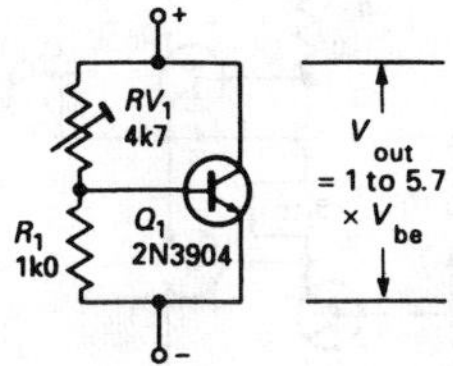

Figure 12.12. Adjustable amplified diode circuit.

Another useful modification that can be made to the basic *Figure 12.7* circuit is to add bootstrapping to its R_1 collector load, to boost its effective impedance and thus raise the circuit's overall voltage gain (the 'bootstrapping' technique is fully described in Chapter 10). *Figures 12.13* and *12.14* show examples of bootstrapped class-AB power amplifier circuits.

In *Figure 12.13* the Q_1 collector load comprises R_1 and R_2 in series, and the circuit's output signal (which also appears across SPKR) is fed back to the R_1–R_2 junction via C_2, thus 'bootstrapping' R_2's value so that its a.c. impedance is boosted by (typically) a factor of about twenty, and the circuit's voltage gain is boosted by a similar amount.

Figure 12.14 shows a version of the circuit that saves two components; in this case the SPKR forms part of Q_1's collector load, and directly bootstraps R_1.

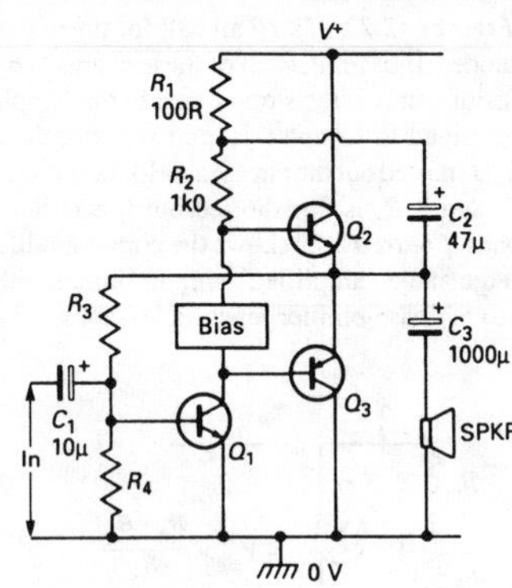

Figure 12.13. Amplifier with bootstrapped driver stage.

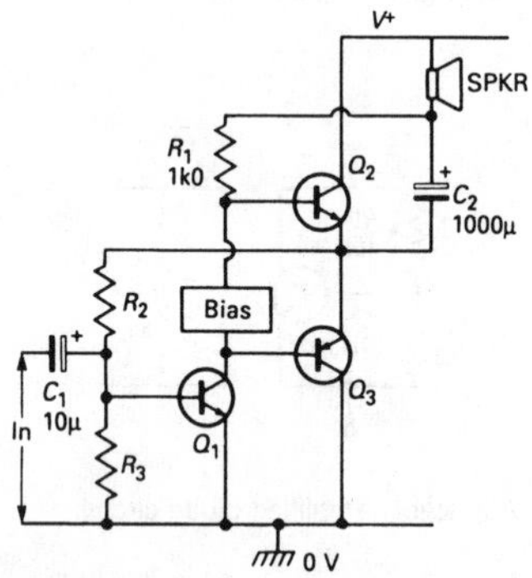

Figure 12.14. Alternative amplifier with bootstrapped driver stage.

Practical class-AB amplifiers

The easiest way to build a class-AB audio amplifier is to do so using one of the many readily-available audio ICs of this type. In some cases, however, particularly when making 'one off' projects, it may be cheaper or more convenient to use a discrete transistor design, such as one of those shown in *Figures 12.15* or *12.16*.

Figure 12.15 shows a simple class-AB amplifier that can typically drive 1 watt into a 3Ω speaker. Here, common-emitter amplifier Q_1 uses collector load LS_1–R_1–D_1–RV_2, and drives the Q_2–Q_3 complementary emitter follower stage. The amplifier's output is fed (via C_2) to the LS_1–R_1 junction, thus providing a low impedance drive to the loudspeaker and simultaneously bootstrapping the R_1 value so that the circuit gives high voltage gain. The output is also fed back to Q_1 base

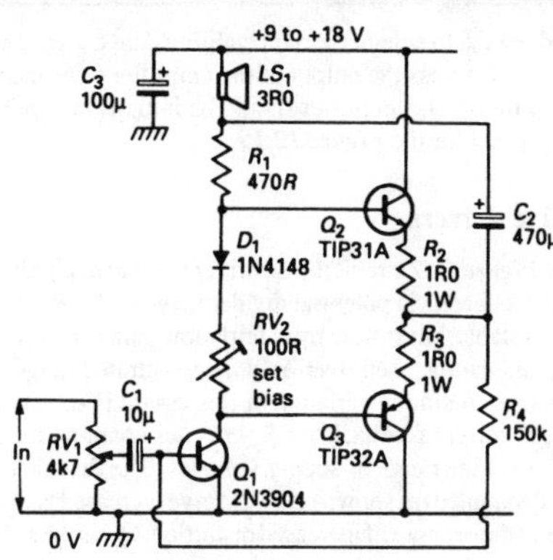

Figure 12.15. Simple 1 watt amplifier.

via R_4, thus providing base bias via a negative feedback loop. In use, RV_2 should be trimmed to give minimal audible signal cross-over distortion consistent with low quiescent current consumption (typically in the range 10mA to 15mA).

Figure 12.16 shows a rather more complex audio power amplifier that can deliver about 10 watts into an 8R0 load when powered from a 30V supply. This circuit uses high-gain quasi-complementary output stages (Q_3 to Q_6) and uses an adjustable 'amplified diode' (Q_1) as an output biasing device. The Q_2 common emitter amplifier stage has its main load resistor (R_2) bootstrapped via C_2, and is d.c. biased via R_3, which should set the quiescent output voltage at about half-supply value (if not, alter the R_3 value). The upper frequency response of the amplifier

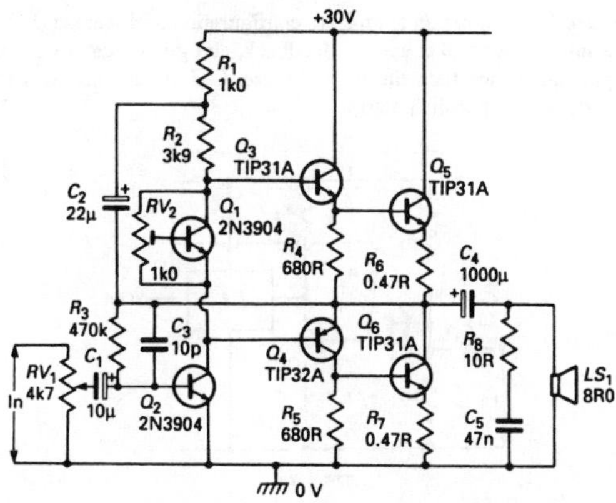

Figure 12.16. 10 watt audio amplifier.

is restricted via C_3, to enhance circuit stability, and C_5–R_8 are wired as a Zobel network across the output of the amplifier to further enhance the stability. In use, the amplifier should be initially set up in the way already described for the *Figure 12.15*.

Alternative drivers

In the basic *Figure 12.7* circuit the Q_1 driver stage uses parallel d.c. and a.c. voltage feedback via potential divider network R_2–R_3. This circuit is simple and stable, but suffers from fairly low gain and very low input resistance, and can be used over only a very limited range of power supply voltages. A simple variation of this circuit is shown in *Figure 12.17*. It uses current feedback via R_1–R_2, thus enabling the circuit to be used over a wide range of supply voltages. The feedback resistors can be a.c.-decoupled (as shown) via C_2 to give increased gain and input impedance, at the expense of increased distortion. Q_1 can be a Darlington type, if a very high input impedance is required.

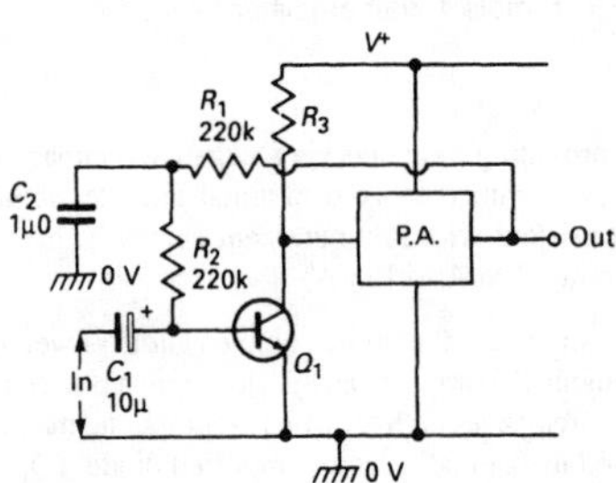

Figure 12.17. Driver stage with decoupled parallel d.c. feedback.

Figure 12.18 shows an alternative configuration of driver stage. This design uses series d.c. and a.c. feedback, and gives greater gain and input impedance than the basic *Figure 12.7* circuit, but uses two transistors of opposite polarities.

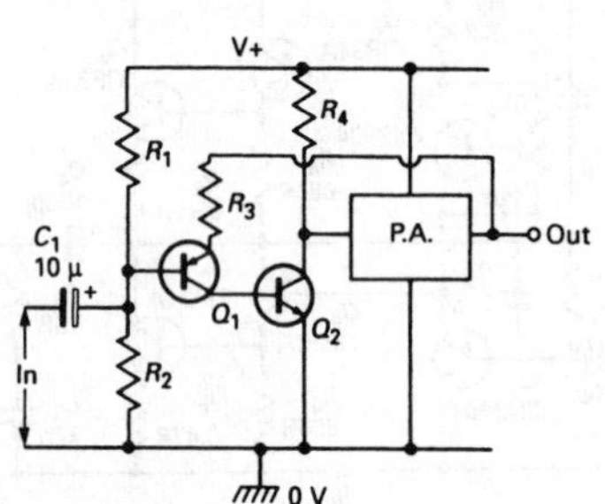

Figure 12.18. Driver stage with series d.c. feedback.

Finally, to complete this look at audio power amplifiers, *Figure 12.19* shows a circuit that has direct-coupled ground-referenced inputs and outputs and uses split power supplies. It has a long-tailed pair input stage, and the input and output both centre on zero volts if R_1 and R_4 have equal values. The circuit can be used with a single ended power supply by grounding one supply line and using a.c. coupling of the input and the output signals. This basic circuit forms the basis of many IC power amplifier designs.

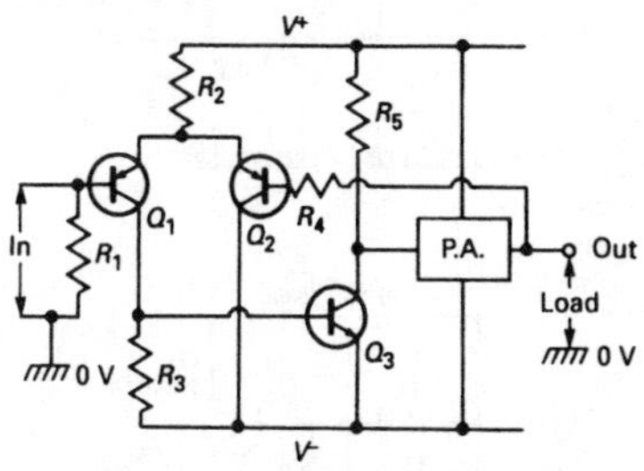

Figure 12.19. Driver stage with long-tailed pair input.

Scratch/rumble filters

A common annoyance when playing old records is that of 'scratch' and/ or 'rumble' sounds. The 'scratch' noises are mainly high-frequency (greater than 10kHz) sounds picked up from the disc surface, and the 'rumbles' are low-frequency (less than 50Hz) sounds that are mostly caused by slow variations in motor-drive speed. Each of these noises can be greatly reduced or eliminated by passing the player's audio signals through a filter that rejects the troublesome parts of the audio spectrum. *Figures 12.20* and *12.21* show suitable circuits.

The high-pass rumble filter of *Figure 12.20* gives unity voltage gain to signals above 50Hz, but gives 12dB per octave rejection to those below this value, i.e., it gives 40dB of attenuation at 5Hz, etc. Emitter-follower Q_1 is biased at half-supply volts from the R_1–R_2–C_3 low-impedance point, but has negative feedback applied via the R_3–C_2–C_1–R_4 filter network. The circuit's frequency turn-over point can be altered by changing the C_1–C_2 values (which must be equal); thus, if the C_1–C_2 values are halved (to 110n), the turn-over frequency doubles (to 100Hz), etc.

The low-pass scratch filter of *Figure 12.21* gives unity voltage gain to signals below 10kHz, but gives 12dB per octave rejection to those above this value. This circuit is similar to that of *Figure 12.20*, except that the positions of the resistors and capacitors are transposed in the C_2–R_4–C_4–R_5 filter network. The circuit's turn-over frequency can be altered by changing the C_2–C_4 values; e.g., values of 3n3 give a frequency of 7.5kHz.

The *Figures 12.20* and *12.21* circuits can be combined, to make a composite 'scratch and rumble' filter, by connecting the output of the high-pass filter to the input of the low-pass filter; if desired, the filters

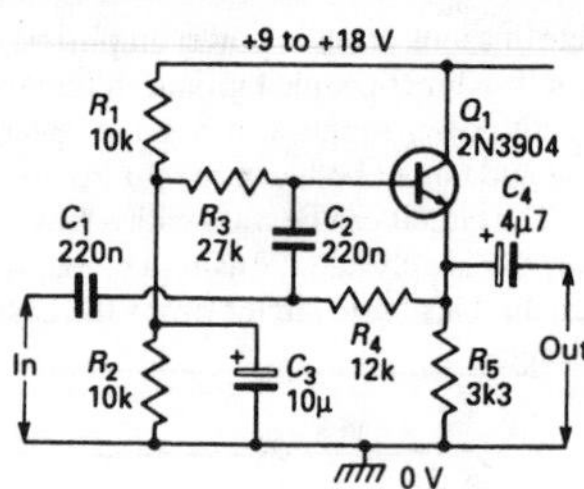

Figure 12.20. 50Hz rumble or hi-pass filter.

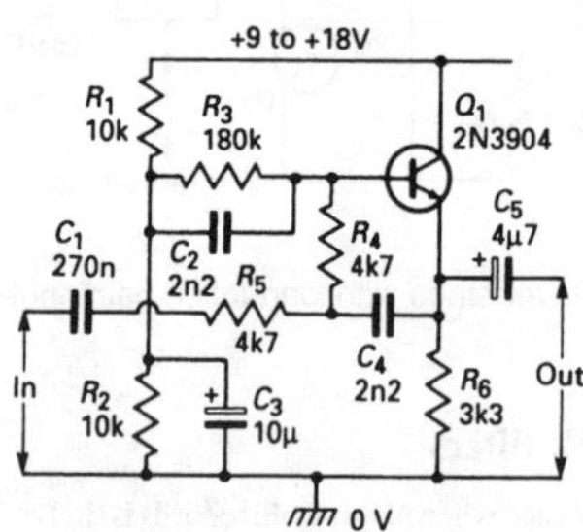

Figure 12.21. 10kHz scratch or low-pass filter.

can be provided with bypass switches, enabling them to be easily switched in and out of circuit, by using the connections of *Figure 12.22*. Note that if the *Figures12.20* and *12.21* designs are to be built as a single unit, a few components can be saved by making the R_1–R_2–C_3 biasing network common to both circuits.

A noise limiter

Unwanted noise can be a great nuisance; when listening to very weak broadcast signals, for example, it is often found that peaks of background noise completely swamp the broadcast signal, making it unintelligible. This problem can often be overcome by using the 'noise limiter' circuit of *Figure 12.23*. Here, the signal-plus-noise waveform

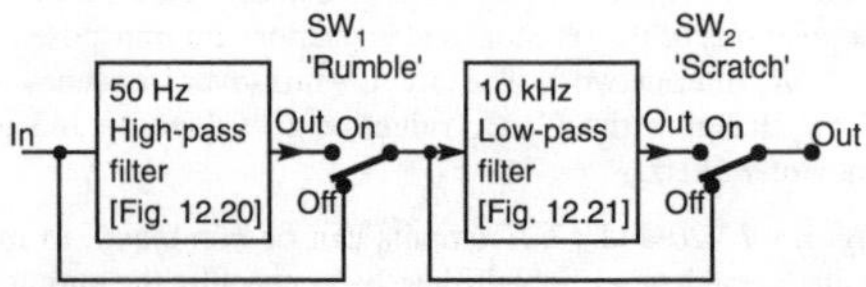

Figure 12.22. Complete scratch/rumble filter, with switching.

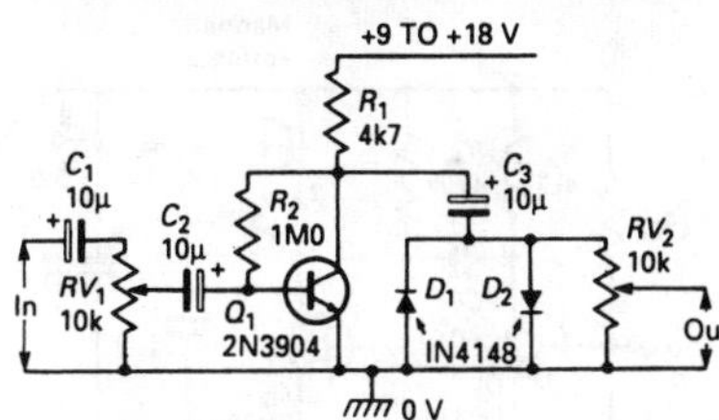

Figure 12.23. Noise limiter.

is fed to amplifier Q_1 via RV_1. Q_1 amplifies both waveforms equally, but D_1 and D_2 automatically limit the peak-to-peak output swing of Q_1 to about 1.2V. Thus, if RV_1 is adjusted so that the 'signal' output is amplified to this peak level, the 'noise' peaks will not be able to greatly exceed the signal output, and intelligibility is greatly improved.

Astable multivibrators

The astable multivibrator circuit has many uses. *Figure 12.24* shows it used to generate a non-symmetrical 800Hz waveform that produces a monotone audio signal in the loudspeaker when S_1 is closed. The circuit can be used as a morse-code practice oscillator by using a morse key as S_1; the tone frequency can be changed by altering the C_1 and/or C_2 values.

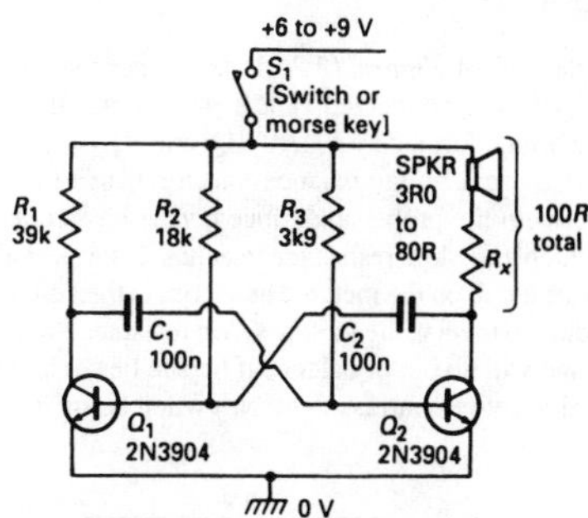

Figure 12.24. Morse-code practice oscillator.

Figure 12.25 shows an astable multivibrator used as the basis of a 'signal injector-tracer' item of test gear. When SW_1 is in INJECT position '1', Q_1 and Q_2 are configured as a 1kHz astable, and feed a good square-wave into the probe terminal via R_1–C_1. This waveform is rich in harmonics, so if it is injected into any AF or RF stage of an AM radio it produces an audible output via the radio's loudspeaker, unless one of the radio's stages is faulty. By choosing a suitable injection point, the 'injector' can thus be used to trouble-shoot a defective radio.

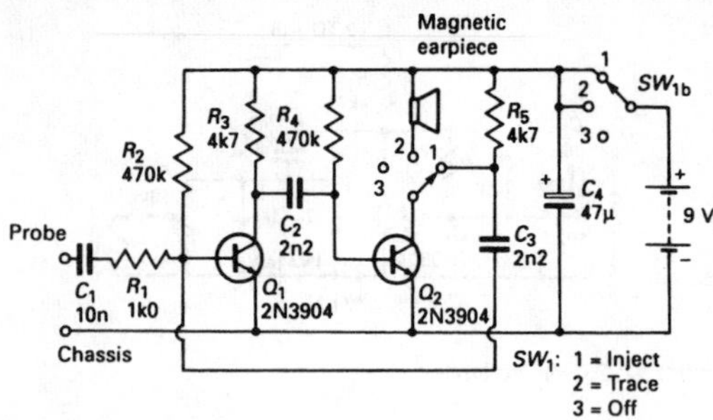

Figure 12.25. Signal injector–tracer.

When SW_1 is switched to TRACE position '2', the *Figure 12.25* circuit is configured as a cascaded pair of common-emitter amplifiers, with the probe input feeding to Q_1 base, and Q_2 output feeding into an earpiece or head-set. Any weak audio signals fed to the probe are directly amplified and heard in the earpiece, and any amplitude-modulated RF signals fed to the probe are demodulated by the non-linear action of Q_1 and the resulting audio signals are then amplified and heard in the earpiece. By connecting the probe to suitable points in a radio, the 'tracer' can thus be used to trouble-shoot a faulty radio, etc.

Lie detector

The 'lie detector' of *Figure 12.26* is an 'experimenters' circuit, in which the victim is connected (via a pair of metal probes) into a Wheatstone bridge formed by R_1–RV_1–Q_1 and R_3–R_4; the 1mA centre-zero meter is used as a bridge-balance detector. In use, the victim makes firm contact with the probes and, once he/she has attained a relaxed state (in which the skin resistance reaches a stable value), RV_1 is adjusted to set a null on the meter. The victim is then cross-questioned and, according to theory, the victim's skin resistance will then change and the bridge will go out of balance if he/she lies or shows any signs of emotional upset (embarrassment, etc) when being questioned.

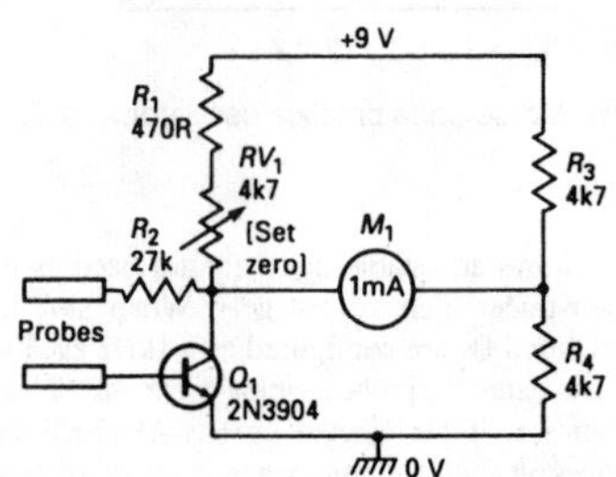

Figure 12.26. Simple lie detector.

Current mirrors

A current mirror is a constant-current generator in which the output current magnitude is virtually identical to that of an independent input 'control' current. This type of circuit is widely used in modern linear IC design. *Figure 12.27* shows a simple current mirror using ordinary npn transistors; Q_1 and Q_2 are a matched pair and share a common thermal environment. When input current I_{IN} is fed into diode-connected Q_1 it generates a proportionate forward base-emitter voltage, which is applied directly to the base-emitter junction of matched transistor Q_2, causing it to sink an almost identical ('mirror') value of collector current, I_{SINK}. Q_2 thus acts as a constant current sink that is controlled by I_{IN}, even at collector voltages as low as a few hundred millivolts.

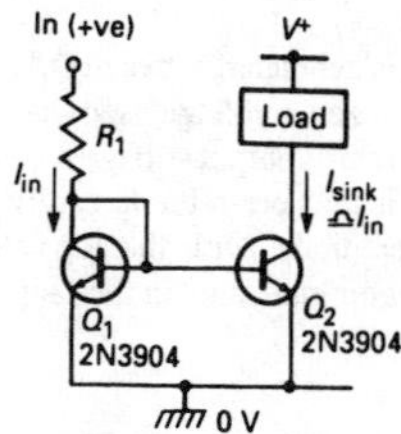

Figure 12.27. An npn current mirror.

Figure 12.28 shows a pnp version of the simple current mirror circuit. This works in the same basic way as already described, except that Q_2's collector acts as a constant current source that has its amplitude controlled by I_{IN}. Note that both of these circuits still work quite well as current-controlled constant current sinks or sources even if Q_1 and Q_2 have badly matched characteristics, but in this case may not act as true current mirrors, since their I_{SINK} and I_{IN} values may be very different.

An adjustable zener

Figure 12.29 shows the circuit of an 'adjustable zener' that can have its output voltage pre-set over the range 6.8V to 21V via RV_1. The circuit action is such that a fixed reference voltage (equal to the sum of the

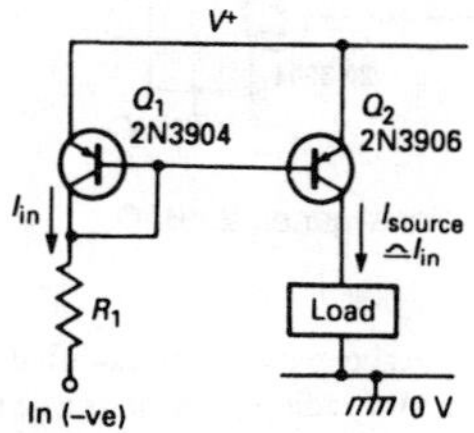

Figure 12.28. A pnp current mirror.

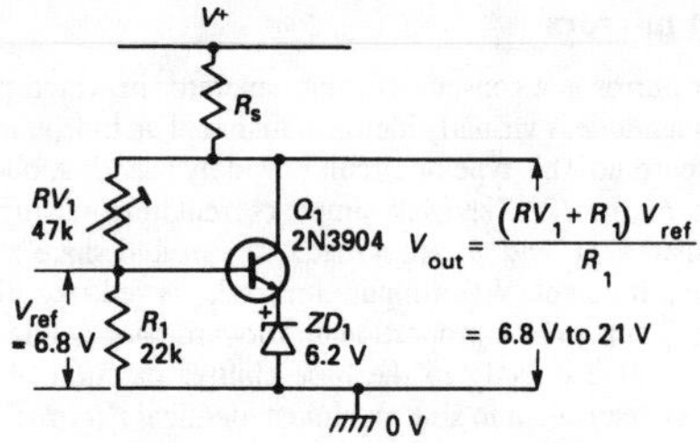

Figure 12.29. Adjustable zener.

zener and V_{BE} values) is generated between Q_1's base and ground, and (because of the value of zener voltage used) has a near-zero temperature coefficient. The circuit's output voltage is equal to V_{REF} multiplied by $(RV_1 + R_1)/R_1$, and is thus pre-settable via RV_1. This circuit is used like an ordinary zener diode, with the R_S value chosen to set its operating current at a nominal value in the range 5 to 20mA.

L–C oscillators

L–C oscillators have many applications in test gear and gadgets, etc. *Figure 12.30* shows an L–C medium-wave (MW) signal generator or beat-frequency oscillator (BFO), with Q_1 wired as a Hartley oscillator that uses a modified 465kHz IF transformer as its collector load. The IF transformer's internal tuning capacitor is removed, and variable oscillator tuning is available via VC_1, which enables the output frequency (on either fundamentals or harmonics), to be varied from well below 465kHz to well above 1.7MHz. Any MW radio will detect the oscillation frequency if placed near the circuit; if the unit is tuned to the radio's IF value, a beat note will be heard, enabling c.w. and s.s.b. transmissions to be clearly detected.

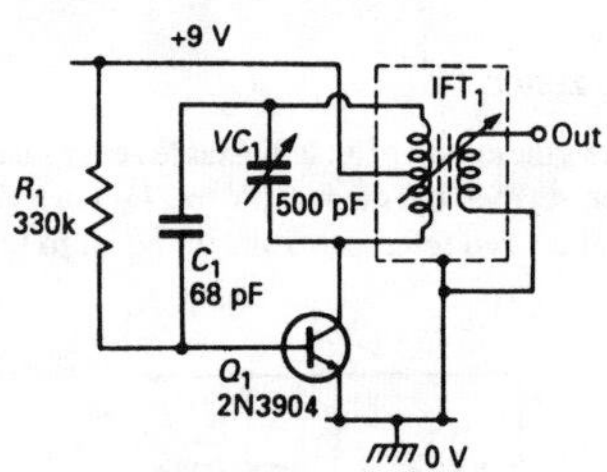

Figure 12.30. MW signal generator/BFO

Figure 12.31 shows the above oscillator modified so that, when used in conjunction with a MW radio, it functions as a simple 'metal/pipe locator'. Oscillator coil L_1 is hand-wound and comprises 30 centre-tapped turns of wire, firmly wound over about 25mm length of a 75 to

100mm diameter non-metallic former or 'search head' and connected to the main circuit via 3-core cable. The search head can be fixed to the end of a long non-metallic handle if the circuit is to be used in the classic 'metal detector' mode, or can be hand-held if used to locate metal pipes or wiring that are hidden behind plasterwork, etc. Circuit operation relies on the fact that L_1's electromagnetic field is disturbed by the presence of metal, causing the inductance of L_1 and the frequency of the oscillator to alter. This frequency shift can be detected on a portable MW radio placed near L_1 by tuning the radio to a local station and then adjusting VC_1 so that a low frequency 'beat' or 'whistle' note is heard from the radio. This beat note changes if L_1 (the search head) is placed near metal.

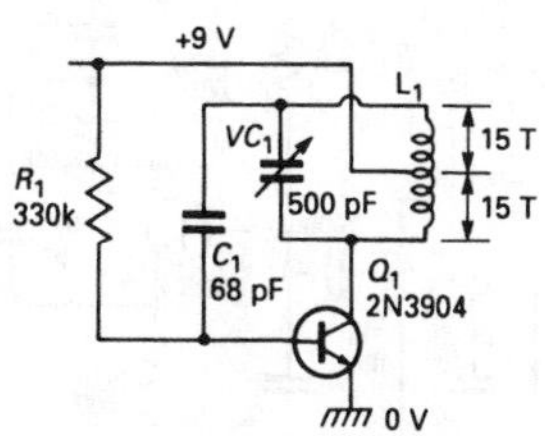

Figure 12.31. Metal/pipe locator.

Figure 12.32 shows another application of the Hartley oscillator. In this case the circuit functions as a D.C.-to-D.C. converter, which converts a 9V battery supply into a 300V D.C. output. T_1 is a 9V–0–9V to 250V transformer, with its primary forming the 'L' part of the oscillator. The supply voltage is stepped up to about 350V peak at T_1 secondary, and is half-wave rectified by D_1 and used to charge C_3. With no permanent load on C_3, the capacitor can deliver a powerful but non-lethal 'belt'. With a permanent load on the output, the output falls to about 300V at a load current of a few mA.

FM transmitters

Figures 12.33 and *12.34* show a pair of low-power FM transmitters that generate signals that can be picked up at a 'respectable' range on any 88 to 108MHz FM-band receiver. The *Figure 12.33* circuit uses IC_1 as

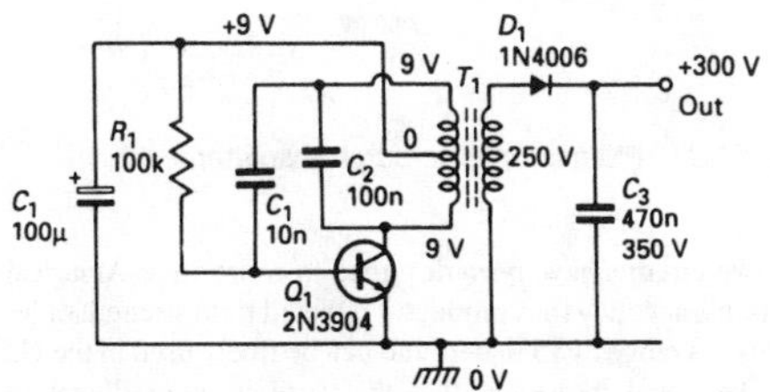

Figure 12.32. 9V to 300V D.C.-to-D.C. converter.

a 1kHz square wave generator that modulates the Q_1 VHF oscillator, and produces a harsh 1kHz tone signal in the receiver; this circuit thus acts as a simple alarm-signal transmitter.

The *Figure 12.34* circuit uses a 2-wire electret microphone insert to pick up voice sounds, etc., which are amplified by Q_1 and used to modulate the Q_2 VHF oscillator; this circuit thus acts as an FM microphone or 'bug'. In both circuits the VHF oscillator is a Colpitts type but with the transistor used in the common-base mode, with C_7 giving feedback from the tank output back to the emitter 'input'.

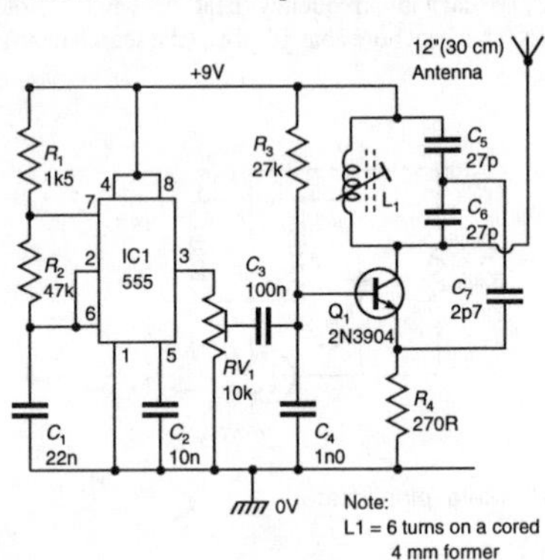

Figure 12.33. FM radio transmitter alarm.

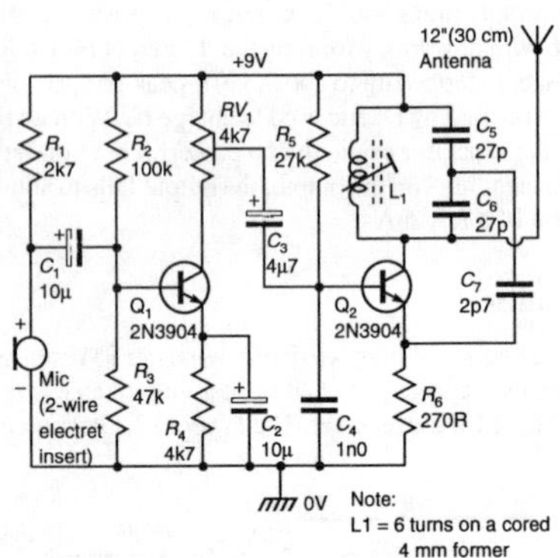

Figure 12.34. FM microphone/bug transmitter.

These two circuits have been designed to conform to American FCC regulations, and they thus produce a radiated field strength of less than 50µV/m at a range of 15 meters and can be freely used in the U.S.A. It should be noted, however, that their use is quite illegal in most countries, including the U.K.

To set up these circuits, set the coil slug at its middle position, connect the battery, and tune the FM receiver to locate the transmitter frequency. If necessary, trim the slug to tune the transmitter to a clear spot in the FM band. RV_1 should then be trimmed to set the modulation at a 'clean' level.

Transistor a.c. voltmeters

An ordinary moving-coil meter can be made to read a.c. voltages by feeding them to it via a rectifier and suitable 'multiplier' resistor, but produces grossly non-linear scale readings if used to give f.s.d. values below a few volts. This non-linearity problem can be overcome by connecting the meter circuitry into the feedback loop of a transistor common-emitter amplifier, as shown in the circuits of *Figures 12.35* to *12.37*, which (with the R_m value shown) each read 1V f.s.d.

The *Figure 12.35* circuit uses a bridge rectifier type of meter network, and draws a quiescent current of 0.3mA, has an f.s.d. frequency response that is flat from below 15Hz to above 150kHz, and has superb linearity up to 100kHz when using IN4148 silicon diodes or to above 150kHz when using BAT85 Schottky types. R_1 sets Q_1's quiescent current at about treble the meter's f.s.d. value, and thus gives the meter automatic overload protection.

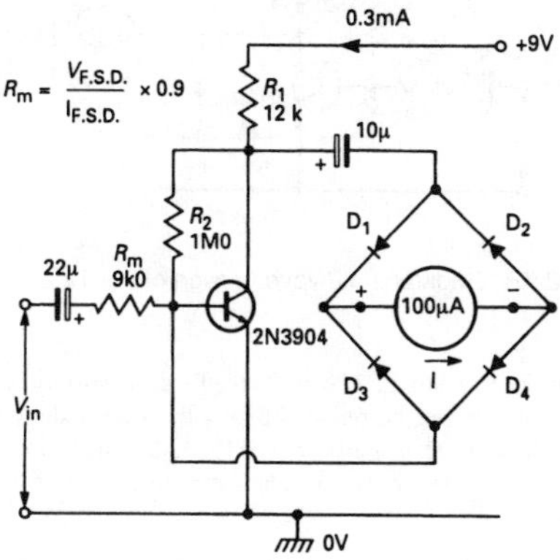

Figure 12.35. The frequency response of this 1V a.c. meter is flat to above 150kHz.

Figures 12.36 and *12.37* show 'pseudo full-wave' and 'ghosted half-wave' versions of the above circuit. These have a performance similar to that of *Figure 12.35*, but with better linearity and lower sensitivity. D_3 is sometimes used in these circuits to apply slight forward bias to D_1 and D_2 and thus enhance linearity, but this makes the meter pass a 'standing' current when no a.c. input is applied. The diodes used in these and all other electronic 'a.c. meter' circuits shown in this chapter should be either silicon (IN4148, etc.) or (for an exceptionally good performance) Schottky types; germanium types should not be used.

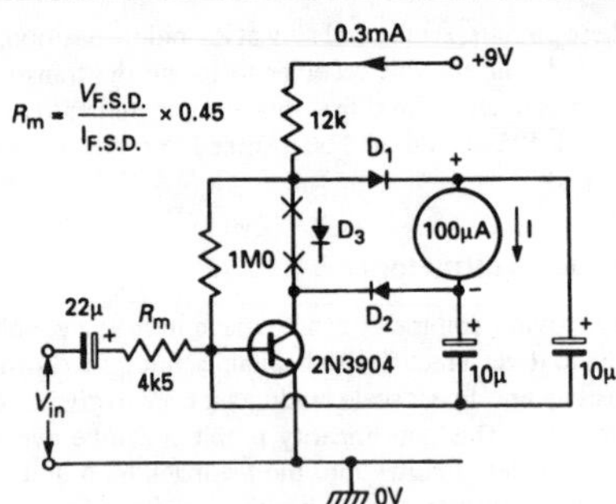

Figure 12.36. Pseudo full-wave version of the 1V a.c. meter.

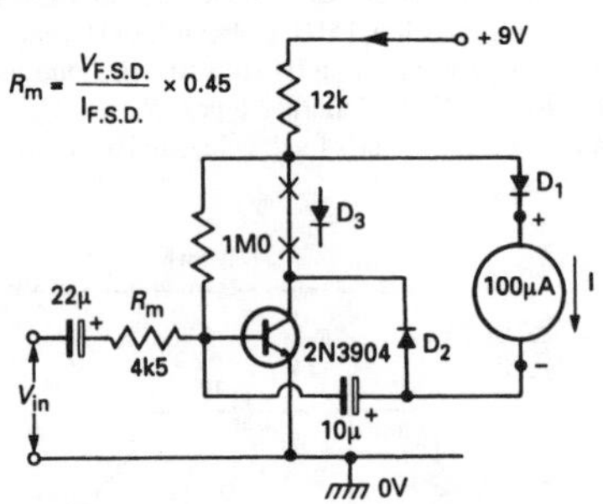

Figure 12.37. Ghosted half-wave version of the 1V a.c. meter.

In the *Figures 12.35* to *12.37* circuits the f.s.d. sensitivity is set at 1V by R_m, which can not be reduced below the values shown without incurring a loss of meter linearity. The R_m value can, however, safely be increased, to give higher f.s.d. values, e.g., by a factor of ten for 10V f.s.d., etc.

If greater f.s.d. sensitivity is wanted from the above circuits it can be obtained by applying the input signal via a suitable pre-amplifier, i.e., via a +60dB amplifier for 1mV sensitivity, etc. *Figure 12.38* shows this technique applied to the *Figure 12.35* circuit, to give a f.s.d. sensitivity variable between 20mV and 200mV via RV_1. With the sensitivity set at 100mV f.s.d. this circuit has an input impedance of 25k and a bandwidth that is flat within 0.5dB to 150kHz.

A.c. millivoltmeter circuits

A one-transistor 'a.c. meter' can not be given an f.s.d. sensitivity greater than 1V without loss of linearity. If greater sensitivity is needed, two or more stages of transistor amplification must be used. The highest useful f.s.d. sensitivity that can be obtained (with good linearity and gain stability) from a two-transistor circuit is 10mV, and *Figure 12.39*

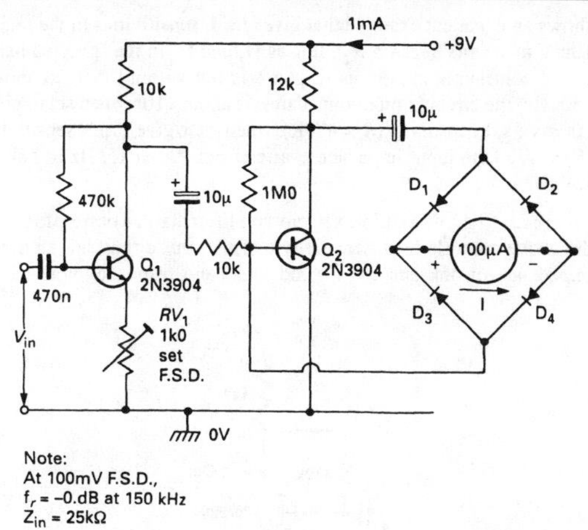

Figure 12.38. This a.c. voltmeter can be set to give f.s.d. sensitivities in the range 20mV to 200mV.

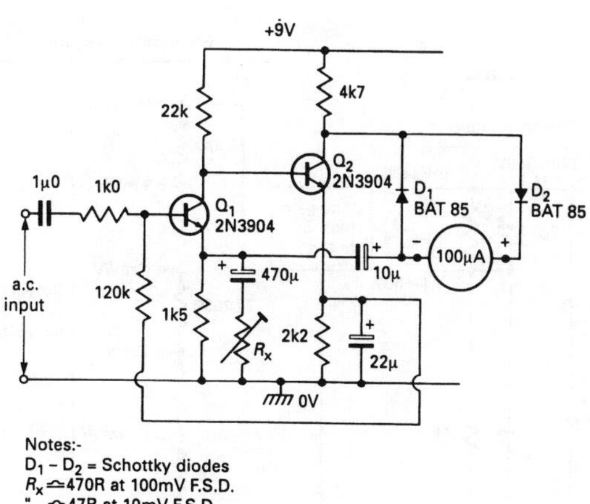

Figure 12.39. Wideband a.c. millivoltmeter with f.s.d. sensitivity variable from 10mV to 100mV via R_x.

shows an excellent example that gives f.s.d. sensitivities in the range 10mV to 100mV (set via R_x). It uses D_1 and D_2 in the 'ghosted half-wave' configuration, and its response is flat within 0.5dB to above 150kHz; the circuit's input impedance is about 110k when set to give 100mV f.s.d. sensitivity (R_x=470R); when set to give 10mV sensitivity (R_x = 47R) the input impedance varies from 90k at 15kHz to 56k at 150kHz.

Figure 12.40 shows a simple x10 pre-amplifier that can be used to boost the above circuit's f.s.d. sensitivity to 1mV; this circuit has an input impedance of 45k and has a good wideband response. Note when

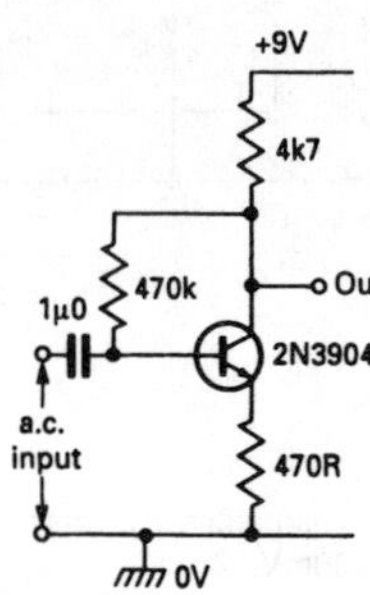

Figure 12.40. This x10 wideband pre-amplifier can be used to boost an a.c. millivoltmeter's sensitivity.

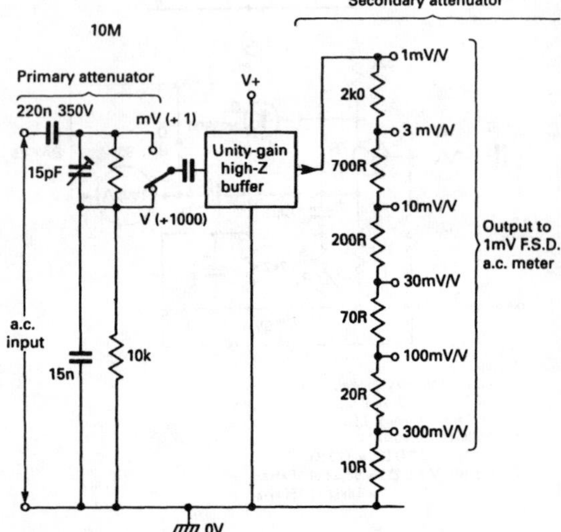

Figure 12.41. Basic multi-range a.c. volt/millivolt meter circuit.

building highly sensitive a.c. millivoltmeters that great care must be taken to keep all connecting leads short, to prevent unwanted RF pickup.

A wide-range a.c. volt/millivoltmeter can be made by feeding the input signals to a sensitive a.c. meter via suitable attenuator circuitry. To avoid excessive attenuator complexity, the technique of *Figure 12.41* is often adopted; the input is fed to a high-impedance unity-gain buffer, either directly (on 'mV' ranges) or via a compensated 60dB attenuator (on 'V' ranges), and the buffer's output is fed to a basic 1mV f.s.d. meter via a simple low-impedance attenuator, which in this example has 1 – 3 – 10, etc., ranging.

Figure 12.42 shows a useful variation of the above technique. In this case the input buffer also serves as a x10 amplifier, and the secondary attenuator's output is fed to a meter with 10mV f.s.d. sensitivity, the net effect being that a maximum overall sensitivity of 1mV is obtained with a minimum of complexity.

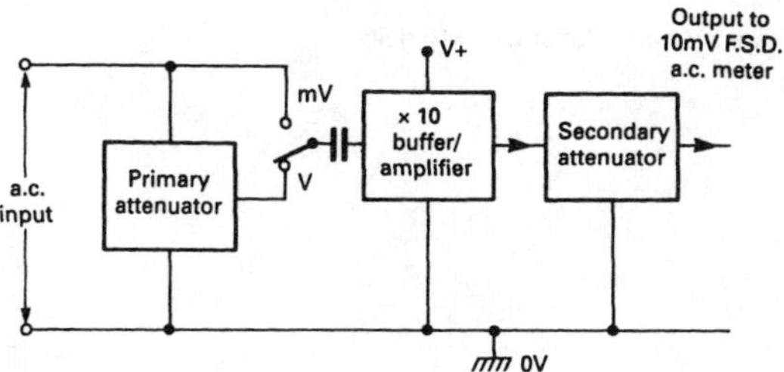

Figure 12.42. A useful a.c. volt/millivolt meter circuit variation.

Figures 12.43 and *12.44* show input buffers suitable for use with the above types of multi-range circuit. The *Figure 12.43* design is that of a unity-gain buffer; it gives an input impedance of about 4M0. The *Figure 12.44* buffer gives a x10 voltage gain (set by the R_1/R_x ratio) and has an input impedance of 1M0.

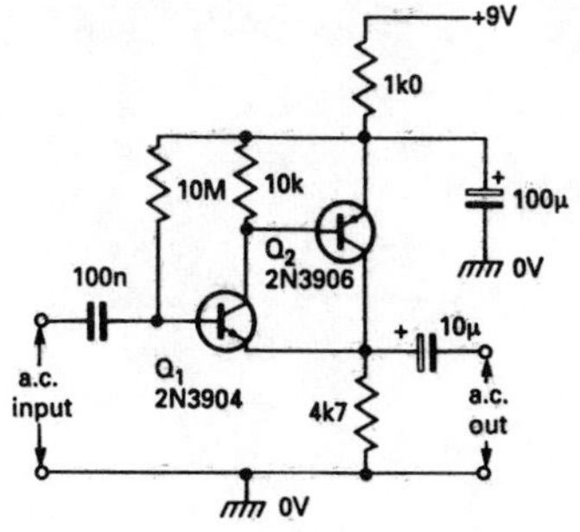

Figure 12.43. Unity-gain input buffer.

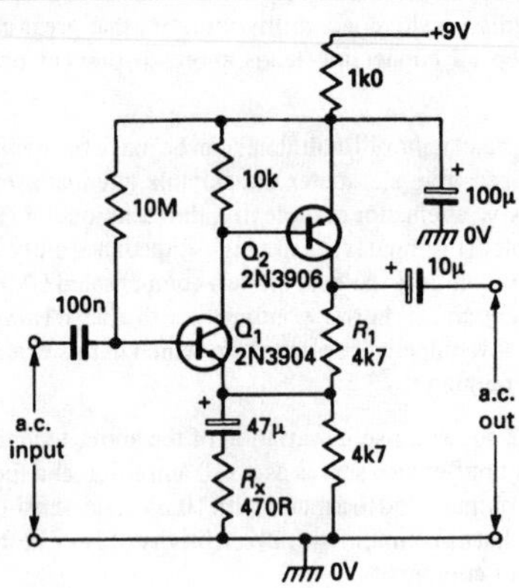

Figure 12.44. Buffer with x10 gain.

13 Optoelectronic circuits

Chapter 7 of this volume, in describing the basic characteristics of the junction diode and associated devices, makes brief mention of photodiodes and LEDs. The present chapter expands on this theme and looks at ways of using simple optoelectronic devices such as LEDs, photodiodes, phototransistor, and optocouplers.

LED basic circuits

A LED (light emitting diode) is a special type of junction diode, made of gallium phosphide or gallium arsenide, etc., that emits a fairly narrow bandwidth of visible (usually red, orange, yellow or green) or invisible (infra-red) light when stimulated by a forward current. LEDs have power-to-light energy conversion efficiencies some ten to fifty times better than a tungsten lamp and have very fast response times (about 0.1µS, compared with tens or hundreds of mS for a tungsten lamp), and are thus widely used as visual indicators in moving-light displays, etc.

A significant voltage is developed across a LED when it is passing a forward current, and *Figure 13.1* shows the typical forward voltages of different coloured LEDs at forward currents of 20mA. If a LED is reverse biased it avalanches or 'zeners' at a fairly low voltage value, as shown in *Figure 13.2*; most LEDs have maximum reverse voltage ratings of about 5 volts.

In use, a LED must be wired in series with a current-limiting device such as a resistor; *Figure 13.3* shows how to work out the *'R'* value needed to give a particular current from a particular supply voltage. In practice, *R* can be connected to either the anode or the cathode of the

Colour	Red	Orange	Yellow	Green
V_f (typical)	1V8	2V0	2V1	2V2

Figure 13.1. Typical forward voltages of standard LEDs at I_f = 20mA.

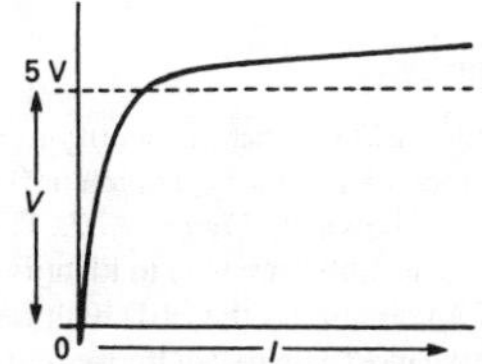

Figure 13.2. A reverse biased LED acts like a zener diode.

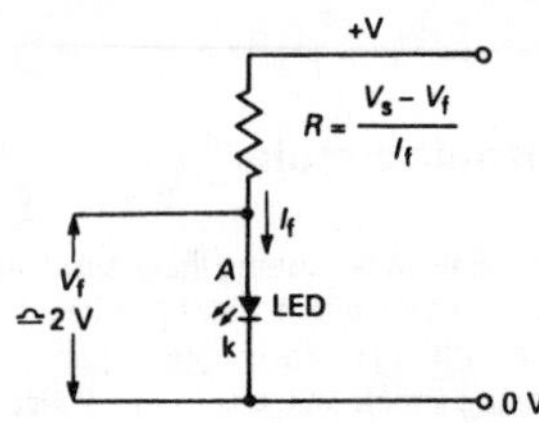

Figure 13.3. Method of finding the *R* value for a given V_s and I_f.

LED. The LED brightness is proportional to the LED current; most LEDs will operate safely up to absolute maximum currents of 30 to 40mA.

A LED can be used as an A.C. indicator by reverse-shunting it with a normal diode, as shown in *Figure 13.4*; for a given brightness, the *'R'* value should be halved relative to that of the D.C. circuit. If this circuit is used with high-value A.C. supplies, *'R'* may need a fairly high power rating; thus, if used with a 250V supply it will need a minimum rating of 2.5W at a mean LED current of 10mA; this snag can be overcome by replacing *'R'* with a current-limiting capacitor, as shown in *Figure 13.5*. Here, the C_S impedance limits the LED current to the desired value, but C_S dissipates near-zero power, since its current and voltage are ninety degrees out of phase. C_S values of 100nF and 220nF are usually adequate on 250V and 125V 50–60Hz A.C. lines respectively.

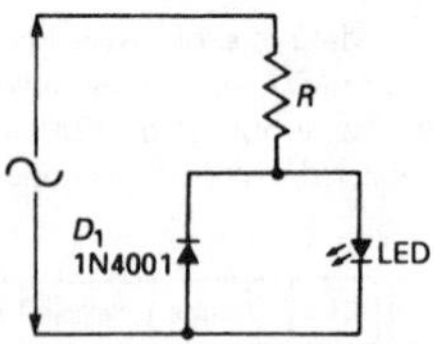

Figure 13.4. Using a LED as an indicator in a low-voltage A.C. circuit.

The basic *Figure 13.5* circuit can be used as a 'blown fuse' indicator by wiring it as shown in *Figure 13.6*. The circuit is shorted out by the fuse, but becomes enabled when the fuse is 'blown'.

Practical usage notes

One problem in using a LED is that of identifying its polarity. Most LEDs have their cathode identified by a notch or flat on the package, or by a short lead, as shown in *Figure 13.7*. This practice is not universal, however, so the only sure way to identify a LED is to test it in the basic *Figure 13.3* circuit: try the LED both ways round: when it glows, the cathode is the most negative of the two terminals; always test a LED before wiring it into circuit.

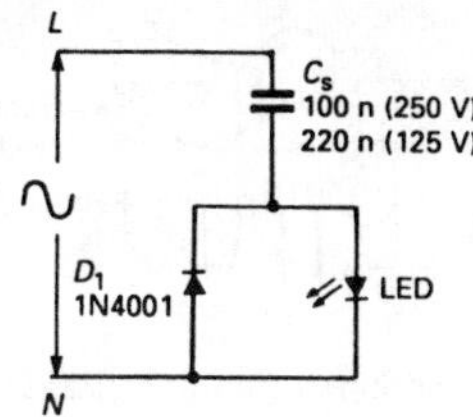

Figure 13.5. Using a LED as an indicator in an A.C. power line circuit.

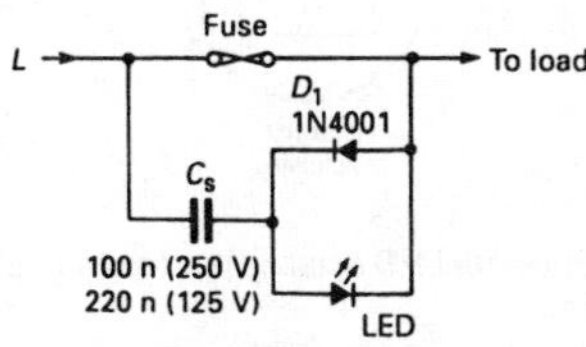

Figure 13.6. A.C. power line fuse blown indicator.

Special mounting kits, comprising a plastic clip and ring, are available for fixing LEDs into PC boards and front panels, etc. *Figure 13.8* illustrates the functioning of such a kit.

Most LEDs come in a single-LED package of the type shown in *Figure 13.7*. Multi-LED packages are also available. The best known of these are the 7-segment displays, comprising seven (or eight) LEDs packaged in a form suitable for displaying alpha-numeric characters. So-called 'bar-graph' displays, comprising 10 to 30 linearly-mounted LEDs in a single package, are also available.

Most LEDs generate a single output colour, but a few special devices provide 'multicolour' outputs; these are actually 2-LED devices, and *Figure 13.9* shows one that comprises a pair of LEDs connected in inverse parallel, so that green is emitted when it is biased in one direction, and red (or yellow) when it is biased in the reverse direction. This device is useful for giving polarity indication or null detection.

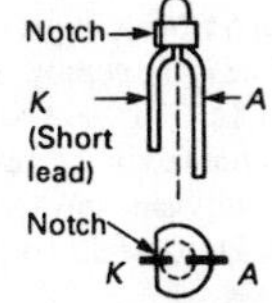

Figure 13.7. Typical outline and method of recognising the polarity of a LED.

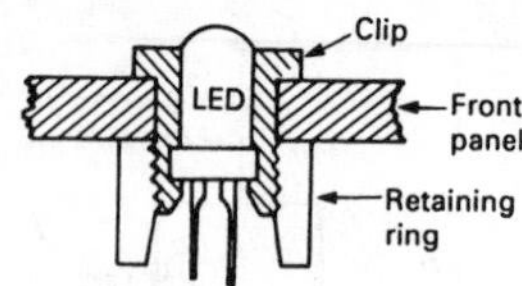

Figure 13.8. Clip and ring kit used to secure a LED to a front panel.

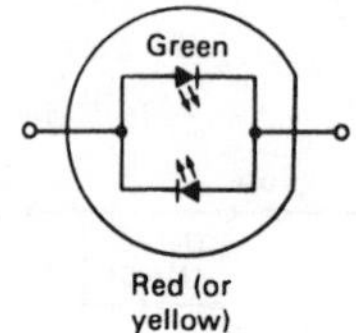

Figure 13.9. Bi-colour LED actually houses two LEDs connected in inverse parallel.

Another type of 'multicolour' LED is shown in *Figure 13.10*. This comprises a green and a red LED mounted in a 3-pin common-cathode package, and can generate green or red colours by turning on only one LED at a time, or orange or yellow ones by turning on the two LEDs in the ratios shown in the table.

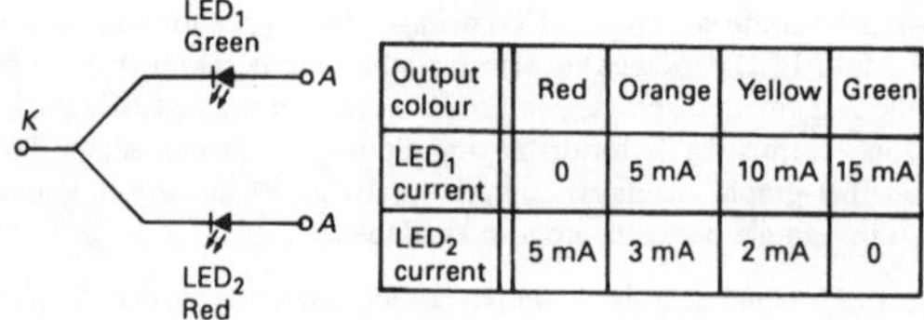

Output colour	Red	Orange	Yellow	Green
LED_1 current	0	5 mA	10 mA	15 mA
LED_2 current	5 mA	3 mA	2 mA	0

Figure 13.10. Multicolour LED giving four colours from two junctions.

Multi-LED circuits

Several LEDs can be powered from a single source by wiring them in series as shown in *Figure 13.11*; the supply voltage must be far larger than the sum of the individual LED forward voltages. This circuit draws minimal total current, but is limited in the number of LEDs that it can drive. A number of these circuits can, however, be wired in parallel, to enable a large number of LEDs to be driven from a single source.

Another way of powering several LEDs is by using several *Figure 13.3* circuits in parallel, as shown in *Figure 13.12*, but this technique is wasteful of current.

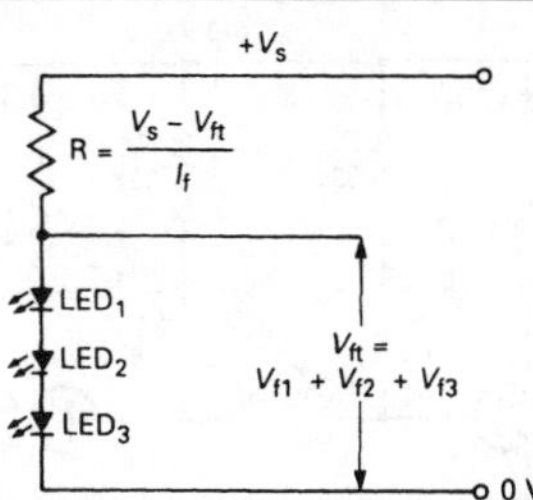

Figure 13.11. LEDs wired in series and driven via a single current-limiting resistor.

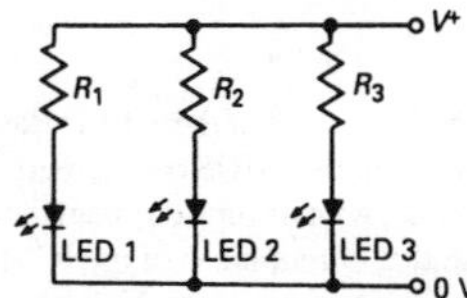

Figure 13.12. This circuit can drive a large number of LEDs, but at the expense of high current.

Figure 13.13 shows a 'what not to do' circuit. This design will not work correctly because inevitably differences in LED forward voltage characteristics usually cause one LED to 'hog' most of the available current, leaving little or none for the remaining LEDs.

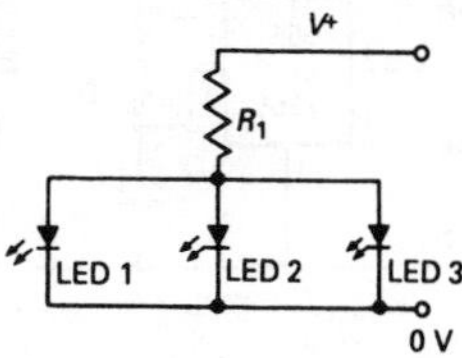

Figure 13.13. This LED-driving circuit will not work. One LED will hog all the current.

LED 'flasher' circuits

A LED 'flasher' is a circuit that turns a LED repeatedly on and off, usually at a rate of one or two flashes per second, to give an eye-catching display action; it may control a single LED or may control two LEDs in such a way that one turns on as the other turns off, and *vice versa*. *Figure 13.14* shows the practical circuit of a 2-transistor 2-LED flasher, which can be converted to single-LED operation by simply replacing the unwanted LED with a short circuit. Here, Q_1 and Q_2 are wired as a 1 cycle-per-second astable multivibrator, with switching rates controlled via C_1–R_3 and C_2–R_4.

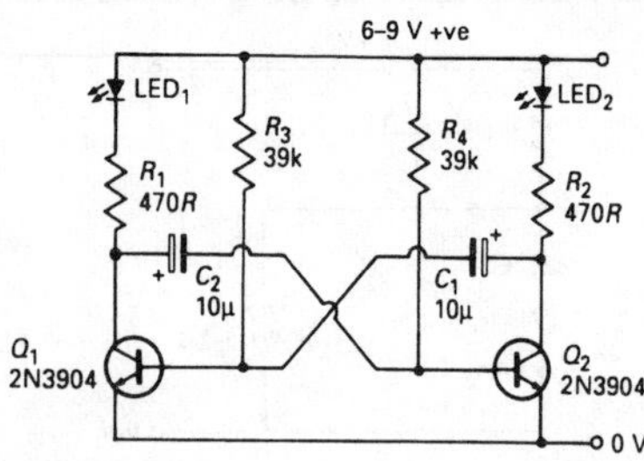

Figure 13.14. Transistor two-LED flasher circuit operates at one flash per second.

Figure 13.15 shows an IC version of the 2-LED flasher. This design is based on the 555 timer IC or its CMOS counterpart, the 7555, which is wired in the astable mode, with its time constant determined by C_1 and R_4. The action is such that output pin-3 of the IC alternately switches between the ground and the positive supply voltage levels, alternately shorting out (disabling) one or other of the two LEDs. The circuit can be converted to single-LED operation by omitting the unwanted LED and its associated current-limiting resistor.

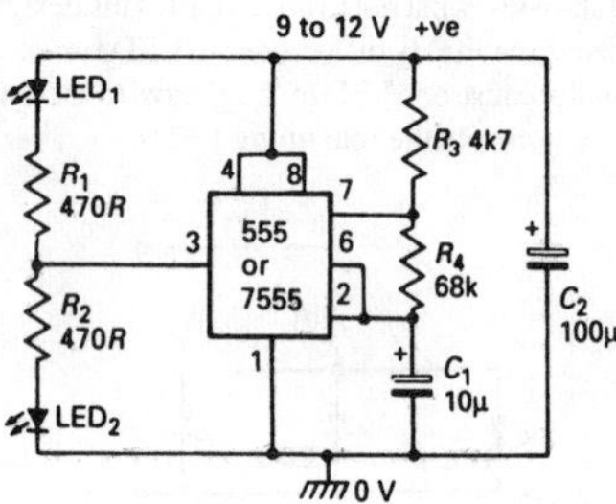

Figure 13.15. IC two-LED flasher circuit operates at about one flash per second.

Another widely used type of LED-display circuit is that used for LED 'sequencing'; these drive a chain of LEDs in such a way that each LED is switched on and off in a time-controlled sequence, so that a ripple of light seems to run along the chain. Another type is that used to give 'bar-graph' type analogue-value indication; these drive a chain of linearly-spaced LEDs in such a way that the illuminated chain length is proportional to the analogue value of a voltage applied to the input of the driver circuit, e.g., so that the circuit acts like an analogue voltmeter. Lots of LED display circuits of these types are presented in the author's 'Optoelectronics Circuits Manual'.

Photodiodes

When p–n silicon junctions are reverse biased their leakage currents and impedances are inherently photo-sensitive; they act as very high impedances under dark conditions and as low impedances under bright ones. Normal diodes have their junctions shrouded in opaque material to stop this effect, but photodiodes are made to exploit it and have their junctions encased in translucent material. Some photodiodes are made to respond to visible light, and some to infra-red (IR) light. In use, the photodiode is simply reverse biased and the output voltage is taken from across a series resistor, which may be connected between the diode and ground, as in *Figure 13.16(a)*, or between the diode and the positive supply line, as in *Figure 13.16(b)*.

Photodiodes have a far lower light-sensitivity than cadmium-sulphide LDRs, but give a far quicker response to changes in light level. Generally, LDRs are ideal for use in slow-acting direct-coupled 'light-level' sensing applications, while photodiodes are ideal for use in fast a.c.-coupled 'signalling' applications. Typical photodiode applications include IR remote-control circuits, IR 'beam' switches and alarm circuits, and photographic 'flash' slave circuits, etc.

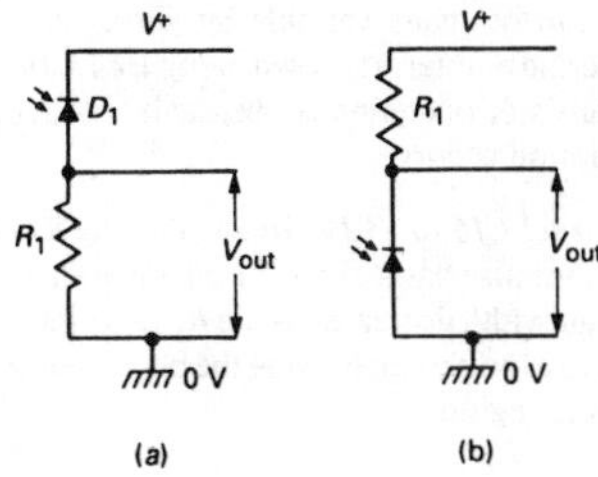

Figure 13.16. Alternative ways of using a photodiode.

Phototransistors

Ordinary silicon transistors are made from an npn or pnp sandwich, and thus inherently contain a pair of photo-sensitive junctions. Some types are available in phototransistor form, and use the *Figure 13.17* symbol.

Figure 13.18 shows three different basic ways of using a phototransistor; in each case the base-collector junction is effectively reverse biased and thus acts as a photodiode. In *(a)* the base is grounded, and the transistor acts as a simple photodiode. In *(b)* and *(c)* the base terminal is open-circuit and the photo-generated currents effectively feed directly into

Figure 13.17. Phototransistor symbol.

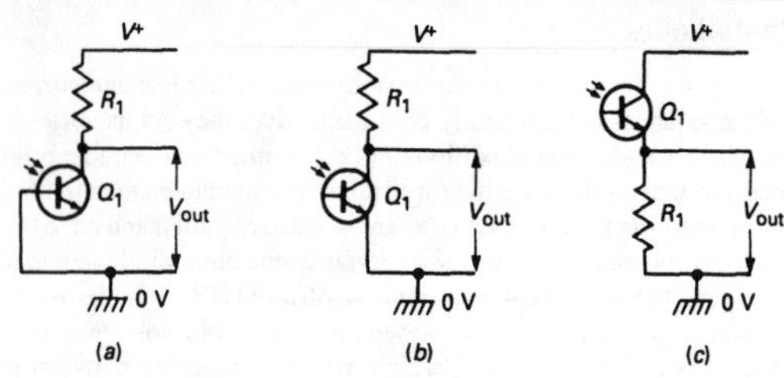

Figure 13.18. Alternative ways of using a phototransistor.

the base and, by normal transistor action, generate a greatly amplified collector-to-emitter current that produces an output voltage across series resistor R_1.

The sensitivity of a phototransistor is typically one hundred times greater than that of a photodiode, but its useful maximum operating frequency (a few hundred kHz) is proportionally lower than that of a photodiode (tens of MHz). The sensitivity (and operating speed) of a phototransistor can be made variable by wiring a variable resistor between the base and emitter, as shown in *Figure 13.19*; with RV_1 open circuit, phototransistor operation is obtained; with RV_1 short circuit, photodiode operation occurs.

Note in the *Figures13.16* to *13.19* circuits that the R_1 value is usually chosen on a compromise basis, since the circuit voltage gain increases but the useful bandwidth decreases as the R_1 value increases. Also, the R_1 value may have to be chosen to bring the photo-sensitive device into its linear operating region.

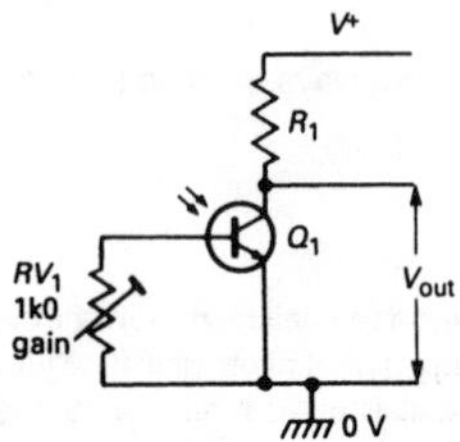

Figure 13.19. Variable-sensitivity phototransistor circuit.

Optocouplers

An optocoupler is a device housing a LED (usually an IR type) and a matching phototransistor (or photodiode); the two devices are closely optically coupled and mounted in a light-excluding housing. *Figure 13.20* shows a basic optocoupler 'usage' circuit. The LED is used as the input side of the circuit, and the photo-transistor as the output. Normally, SW_1 is open, and the LED and Q_1 are thus off. When SW_1 is

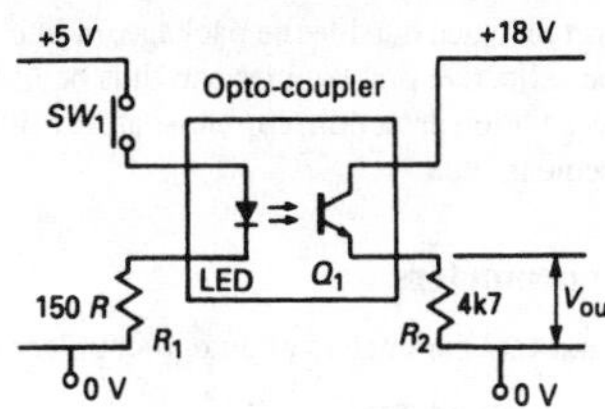

Figure 13.20. Basic optocoupling circuit.

closed a current flows through the LED via R_1, and Q_1 is turned on optically and generates an output voltage across R_2. The output circuit is thus controlled by the input one, but the two circuits are fully isolated electrically ('isolation' is the major feature of this type of optocoupler, which can be used to couple either digital or analogue signals).

The *Figure 13.20* device is a simple 'isolating' optocoupler. *Figures 13.21* and *13.22* show two other types of optocoupler. The *Figure 13.21* 'slotted' one has a slot moulded into the package between the LED light source and the phototransistor light sensor. Light can normally pass from the LED to Q_1 via the slot, but can be completely blocked by placing an opaque object in the slot. The slotted optocoupler can thus be used in a variety of 'presence detecting' applications, including end-of-tape detection, limit switching, and liquid-level detection, etc.

The *Figure 13.22* 'reflective' optocoupler has the LED and Q_1 optically screened from each other within the package, and both face outwards towards an external point. The construction is such that an optocoupled link can be set up by a reflective object (such as metallic paint or tape,

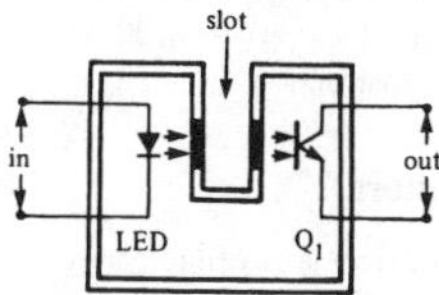

Figure 13.21. Slotted optocoupler device.

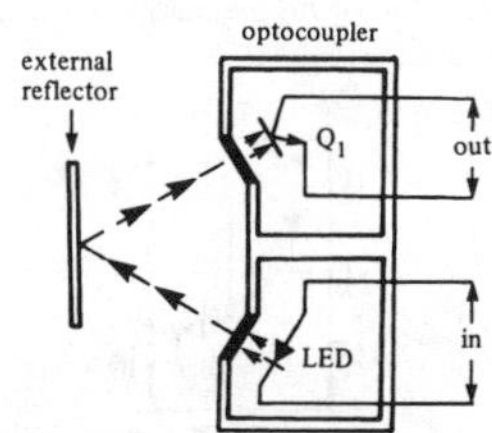

Figure 13.22. Reflective optocoupler.

etc.) placed a short distance outside the package, in line with both the LED and Q_1. The reflective optocoupler can thus be used in applications such as tape-position detection, engine-shaft revolution counting or speed measurement, etc.

Optocoupler parameters

The five most important parameters of an optocoupler are as follows.

Current transfer ratio (CTR)

This is a measure of optocoupling efficiency, and is actually quoted in terms of output-to-input 'current transfer ratio' (CTR), i.e., the ratio of the output current (I_C) measured at the collector of the phototransistor, to the input current (I_F) flowing into the LED. Thus, CTR = I_C/I_F. In practice, CTR may be expressed as a simple figure such as '0.5', or (by multiplying this figure by 100) as a percentage figure such as '50%'. Simple isolating optocouplers with single-transistor output stages have typical CTR values in the range 20% to 100%.

Isolation voltage

This is the maximum permissible D.C. voltage allowed to exist between the input and output circuits. Typical values vary from 500V to 4kV.

$V_{CE(MAX)}$

This is the maximum allowable D.C. voltage that can be applied across the output transistor. Typical values vary from 20V to 80V.

$I_{F(MAX)}$

This is the maximum permissible D.C. current that can be allowed to flow in the input LED. Typical values vary from 40mA to 100mA.

Bandwidth

This is the typical maximum signal frequency (in kHz) that can be usefully passed through the optocoupler when it is operated in its 'normal' mode. Typical values vary from 20kHz to 500kHz, depending on the type of device construction.

Practical optocouplers

Practical optocoupler devices are available in six basic forms, and these are illustrated in *Figure 13.23* to *Figure 13.28*. Four of these devices are 'isolating' types, housing 1, 2 or 4 simple or Darlington-type optocouplers in a single package; the remaining two are the 'slotted' optocoupler *(Figure 13.27)* and the Darlington 'reflective' optocoupler *(Figure 13.28)*. The table of *Figure 13.29* lists the typical parameter values of these six devices.

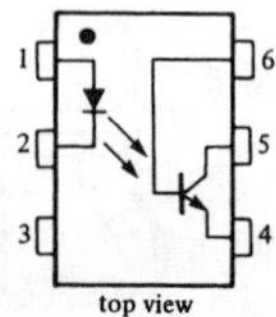

Figure 13.23. Typical simple isolating optocoupler.

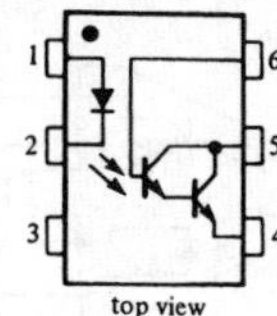

Figure 13.24. Typical Darlington isolating optocoupler.

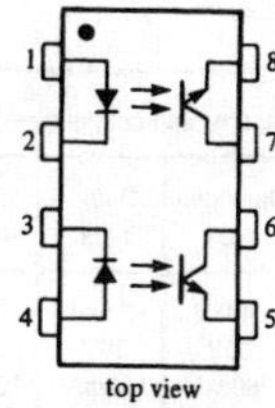

Figure 13.25. Typical dual isolating optocoupler.

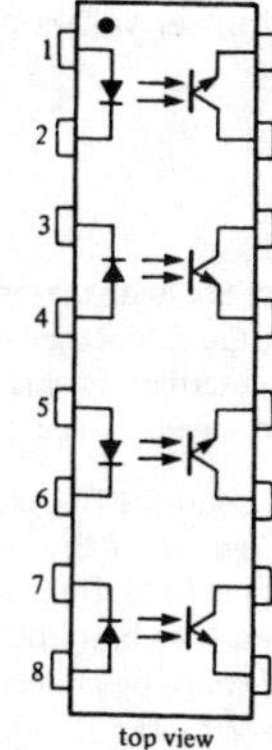

Figure 13.26. Typical quad isolating optocoupler.

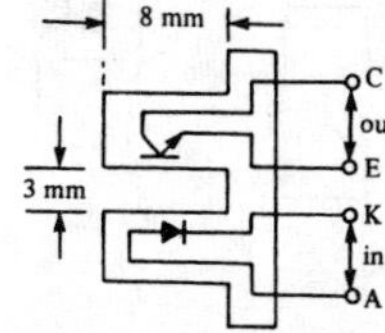

Figure 13.27. Typical slotted optocoup;ler.

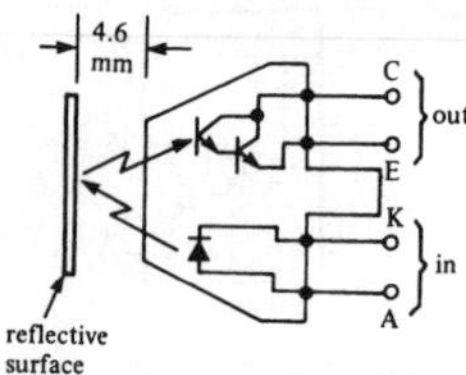

Figure 13.28. Typical reflective optocoupler.

Parameter	Isolating opto-couplers: Simple type	Darlington type	Dual type	Quad type	Slotted opto-coupler	Reflective opto-coupler
Isolating voltage	±4kV	±4kV	±1.5kV	±1.5kV	N.A.	N.A.
V_{CE} (max)	30V	30V	30V	30V	30V	15V
I_F (max)	60mA	60mA	100mA	100mA	50mA	40mA
CTR (min)	20%	300%	12.5%	12.5%	10%	0.5%
Bandwidth	300kHz	30kHz	200kHz	200kHz	300kHz	20kHz
Outline	Fig 13.23	Fig 13.24	Fig 13.25	Fig 13.26	Fig 13.27	Fig 13.28

Figure 13.29. Typical parameter values of the *Figures 13.23* to *13.28* devices.

Usage notes

Optocouplers are very easy to use, with the input side being used in the manner of a normal LED and the output used in the manner of a normal phototransistor as already described in this chapter. The following notes give a summary of the salient 'usage' points.

The input current to the optocoupler LED must be limited via a series resistor which, as shown in *Figure 13.30*, can be connected on either the anode or the cathode side of the LED. If the LED is to be driven from an A.C. source, or there is a possibility of a reverse voltage being applied to the LED, the LED must be protected via an external diode connected as shown in *Figure 13.31*.

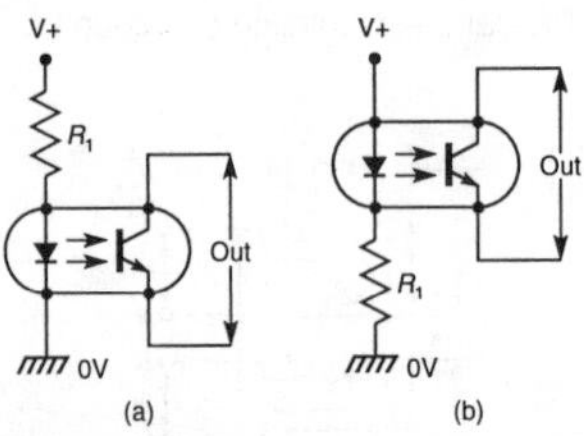

Figure 13.30. The LED current must be limited via a series resistor, which can be connected to either the anode *(a)* or the cathode *(b)*.

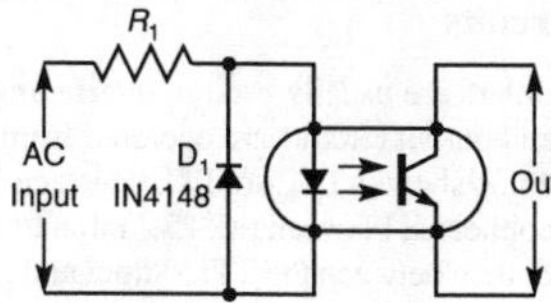

Figure 13.31. The input LED can be protected against reverse voltages via an external diode.

The phototransistor's output current can be converted into a voltage by wiring a resistor in series with its collector or emitter, as shown in *Figure 13.32*. The greater the resistor value, the greater is the circuit's sensitivity but the lower is its bandwidth.

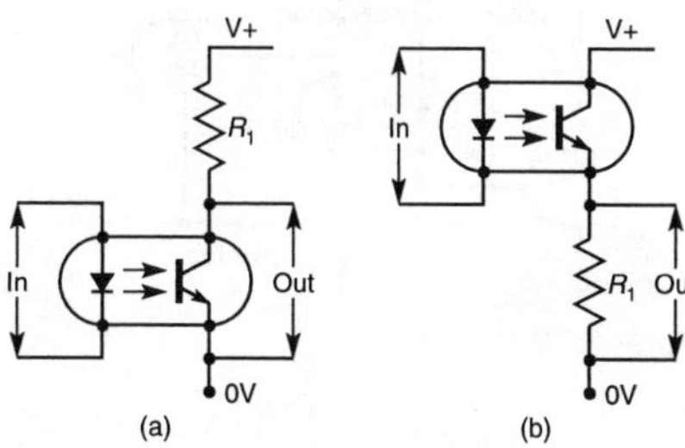

Figure 13.32. An external output resistor, wired in series with the phototransistor, can be connected to either the collector *(a)* or emitter *(b)*.

The phototransistor is normally used with its base terminal open circuit; it can be converted into a photodiode by using the base terminal as shown in *Figure 13.33* and ignoring the emitter terminal (or shorting it to the base). This connection results in a greatly increased bandwidth (typically 30MHz), but a greatly reduced CTR value (typically 0.2%).

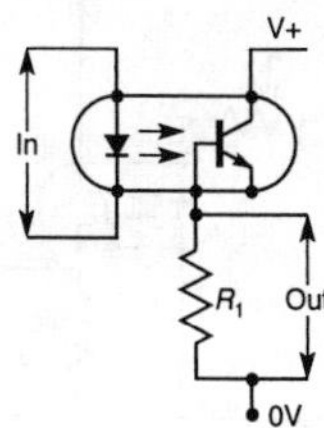

Figure 13.33. If its base is available, the phototransistor can be made to function as a photodiode.

Interfacing circuits

Isolating optocouplers are usually used in interfacing applications in which the input and output circuits are operated from different power supplies. *Figure 13.34* shows a TTL-to-TTL interface. The optocoupler LED and R_1 are connected between the +5V rail and the output of the TTL driver (rather than between the TTL output and ground), because TTL outputs can typically sink 16mA, but can source only 400μA. R_3 is used as a 'pull-up' resistor, to ensure that the LED turns fully off when the TTL output is in the logic-1 state. The phototransistor is wired between the input and ground of the 'driven' TTL gate, since a TTL input needs to be pulled down to below 800mV at 1.6mA to ensure correct logic-0 operation. This circuit provides non-inverting optocoupler action.

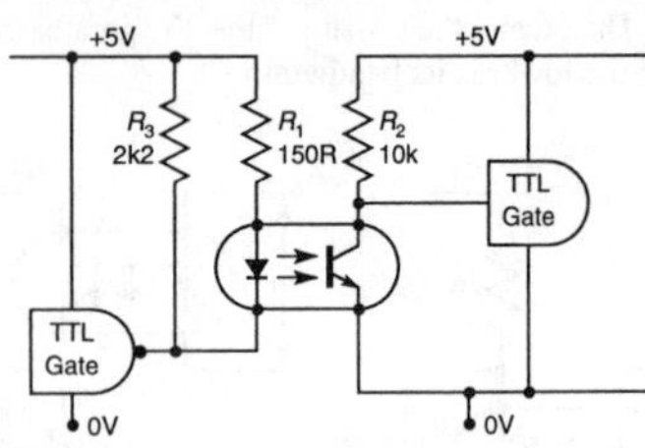

Figure 13.34. TTL interface.

CMOS IC outputs can source or sink currents up to several mA, and can thus be interfaced by using a 'sink' configuration similar to that of *Figure 13.34*, or using the 'source' configuration shown in *Figure 13.35*. In either case, the R_2 value must be large enough to provide an output voltage swing that switches fully between the CMOS logic-0 and logic-1 states.

Figure 13.36 shows an optocoupler used to interface a computer's digital output signal (5V, 5mA) to a 12V D.C. motor that draws an operating current of less than 1 amp. With the computer output high, the

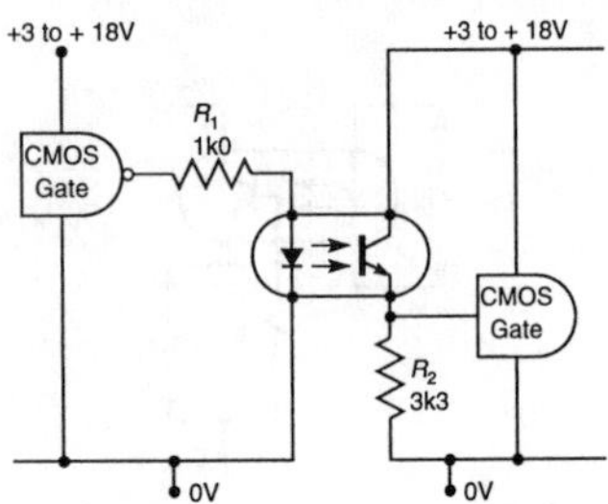

Figure 13.35. CMOS interface.

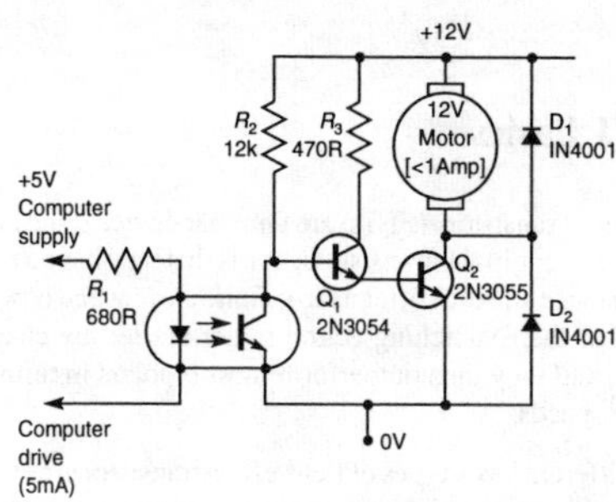

Figure 13.36. Computer-driven D.C. motor.

optocoupler LED and phototransistor are both off, so the motor is driven on via Q_1 and Q_2. When the computer output goes low, the LED and phototransistor are driven on, so Q_1–Q_2 and the motor are cut off. Finally, *Figure 13.37* shows how an optocoupler can be used to interface audio analogue signals from one circuit to another. Here, the op-amp is used as a unity-gain voltage follower with the optocoupler LED wired into its negative feedback loop, and is d.c. biased at half-supply volts via R_1–R_2 and can be a.c.-modulated by an audio signal applied via C_1. A standing current of 1mA to 2mA (set by R_3) thus flows through the LED and is modulated by the a.c. signal. On the output side of the circuit, a quiescent current is set up (by the optocoupler action) in the phototransistor, and causes a quiescent voltage to be set up across RV_1, which should have its value adjusted to give a quiescent output value of half-supply voltage. The audio output signal appears across RV_1 and is decoupled via C_2.

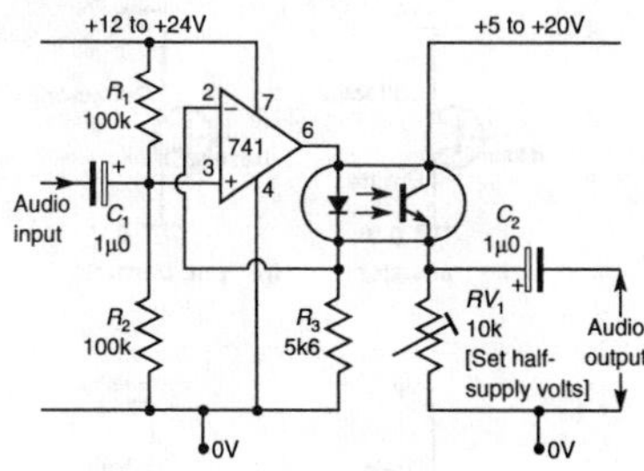

Figure 13.37. Audio-coupling circuit.

14 FET basics

Field-Effect Transistors (FETs) are unipolar devices, and have two big advantages over bipolar transistors; one is that they have a near-infinite input resistance and thus offer near-infinite current and power gain; the other is that their switching action is not marred by charge-storage problems, and they thus outperform most bipolars in terms of digital switching speeds.

Several different basic types of field-effect transistor are available, and this chapter looks at their basic operating principles; Chapters 15 to 17 show practical ways of using FETs.

FET basics

A FET is a three-terminal amplifying device; its terminals are known as the source, gate, and drain, and correspond respectively to the emitter, base, and collector of a normal transistor. Two distinct families of FET are in general use. The first of these is known as the 'junction-gate' type of FET, this term generally being abbreviated to either JUGFET or (more usually) JFET. The second family is known as either 'insulated-gate' FETs or Metal Oxide Semiconductor FETs, and these terms are generally abbreviated to IGFET or MOSFET respectively. 'N-channel' and 'p-channel' versions of both types of FET are available, just as normal transistors are available in npn and pnp versions, and *Figure 14.1* shows the symbols and supply polarities of both types of bipolar transistor and compares them with both JFET versions.

Figure 14.2 illustrates the basic construction and operating principle of a simple n-channel JFET. It consists of a bar of n- type semiconductor material with a drain terminal at one end and a source terminal at the

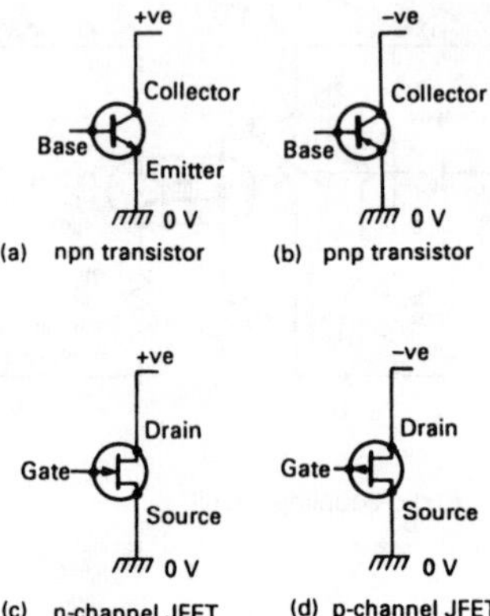

Figure 14.1. Comparison of transistor and JFET symbols, notations, and supply polarities.

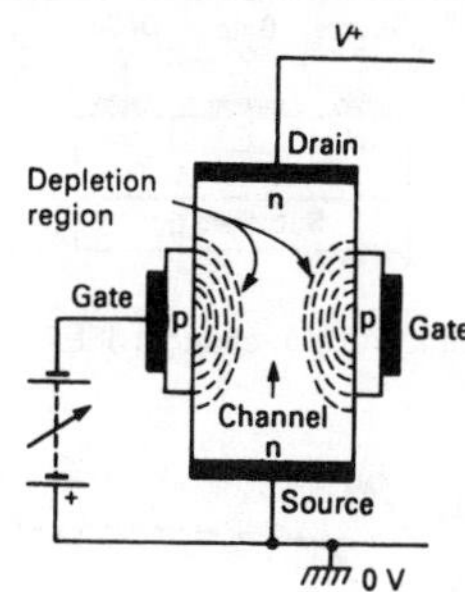

Figure 14.2. Basic structure of a simple n-channel JFET, showing how channel width is controlled via the gate bias.

other: a p-type control electrode or gate surrounds (and is joined to the surface of) the middle section of the n-type bar, thus forming a p-n junction.

In normal use the drain terminal is connected to a positive supply and the gate is biased at a value that is negative (or equal) to the source voltage, thus reverse biasing the JFET's internal p-n junction and accounting for its very high input impedance. With zero gate bias applied, a current flows from drain to source via a conductive 'channel' in the n-type bar. When negative gate bias is applied, a high resistance region is formed within the junction, and reduces the width of the n-type conduction channel and thus reduces the magnitude of the drain-to-source current. As the gate bias is increased, the 'depletion' region spreads deeper into the n-type channel, until eventually, at some 'pinch-off' voltage value, the depletion layer becomes so deep that conduction ceases completely.

Thus, the basic JFET of *Figure 14.2* passes maximum current when its gate bias is zero, and its current is reduced or 'depleted' when the gate bias is increased. It is thus known as a 'depletion-type' n-channel JFET. A p-channel version of the device can be made by simply transposing the p and n materials.

JFET details

Figure 14.3 shows the basic form of construction of a practical n-channel JFET; a p-channel JFET can be made by transposing the p and n materials. All JFETs operate in the depletion mode, as already described. *Figure 14.4* shows the typical transfer characteristics of a low-power n-channel JFET, and illustrates some important features of this type of device. The most important characteristics of the JFET are as follows.

(1) When a JFET is connected to a supply with the polarity shown in *Figure 14.1* (drain +ve for a n-channel FET, -ve for a p-channel FET), a drain current (I_D) flows and can be controlled via gate-to-source bias voltage V_{GS}.

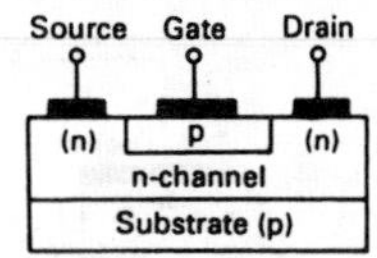

Figure 14.3. Construction of n-channel JFET.

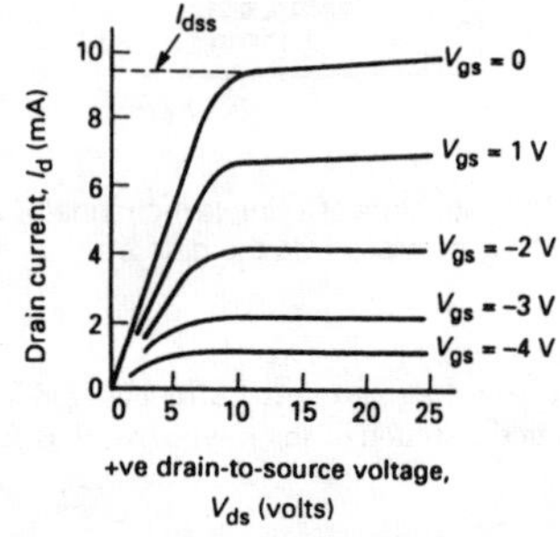

Figure 14.4. Idealised transfer characteristics of n-channel JFET.

(2) I_D is greatest when $V_{GS} = 0$, and is reduced by applying a reverse bias to the gate (negative bias in a n-channel device, positive bias in a p-type). The magnitude of V_{GS} needed to reduce I_D to zero is called the 'pinch-off' voltage, V_P, and typically has a value between 2 and 10 volts. The magnitude of I_D when $V_{GS} = 0$ is denoted I_{DSS}, and typically has a value in the range 2 to 20mA.

(3) The JFET's gate-to-source junction has the characteristics of a silicon diode. When reverse biased, gate leakage currents (I_{GSS}) are only a couple of nA (1nA = .001μA) at room temperature. Actual gate signal currents are only a fraction of a nA, and the input impedance of the gate is typically thousands of megohms at low frequencies. The gate junction is shunted by a few pF, so the input impedance falls as frequency is increased.

If the JFET's gate-to-source junction is forward biased, it conducts like a normal silicon diode, and if excessively reverse biased avalanches like a zener diode; in either case, the JFET suffers no damage if gate currents are limited to a few mA.

(4) Note in *Figure 14.4* that, for each V_{GS} value, drain current I_D rises linearly from zero as the drain-to-source voltage (V_{DS}) is increased from zero up to some value at which a 'knee' occurs on each curve, and that I_D then remains virtually constant as V_{DS} is increased beyond this knee value. Thus, when V_{DS} is below the JFET's knee value the drain-to-source terminals act as a resistor, R_{DS}, with a value dictated by V_{GS}, and can thus be used as a voltage-variable resistor, as in *Figure 14.5*. Typically, R_{DS} can be varied from a few hundred ohms (at $V_{GS} = 0$) to

thousands of megohms (at $V_{GS} = V_P$), enabling the JFET to be used as a voltage-controlled switch *(Figure 14.6)* or as an efficient 'chopper' *(Figure 14.7)* that does not suffer from offset-voltage or saturation-voltage problems.

Also note in *Figure 14.4* that when V_{DS} is above the knee value the I_D value is controlled by the V_{GS} value and is almost independent of V_{DS}, i.e., the JFET acts as a voltage-controlled current generator. The JFET can be used as a fixed-value constant-current generator by either tying the gate to the source, as in *Figure 14.8(a)*, or by applying a fixed negative bias to the gate, as in *Figure 14.8(b)*. Alternatively, it can (when suitably biased) be used as a voltage-to-current signal amplifier.

(5) FET 'gain' is specified as transconductance, g_m, and denotes the magnitude of change of drain current with gate voltage, i.e., a g_m of 5mA/V signifies that a V_{GS} variation of one volt produces a 5mA

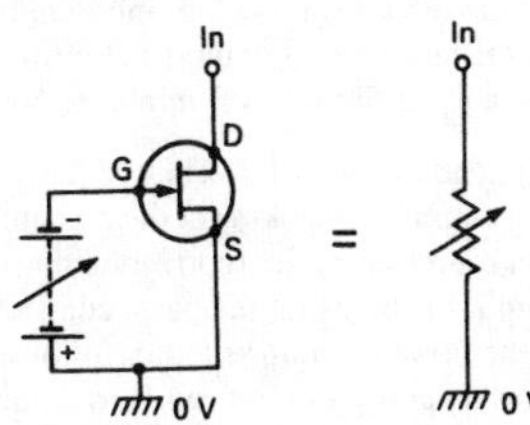

Figure 14.5. An n-channel JFET can be used as a voltage-controlled resistor.

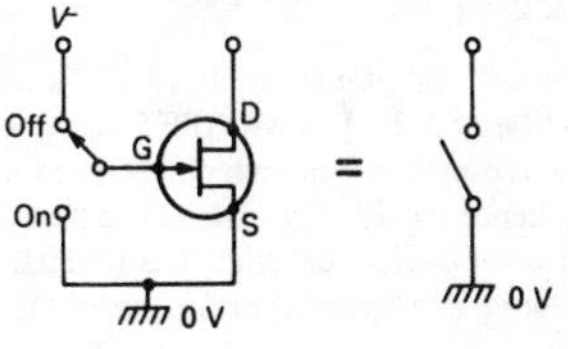

Figure 14.6. An n-channel JFET can be used as a voltage-controlled switch.

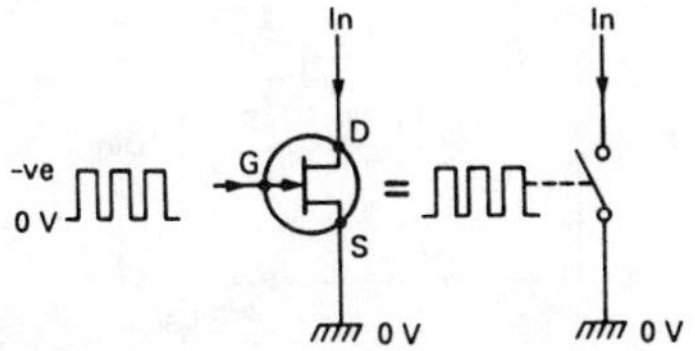

Figure 14.7. An n-channel JFET can be used as an electronic chopper.

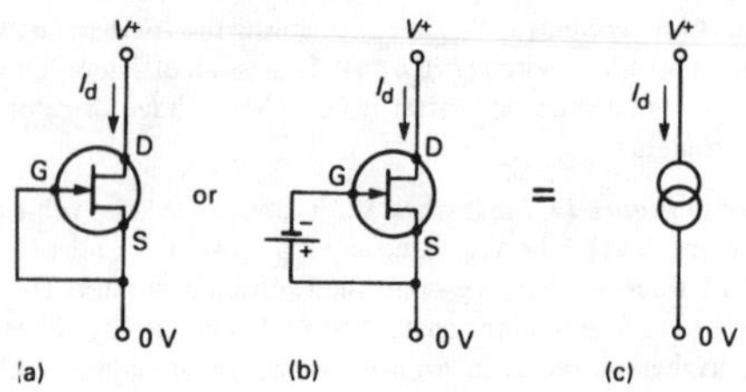

Figure 14.8. An n-channel JFET can be used as a constant-current generator.

change in I_D. Note that the form I/V is the inverse of the ohms formula, so g_m measurements are often expressed in 'mho' units; usually, g_m is specified in FET data sheets in terms of mmhos (milli-mhos) or µmhos (micro-mhos). Thus, a g_m of 5mA/V = 5-mmho or 5000-µmho.

In most practical applications the JFET is biased into the linear region and used as a voltage amplifier. Looking at the n-channel JFET, it can be used as a common source amplifier (corresponding to a bipolar npn common emitter amplifier) by using the basic connections of *Figure 14.9*. Alternatively, the common drain or source follower (similar to the bipolar emitter follower) configuration can be obtained by using the connections of *Figure 14.10*, or the common gate (similar to common base) configuration can be obtained by using the basic *Figure 14.11* circuit. In practice, fairly accurate biasing techniques (discussed in the next chapter) must be used in these circuits.

The IGFET/MOSFET

The second (and most important) family of FETs are those known under the general title of IGFET or MOSFET. In these FETs the gate terminal is insulated from the semiconductor body by a very thin layer of silicon dioxide, hence the title 'Insulated Gate Field Effect Transistor', or IGFET. Also, the devices generally use a 'Metal-Oxide Silicon' semiconductor material in their construction, hence the alternative title of MOSFET.

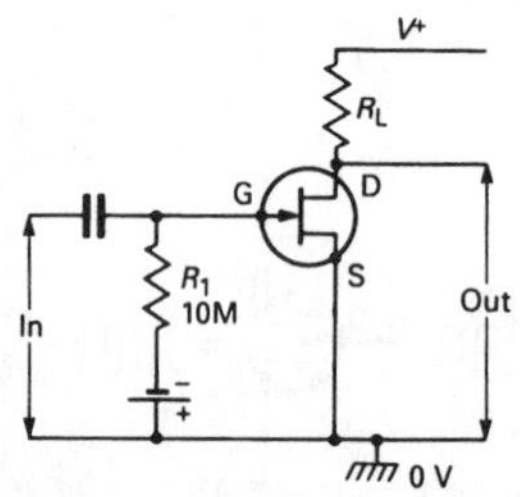

Figure 14.9. Basic n-channel common-source amplifier JFET circuit.

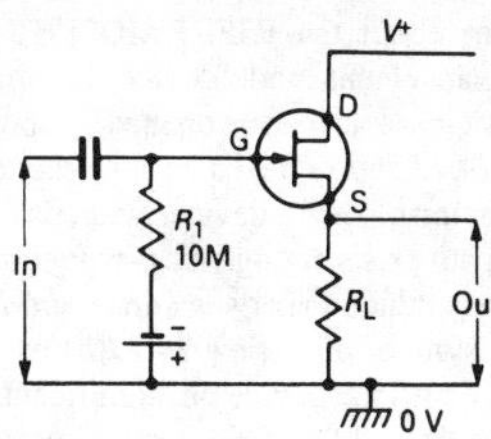

Figure 14.10. Basic n-channel common-drain (source-follower) JFET circuit.

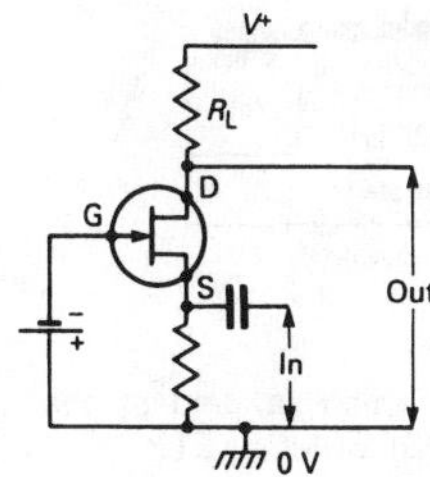

Figure 14.11. Basic n-channel common-gate JFET circuit.

Figure 14.12 shows the basic construction and the standard symbol of the n-channel depletion-mode FET. It resembles the JFET, except that its gate is fully insulated from the body of the FET (as indicated by the *Figure 14.12(b)* symbol), but in fact operates on a slightly different principle to the JFET. It has a normally-open n-type channel between drain and source, but the channel width is controlled by the electrostatic field of the gate bias. The channel can be closed by applying suitable negative bias, or can be increased by applying positive bias. In practice, the FET substrate may be externally available, making a 4-terminal device, or it may be internally connected to the source, making a 3-terminal device.

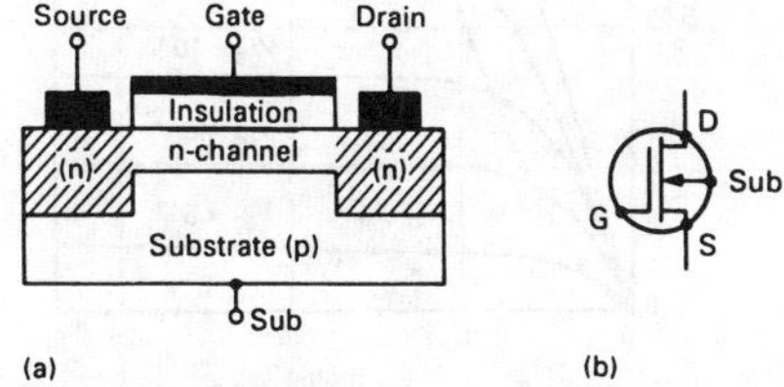

Figure 14.12. Construction *(a)* and symbol *(b)* of n-channel depletion-mode IGFET/MOSFET.

An important point about the IGFET/MOSFET is that it is also available as an enhancement-mode device, in which its conduction channel is normally closed but can be opened by applying forward bias to its gate. *Figure 14.13* shows the basic construction and the symbol of the n-channel version of such a device. Here, no n-channel drain-to-source conduction path exists through the p-type substrate, so with zero gate bias there is no conduction between drain and source; this feature is indicated in the symbol of *Figure 14.13(b)* by the gaps between source and drain. To turn the device on, significant positive gate bias is needed, and when this is of sufficient magnitude it starts to convert the p-type substrate material under the gate into an n-channel, enabling conduction to take place.

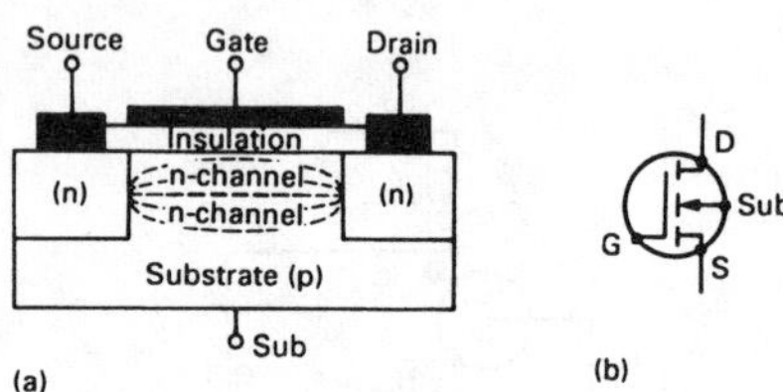

Figure 14.13. Construction *(a)* and symbol *(b)* of n-channel enhancement-mode IGFET/MOSFET.

Figure 14.14 shows the typical transfer characteristics of an n-channel enhancement-mode IGFET/MOSFET, and *Figure 14.15* shows the V_{GS}/I_D curves of the same device when powered from a 15V supply. Note that no I_D current flows until the gate voltage reaches a 'threshold' (V_{TH}) value of a few volts, but that beyond this value the drain current rises in a non-linear fashion. Also note that the transfer graph is divided

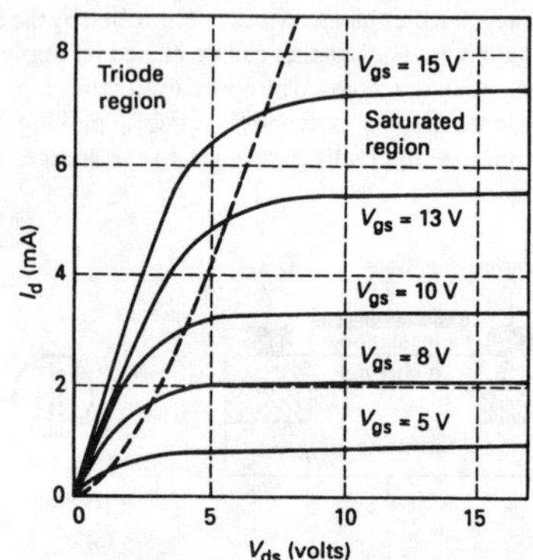

Figure 14.14. Typical transfer characteristics of n-channel enhancement-mode IGFET/MOSFET.

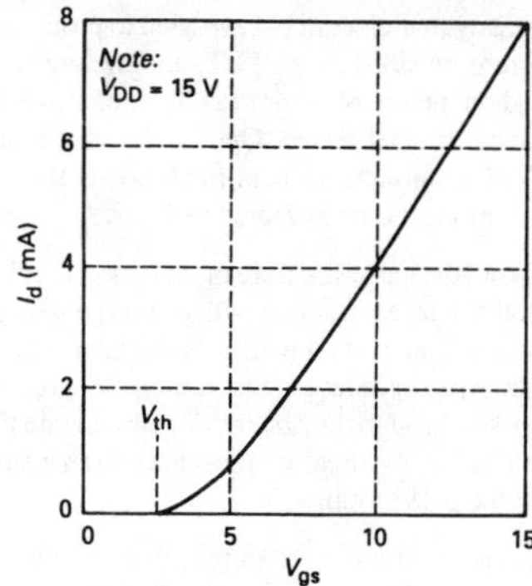

Figure 14.15. Typical V_{GS}/I_d characteristics of n-channel enhancement-mode IGFET/MOSFET.

into two characteristic regions, as indicated by the dotted line, these being the 'triode' region and the 'saturated' region. In the triode region, the device acts like a voltage-controlled resistor; in the saturated region it acts like a voltage-controlled constant-current generator.

The n-channel IGFETs/MOSFETs of *Figures 14.12* and *14.13* can be converted to p-channel devices by simply transposing their p and n materials, in which case their symbols must be changed by reversing the directions of the substrate arrows.

The very high gate impedance of IGFET/MOSFET devices makes them liable to damage from electrostatic discharges, and for this reason they are often provided with internal protection via integral diodes or zeners, as shown in the example of *Figure 14.16*.

VFET devices

In a normal small-signal JFET or IGFET/MOSFET, the main signal current flows 'horizontally' (see *Figures 14.3*, *14.12*, and *14.13*) through the device's conductive channel. This channel is very thin, and maximum operating currents are consequently very limited (typically to maximum values in the range 2 to 40mA).

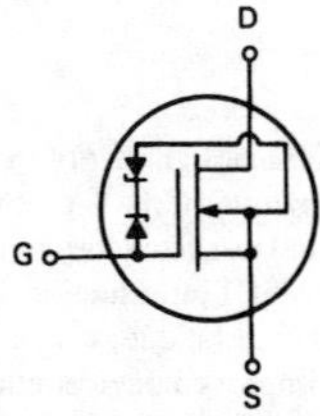

Figure 14.16. Internally-protected n-channel depletion-mode IGFET/MOSFET.

In recent years many manufacturers have tried to produce viable high-power/high-current versions of the FET, and the most successful of these have relied on the use of a 'vertical' flow of current through the conductive channel of the device. One of the best known of these devices is the 'VFET', an enhancement-mode power IGFET/MOSFET which was first introduced by Siliconix in 1976.

Figure 14.17 shows the basic structure of the original Siliconix VFET. It has an essentially 4-layer structure with an n-type source layer at the top, followed by a p-type 'body' layer, an epitaxial n-type layer, and (at the bottom) an n-type drain layer. Note that a 'V' groove (hence the 'VFET' title) passes through the first two layers and into the third layer of the device and is electrostatically connected (via an insulating silicon dioxide film) to the gate terminal.

If the gate is shorted to the source, and the drain is made positive, no drain-to-source current flows, because the diode formed by the p and n materials is reverse biased. But if the gate is made positive to the source the resulting electrostatic field converts the area of p-type material adjacent to the gate into n-type material, thus creating a conduction channel in the position shown in *Figure 14.17* and enabling current to flow vertically from the drain to the source. As the gate becomes more positive, the channel width increases, enabling the drain-to-source current to increase as the drain-to-source resistance decreases. This basic VFET can thus pass high currents (typically up to 2 amps) without creating excessive current density within its channel regions.

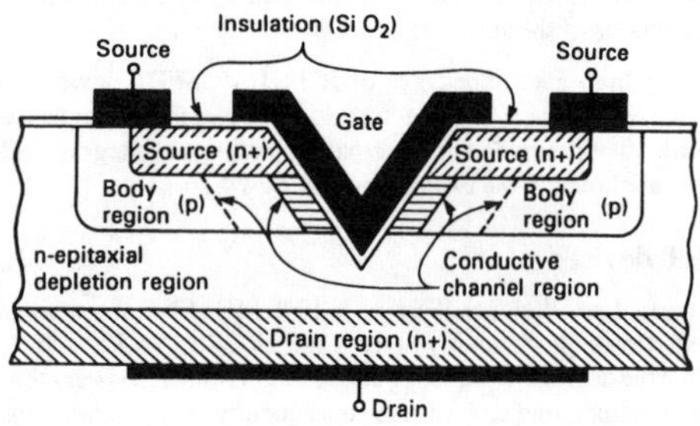

Figure 14.17. Basic structure of the VFET power device.

The original Siliconix VFET design of *Figure 14.17* was successful, but imperfect. The sharp bottom of its V-groove caused an excessive electric field at this point and restricted the device's operating voltage. Subsequent to the original VFET introduction, Intersil introduced their own version of the 'VMOS' technique, with a U-shaped groove (plus other modifications) that improved device reliability and gave higher maximum operating voltages. In 1980, Siliconix added these and other modifications to their own VFET devices, resulting in further improvements in performance.

Other power FETs

Several manufacturers have produced viable power FETs without using the 'V'- or 'U'-groove technique, but still relying on the vertical flow of current between drain and source. Hitachi produce both p-channel and n-channel power MOSFET devices with ratings up to 8A and 200V, but these are suitable for use only in audio and low-RF applications.

Supertex of California and Ferranti of England produce a range of power MOSFETs with the general title of 'vertical DMOS'. These feature high operating voltages (up to 650V), high current rating (up to 16A), low on resistance (down to 50 milliohms) and very fast operating speeds (up to 2GHz at 1A, 500MHz at 10A). Siemens of West Germany use a modified version of DMOS, known as SIPMOS, to produce a range of n-channel devices with voltage ratings as high as 1kV and with current ratings as high as 30A.

The International Rectifier solution to the power FET problem is a device which, in effect, houses a vast array of parallel-connected low-power vertical MOSFETs or 'cells' which share the total current equally between them and thus act like a single power FET, as indicated in *Figure 14.18*. These devices are named HEXFET, after the hexagonal structure of these cells, which have a density of about 100000 per square centimetre of semiconductor material.

In parallel-connected FETs (as in the HEXFET), equal current sharing is ensured by the conduction channel's positive temperature coefficient; if the current of one FET becomes excessive, the resultant heating of its channel raises its resistance, thus reducing its current flow and tending to equalise it with the other parallel-connected FETs. This feature makes power FETs immune to thermal runaway problems.

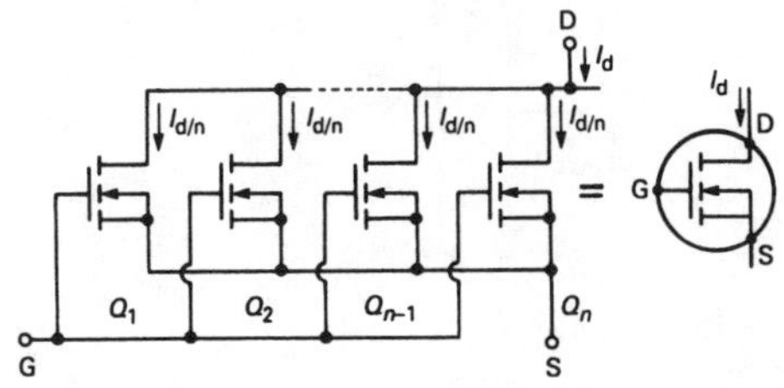

Figure 14.18. The IR HEXFET comprises a balanced matrix of parallel-connected low-power MOSFETs, which are equivalent to a single high-power MOSFET.

CMOS basics

A major FET application is in digital ICs. The best known range of such devices use the technology known as CMOS, and rely on the use of *Complementary* pairs of *MOSFETs*. *Figure 14.19* illustrates basic CMOS principles. The basic CMOS device comprises a p-type and n-type pair of enhancement-mode MOSFETs, wired in series, with their gates shorted together at the input and their drains tied together at the output, as shown in *Figure 14.19(a)*. The pair are meant to use logic-0 or logic-1 digital signals, and *Figures 14.19(b)* and *14.19(c)* respectively show the device's equivalent circuits under these conditions.

When the input is at logic-0, the upper (p-type) MOSFET is biased fully on and acts like a closed switch, and the lower (n-type) MOSFET is biased off and acts like an open switch; the output is thus effectively connected to the positive supply line (logic-1) via a series resistance of about 100Ω. When the input is at logic-1, the MOSFET states are reversed, with Q_1 acting like an open switch and Q_2 acting like a closed switch, so the output is effectively connected to ground (logic-0) via 100Ω. Note in both cases that the entire signal current is fed to the load, and none is shunted off by the CMOS circuitry; this is a major feature of CMOS technology.

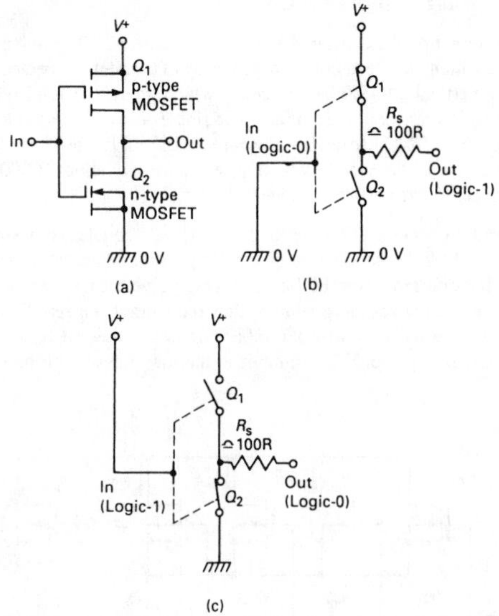

Figure 14.19. *(a)* Basic CMOS circuit, and its equivalent with *(b)* a logic-0 input and *(c)* a logic-1 input.

15 JFET circuits

Chapter 14 explained (amongst other things) the basic operating principles of the JFET. These are low-power devices with a very high input resistance, and invariably operate in the depletion mode, i.e., they pass maximum current when the gate bias is zero, and the current is reduced or 'depleted' by reverse biasing the gate terminal. The two best known real-life JFETs are the 2N3819 n-channel and the 2N3820 p-channel devices, which are usually housed in TO92 plastic packages with the connections shown in *Figure 15.1*. This chapter looks at basic usage information and applications of JFETs, and all the practical circuits shown here are based on the 2N3819; *Figure 15.2* lists the general characteristics of this device.

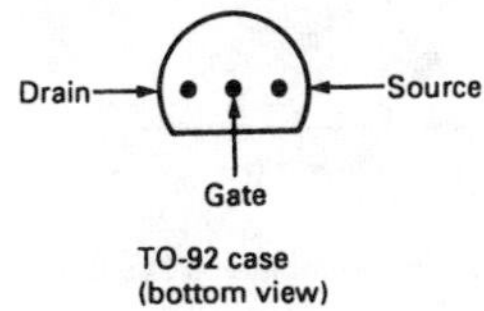

Figure 15.1. Outline and connections of the 2N3919 and 2N3820 JFETs.

V_{ds} = +25 V (= max. drain-to-source voltage)
V_{dg} = +25 V (= max. drain-to-gate voltage)
V_{gs} = −25 V (= max. gate-to-source voltage)
V_p = −8 Vmax (= gate-to-source voltage needed to cut off I_d)
I_{dss} = 2–20 mA (= drain-to-source current with V_{gs} = 0 V)
I_{gss} = −2 nA max. (= gate leakage current at 25°C)
I_g = 10 mA (= max. gate current)
g_m = 2.0 to 6.5 mmho (= small signal transconductance)
C_{iss} = 8 pf max. (= common source input capacitance)
P_T = 200 mW max. (= power dissipation, in free air)
f_T = 100 MHz (= gain-bandwidth product)

Figure 15.2. General characteristics of the 2N3819 n-channel JFET.

JFET biasing

The JFET can be used as a linear amplifier by reverse biasing its gate relative to its source terminal, thus driving it into its linear operating region. Three basic JFET biasing techniques are in common use. The simplest of these is the 'self-biasing' system of *Figure 15.3*, in which the gate is grounded via R_G, and any current flowing in R_S drives the source positive relative to the gate, thus generating reverse bias.

Suppose that an I_D of 1mA is wanted, and that a V_{GS} bias of -2V2 is needed to set this condition; the correct bias can obviously be obtained by giving R_S a value of 2k2; if I_D tends to fall for some reason, V_{GS} naturally falls as well and thus makes I_D increase and counter the original change; the bias is thus self-regulating via negative feedback. In practice, the V_{GS} value needed to set a given I_D varies widely between individual JFETs, and the only sure way of getting a precise I_D value in this system is to make R_S a variable resistor; the system is, however, accurate enough for many applications, and is the most widely used of the three biasing methods.

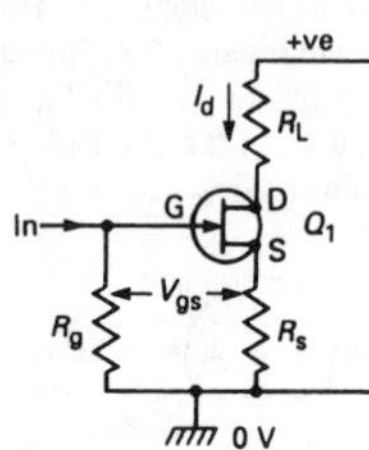

Figure 15.3. Basic JFET self-biasing system.

A more accurate way of biasing the JFET is via the 'offset' system of *Figure 15.4(a)*, in which divider R_1–R_2 applies a fixed positive bias to the gate via R_G, and the source voltage equals this voltage minus V_{GS}. If the gate voltage is large relative to V_{GS}, I_D is set mainly by R_S and is not greatly influenced by V_{GS} variations. This system thus enables I_D values to be set with good accuracy and without need for individual component selection. Similar results can be obtained by grounding the gate and taking the bottom end of R_S to a large negative voltage, as shown in *Figure 15.4(b)*.

The third type of biasing system is shown in *Figure 15.5*, in which constant-current generator Q_2 sets the I_D, irrespective of the JFET characteristics. This system gives excellent biasing stability, but at the expense of increased circuit complexity and cost.

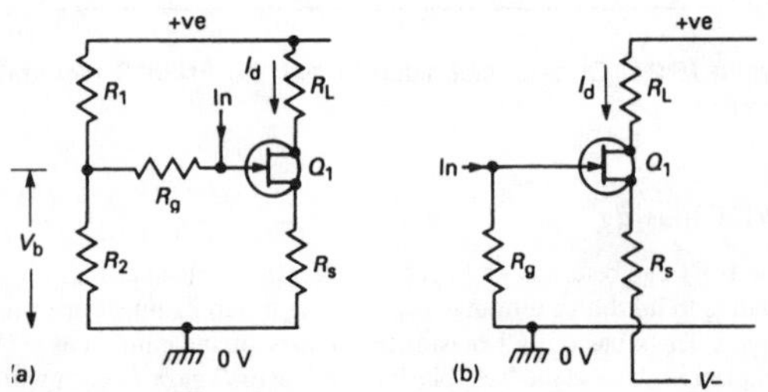

Figure 15.4. Basic JFET offset biasing system.

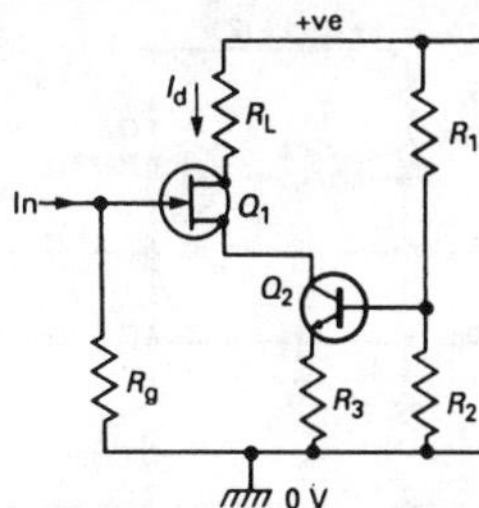

Figure 15.5. Basic JFET constant-current biasing system.

In the three biasing systems described, R_G can have any value up to 10MΩ, the top limit being imposed by the volt drop across R_G caused by gate leakage currents, which may upset the gate bias.

Source follower circuits

When used as linear amplifiers, JFETs are usually used in either the source follower (common-drain) or the common-source modes. The source follower gives a very high input impedance and near-unity voltage gain (hence the alternative title of 'voltage follower'). *Figure 15.6* shows a simple self-biasing (via RV_1) source follower; RV_1 is used to set a quiescent R_2 voltage-drop of 5V6. The circuit's actual input-to-output voltage gain is 0.95. A degree of bootstrapping is applied to R_3 and increases its effective impedance; the circuit's actual input impedance is 10MΩ shunted by 10pF, i.e., it is 10MΩ at very low frequencies, falling to 1MΩ at about 16kHz and 100k at 160kHz, etc.

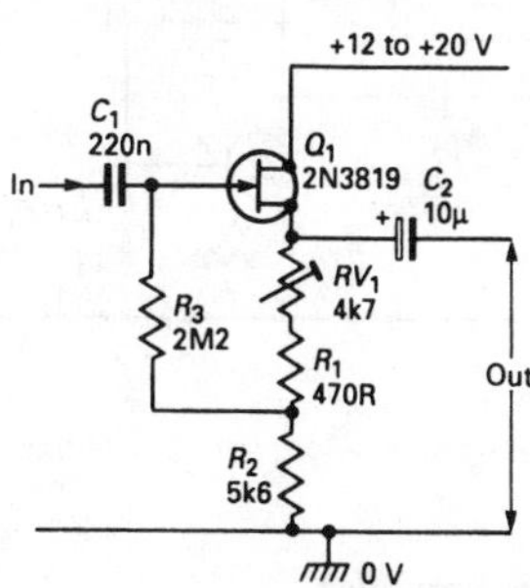

Figure 15.6. Self-biasing source follower. Z_{in} = 10MΩ.

Figure 15.7 shows a source follower with offset gate biasing. Overall voltage gain is about 0.95. C_2 is a bootstrapping capacitor and raises the input impedance to 44MΩ, shunted by 10pF.

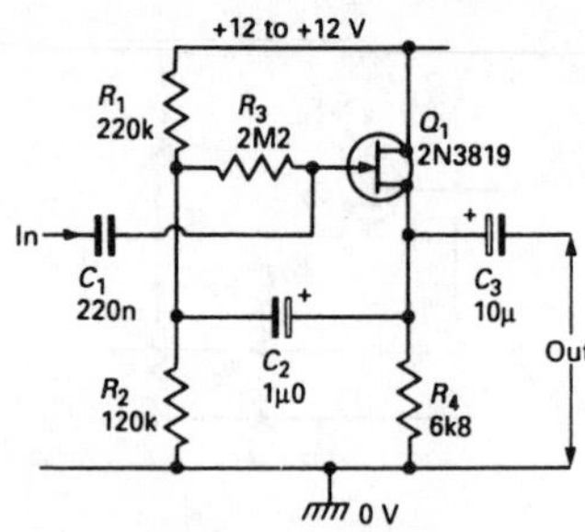

Figure 15.7. Source follower with offset biasing. Z_{in} = 44MΩ.

Figure 15.8 shows a hybrid (JFET plus bipolar) source follower. Offset biasing is applied via R_1–R_2, and constant-current generator Q_2 act as a very-high-impedance source load, giving the circuit an overall voltage gain of 0.99. C_2 bootstraps R_3's effective impedance up to 1000MΩ, which is shunted by the JFET's gate impedance; the input impedance of the complete circuit is 500MΩ, shunted by 10pF. Note that if the high effective value of input impedance of this circuit is to be maintained, the output must either be taken to external loads via an additional emitter follower stage (as shown dotted in the diagram) or must be taken only to fairly high impedance loads.

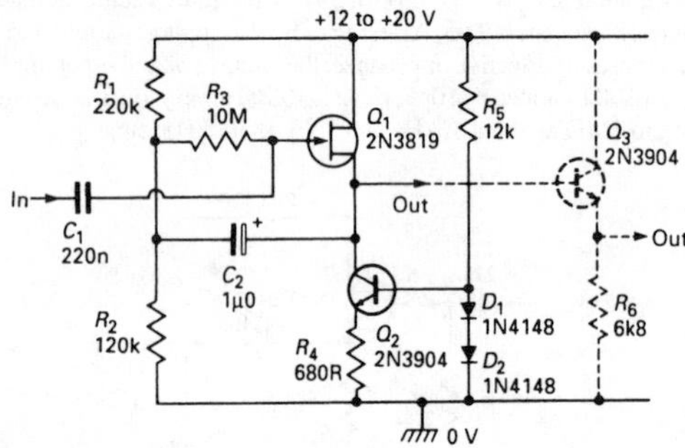

Figure 15.8. Hybrid source follower. Z_{in} = 500MΩ.

Common source amplifiers

Figure 15.9 shows a simple self-biasing common source amplifier; RV_1 is used to set a quiescent 5V6 across R_3. The RV_1–R_2 biasing network is a.c.-decoupled via C_2, and the circuit gives a voltage gain of 21dB (= x12), and has a ±3dB frequency response that spans 15Hz to 250kHz and an input impedance of 2M2 shunted by 50pF (this high shunt value is due to Miller feedback, which multiplies the JFET's effective gate-to-drain capacitance by the circuit's x12 Av value).

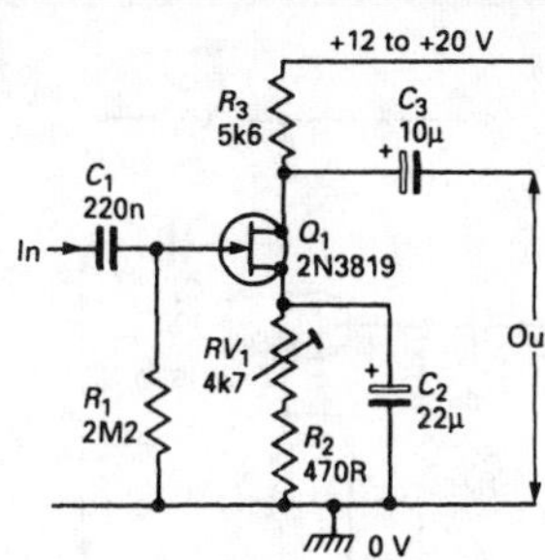

Figure 15.9. Simple self-biasing common-source amplifier.

Figure 15.10 shows a simple self-biasing headphone amplifier that can be used with headphone impedances of 1k0 or greater. It has a built-in volume control (RV_1), has an input impedance of 2M2, and can use any supply in the 9V to 18V range.

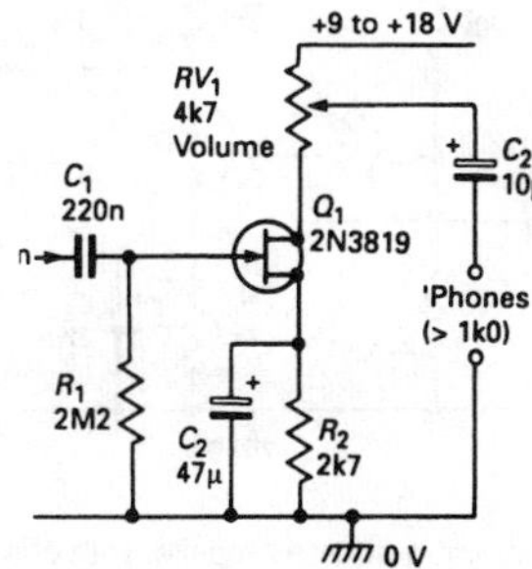

Figure 15.10. Simple headphone amplifier.

Figure 15.11 shows a self-biasing add-on pre-amplifier that gives a voltage gain in excess of 20dB, has a bandwidth that extends beyond 100kHz, and has an input impedance of 2.2MΩ. It can be used with any amplifier that can provide a 9V to 18V power source.

JFET common source amplifiers can, when very high biasing accuracy is needed, be designed using either the 'offset' or 'constant-current' biasing technique. *Figures 15.12* and *15.13* show circuits of these types. Note that the 'offset' circuit of *Figure 15.12* can be used with supplies in the range 16V to 20V only, while the hybrid circuit of *Figure 15.13* can be used with any supply in the 12V to 20V range. Both circuits give a voltage gain of 21dB, a ±3dB bandwidth of 15Hz to 250kHz, and an input impedance of 2M2.

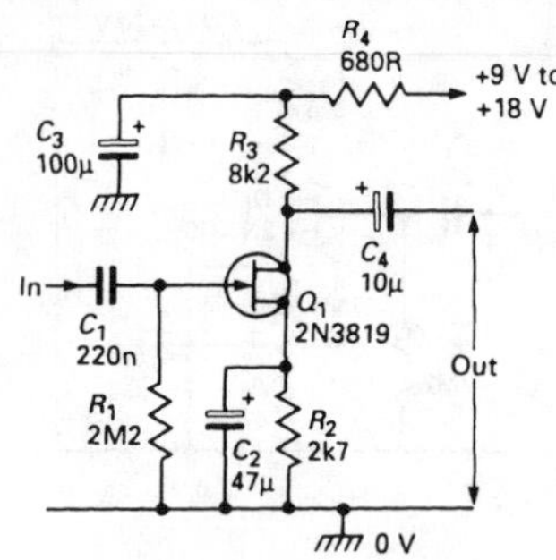

Figure 15.11. General-purpose add-on pre-amplifier.

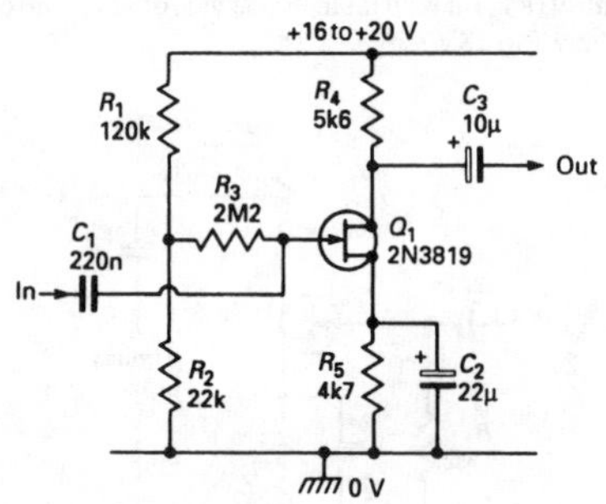

Figure 15.12. Common-source amplifier with offset gate biasing.

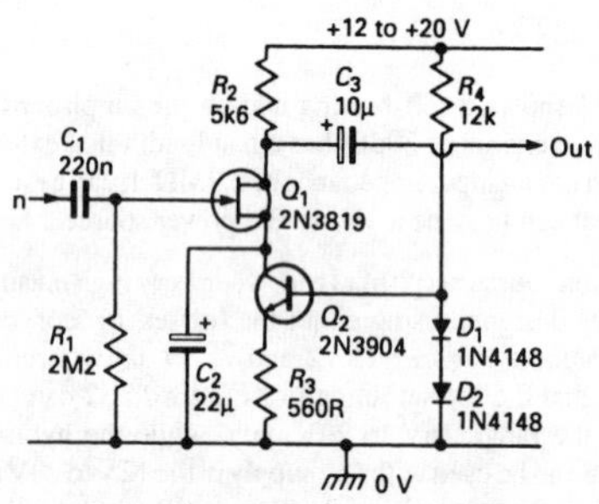

Figure 15.13. Hybrid common-source amplifier.

D.c. voltmeters

Figure 15.14 shows a JFET used to make a simple 3-range d.c. voltmeter with a maximum f.s.d. sensitivity of 0.5V and an input resistance of 11.1MΩ. Here, R_6–RV_2 and R_7 form a potential divider across the 12V supply and, if the R_7–RV_2 junction is used as the circuit's zero-voltage point, sets the top of R_6 at +8V and the bottom of R_7 at -4V. Q_1 is used as a source follower, with its gate grounded via the R_1 to R_4 network and is offset biased by taking its source to -4V via R_5; it consumes about 1mA of drain current.

In *Figure 15.14*, R_6–RV_2 and Q_1–R_5 act as a Wheatstone bridge network, and RV_2 is adjusted so that the bridge is balanced and zero current flows in the meter in the absence of an input voltage at Q_1 gate. Any voltage applied to Q_1 gate then drives the bridge out of balance by a proportional amount, which can be read directly on the meter. R_1 to R_3 form a range multiplier network that, when RV_1 is correctly adjusted, gives f.s.d. ranges of 0.5V, 5.0V, and 50V. R_4 protects Q_1's gate against damage if excessive input voltage is applied to the circuit.

To use the *Figure 15.14* circuit, first trim RV_2 to give zero meter reading in the absence of an input voltage, and then connect an accurate 0.5V d.c. to the input and trim RV_1 to give a precise full-scale meter reading. Repeat these adjustments until consistent zero and full-scale readings are obtained; the unit is then ready for use. In practice, this simple circuit tends to drift with variations in supply voltage and temperature, and fairly frequent trimming of the zero control is needed. Drift can be greatly reduced by using a zener-stabilised 12V supply.

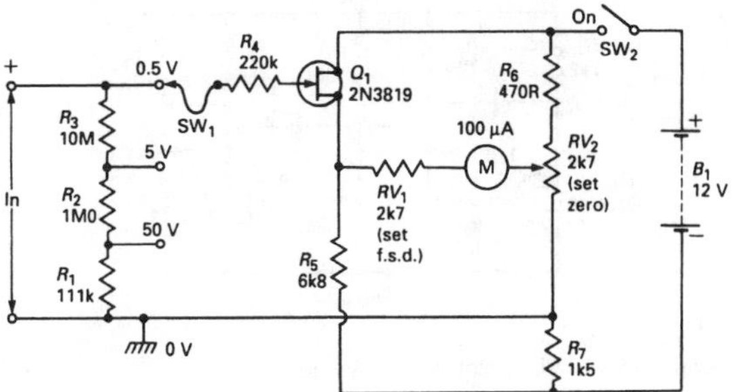

Figure 15.14. Simple three-range d.c. voltmeter.

Figure 15.15 shows a low-drift version of the JFET voltmeter. Q_1 and Q_2 are wired as a differential amplifier, so any drift occuring on one side of the circuit is automatically countered by a similar drift on the other side, and good stability is obtained. The circuit uses the 'bridge' principle, with Q_2–R_5 forming one side of the bridge and Q_2–R_6 forming the other. Q_1 and Q_2 should ideally be a matched pair of JFETs, with I_{dss} values matched within 10%. The circuit is set up in the same way as that of *Figure 15.14*.

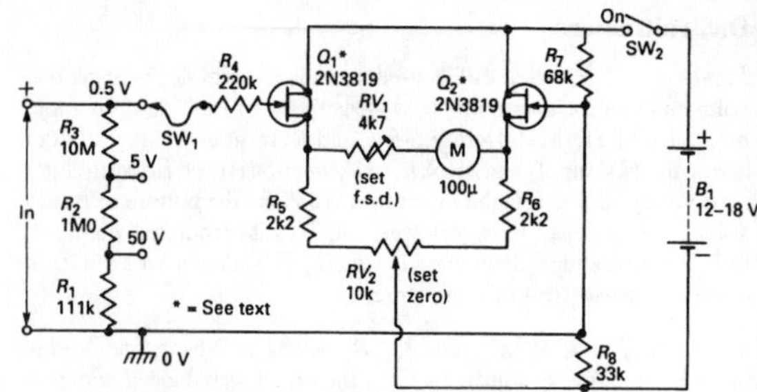

Figure 15.15. Low-drift three-range d.c. voltmeter.

Miscellaneous JFET circuits

To conclude this chapter, *Figures 15.16* to *15.19* show a miscellaneous collection of useful JFET circuits. The *Figure 15.16* design is that of a very-low-frequency (VLF) astable or free-running multivibrator; its on and off periods are controlled by C_1–R_4 and C_2–R_3, and R_3 and R_4 can have values up to 10MΩ. With the values shown, the circuit cycles at a rate of once per 20 seconds, i.e., at a frequency of.05Hz; start button S_1 must be held closed for at least one second to initiate the astable action.

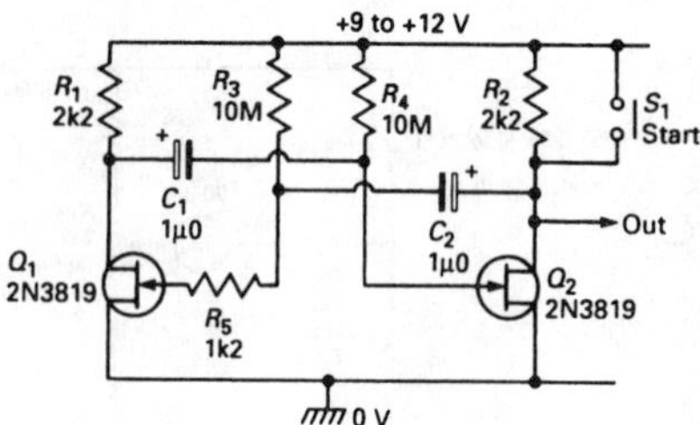

Figure 15.16. VLF astable multivibrator.

Figure 15.17 shows, in basic form, how a JFET and a 741 op-amp can be used to make a voltage-controlled amplifier/attenuator. The op-amp is used in the inverting mode, with its voltage gain set by the R_2/R_3 ratio, and R_1 and the JFET are used as a voltage controlled input attenuator. When a large negative control voltage is fed to Q_1 gate the JFET acts like a near-infinite resistance and causes zero signal attenuation, so the circuit gives high overall gain, but when the gate bias is zero the JFET acts like a low resistance and causes heavy signal attenuation, so the circuit gives an overall signal loss. Intermediate values of signal attenuation and overall gain or loss can be obtained by varying the control voltage value.

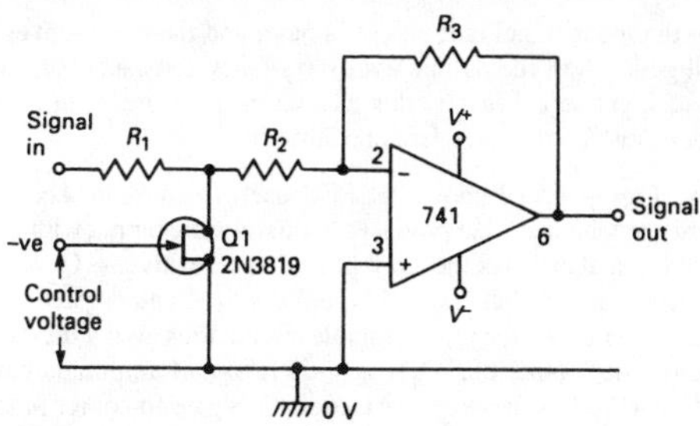

Figure 15.17. Voltage-controlled amplifier/attenuator.

Figure 15.18 shows how this voltage-controlled attenuator technique can be used to make a 'constant volume' amplifier that produces an output signal level change of only 7.5dB when the input signal level is varied over a 40dB range (from 3mV to 300mV r.m.s.). The circuit can accept input signal levels up to a maximum of 500mV r.m.s. Q_1 and R_4 are wired in series to form a voltage-controlled attenuator that controls the input signal level to common emitter amplifier Q_2, which has its output buffered via emitter follower Q_3. Q_3's output is used to generate (via C_5–R_9–D_1–D_2–C_4–R_5) a d.c. control voltage that is fed back to Q_1's gate, thus forming a d.c. negative-feedback loop that automatically adjusts the overall voltage gain so that the output signal level tends to remain constant as the input signal level is varied, as follows.

When a very small input signal is applied to the circuit, Q_3's output signal is also small, so negligible d.c. control voltage is fed to Q_1's gate; Q_1 thus acts as a low resistance under this condition, so almost the full input signal is applied to Q_2 base, and the circuit gives high overall gain. When a large input signal is applied to the circuit, Q_3's output signal tends to be large, so a large d.c. negative control voltage is fed to Q_1's gate; Q_1 thus acts as high resistance under this condition, so only a small

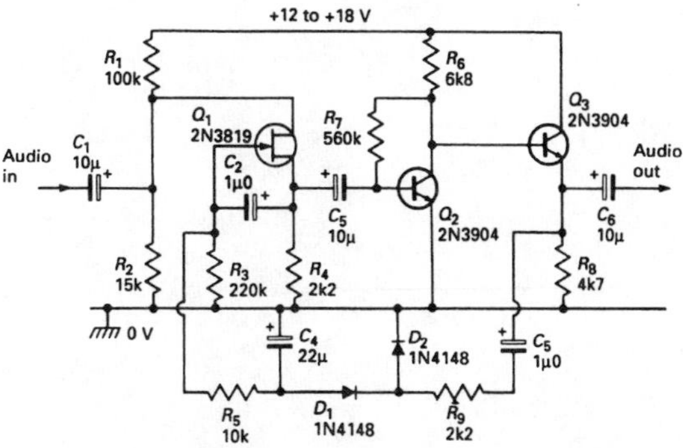

Figure 15.18. Constant-volume amplifier.

part of the input signal is fed to Q_2's base, and the circuit gives low overall gain. Thus, the output level stays fairly constant over a wide range of input signal levels; this characteristic is useful in cassette recorders, intercoms, and telephone amplifiers, etc.

Finally, *Figure 15.19* shows a JFET used to make a d.c.-to-a.c. converter or 'chopper' that produces a square-wave output with a peak amplitude equal to that of the d.c. input voltage. In this case Q_1 acts like an electronic switch that is wired in series with R_1 and is gated on and off at a 1kHz rate via the Q_2–Q_3 astable circuit, thus giving the d.c.-to-a.c. conversion. Note that Q_1's gate-drive signal amplitude can be varied via RV_1; if too large a drive is used, Q_1's gate-to-source junction starts to avalanche, causing a small spike voltage to break through the drain and give an output even when no d.c. input is present. To prevent this, connect a d.c. input and then trim RV_1 until the output is just on the verge of decreasing; once set up this way, the circuit can be reliably used to chop voltages as small as a fraction of a millivolt.

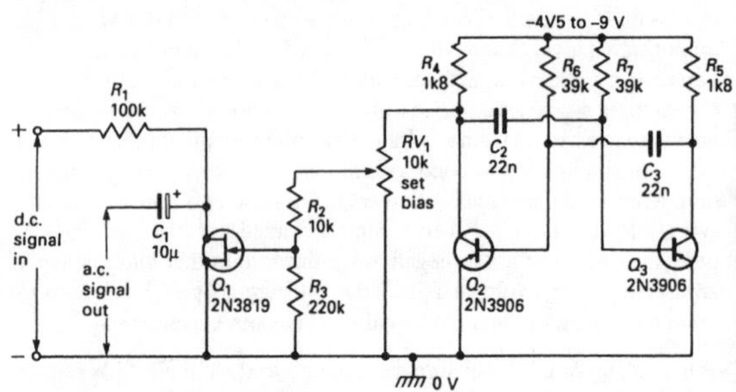

Figure 15.19. D.c.-to-a.c. converter or chopper circuit.

16 MOSFET and CMOS circuits

Chapter 14 explained the basic operating principles of the MOSFET (or IGFET), and pointed out that complementary enhancement-mode pairs of these devices form the basis of the digital technology known as CMOS. The present chapter looks at practical applications of MOSFETs and CMOS-based MOSFET devices.

A MOSFET introduction

MOSFETs are available in both depletion-mode and enhancement-mode versions. Depletion-mode types give a performance similar to a JFET, but with a far higher input impedance. Some depletion mode MOSFETs are equipped with two independent gates, enabling the drain-to-source currents to be controlled via either or both of the gates; these devices are known as dual-gate or tetrode MOSFET; the best known examples of these are the 3N140 and 40673, which use the symbols and TO72 outline shown in *Figure 16.1*.

Figure 16.1. Symbol *(a)* and T072 outline *(b)* of the 3N140 and 40673 dual-gate or tetrode MOSFET.

Most modern MOSFETs are enhancement-mode devices, in which the drain-to-source conduction channel is closed when the gate bias is zero, but can be opened by applying forward gate bias. This 'normally open-circuit' action is implied by the gaps between source and drain in the device's standard symbol, shown in *Figure 16.2(a)*, which depicts an n-channel MOSFET (the arrow head is reversed in a p-channel device). In some devices the semiconductor substrate is made externally available, creating a 'four-terminal' MOSFET, as shown in *Figure 16.2(b)*.

Figure 16.2. Standard symbols of *(a)* three-pin and *(b)* four-pin n-channel enhancement-mode MOSFETs.

Figure 16.3 shows typical transfer characteristics of an n-channel enhancement-mode MOSFET, and *Figure 16.4* shows the V_{GS}/I_D curves of the same device when powered from a 15V supply. Note that no significant I_D current flows until the gate voltage rises to a threshold (V_{TH}) value of a few volts, but that beyond this value the drain current rises in a non-linear fashion. Also note that the graph is divided into two characteristic regions, as indicated by the dotted line, these being the 'triode' region, in which the MOSFET acts like a voltage-controlled resistor, and the 'saturated' region, in which it acts like a voltage-controlled constant-current generator.

Because of their very high input impedances, MOSFETs are vulnerable to damage via electrostatic discharges; for this reason, MOSFETs are sometimes provided with integral protection via diodes or zeners.

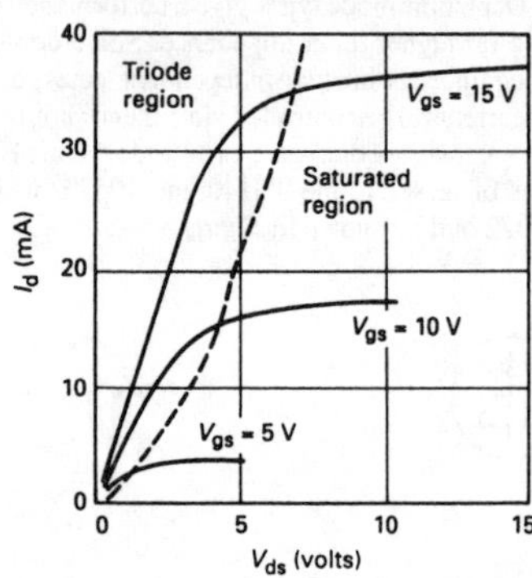

Figure 16.3. Typical transfer characteristics of 4007UB n-channel enhancement-mode MOSFETs.

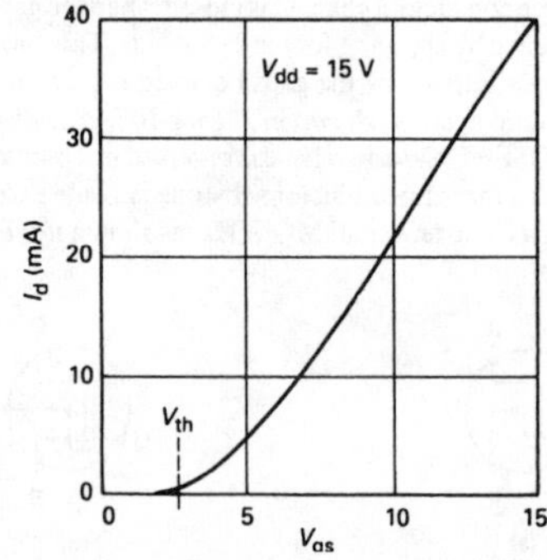

Figure 16.4. Typical V_{GS}/I_D characteristics of 4007UB n-channel enhancement-mode MOSFET.

The 4007UB

The easiest and cheapest practical way of learning about enhancement-mode MOSFETs is via a 4007UB IC, which is the simplest member of the popular CMOS digital IC range and actually houses six useful MOSFETs in a single 14-pin dil package.

Figure 16.5 shows the functional diagram and pin numbers of the 4007UB, which houses two complementary pairs of independently-accessible MOSFETs and a third complementary MOSFET pair that are connected as a standard CMOS inverter stage. Each of the IC's three independent input terminals are internally connected to the standard CMOS protection networks shown in *Figure 16.6*. Within the IC, Q_1, Q_3 and Q_5 are p-channel MOSFETs, and Q_2, Q_4 and Q_6 are n-channel types. Note that the performance graphs of *Figures 16.3* and *16.4* actually apply to the individual n-channel devices within this CMOS IC.

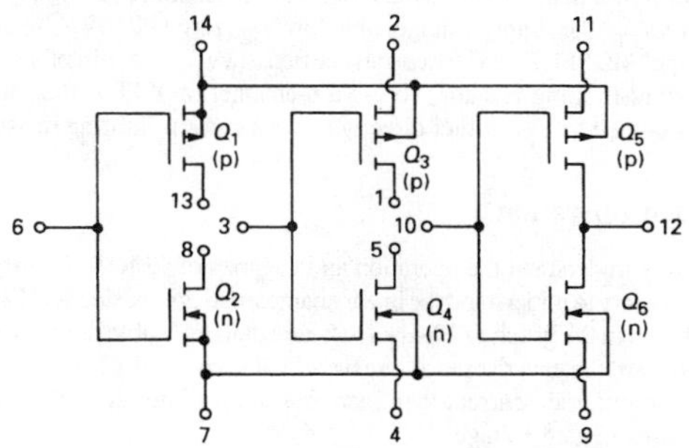

Figure 16.5. Functional diagram of the 4007UB dual CMOS pair plus inverter.

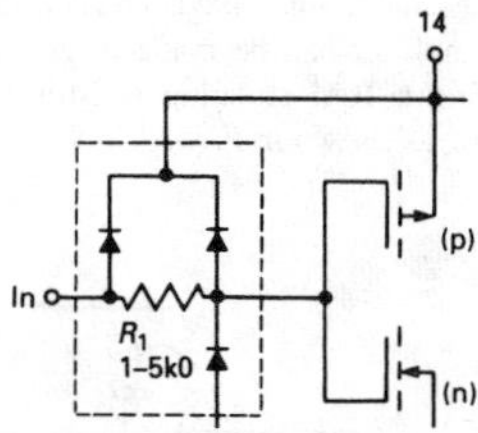

Figure 16.6. Internal input-protection network (within dotted lines) on each input of the 4007UB.

The 4007UB usage rules are simple. In any given application, all unused IC elements must be disabled. Complementary pairs of MOSFETs can be disabled by connecting them as standard CMOS inverters (i.e., gate-to-gate and source-to-source) and tying their inputs to ground, as shown in *Figure 16.7*. Individual MOSFETs can be

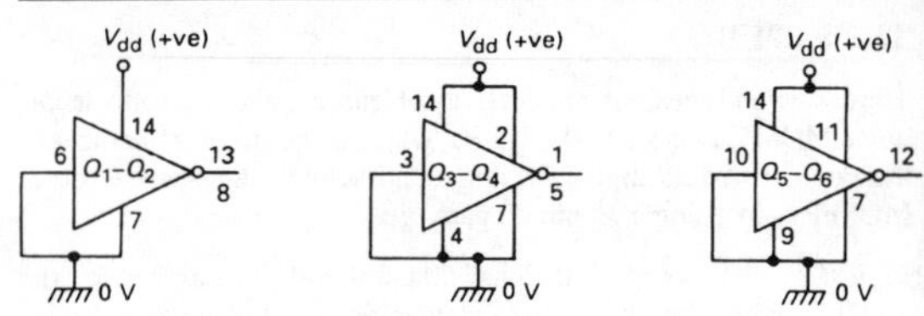

Figure 16.7. Individual 4007UB complementary pairs can be disabled by connecting them as CMOS inverters and grounding their inputs.

disabled by tying their source to their substrate and leaving the drain open circuit. In use, the IC's input terminals must not be allowed to rise above V_{DD} (the supply voltage) or below V_{SS} (zero volts). To use an n-channel MOSFET, the source must be tied to V_{SS}, either directly or via a current-limiting resistor. To use a p-channel MOSFET, the source must be tied to V_{DD}, either directly or via a current-limiting resistor.

Linear operation

To fully understand the operation and vagaries of CMOS circuitry, it is necessary to understand the linear characteristics of basic MOSFETs, as shown in the graph of *Figure 16.4*; note that negligible drain current flows until the gate rises to a 'threshold' value of about 1.5 to 2.5 volts, but that the drain current then increases almost linearly with further increases in gate voltage.

Figure 16.8 shows how to use an n-channel 4007UB MOSFET as a linear inverting amplifier. R_1 acts as Q_2's drain load, and R_2–R_x bias the gate so that Q_2 operates in the linear mode. The R_x value is selected to give the desired quiescent drain voltage, and is normally in the 18k to 100k range. The amplifier can be made to give a very high input impedance by wiring a 10M isolating resistor between the R_2–R_x junction and Q_2 gate, as shown in *Figure 16.9*.

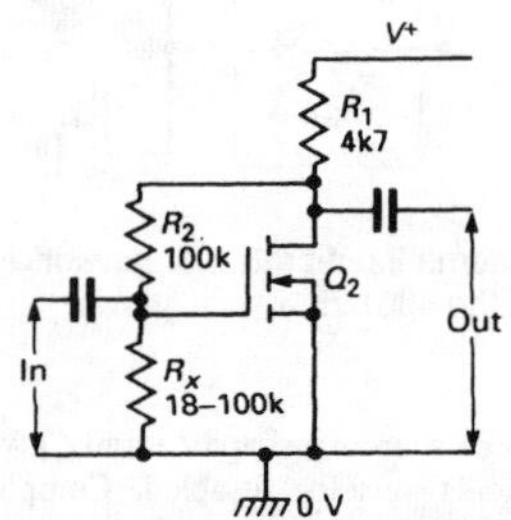

Figure 16.8. Method of biasing n-channel 4007UB MOSFETs for use as a linear inverting amplifier (with medium input impedance).

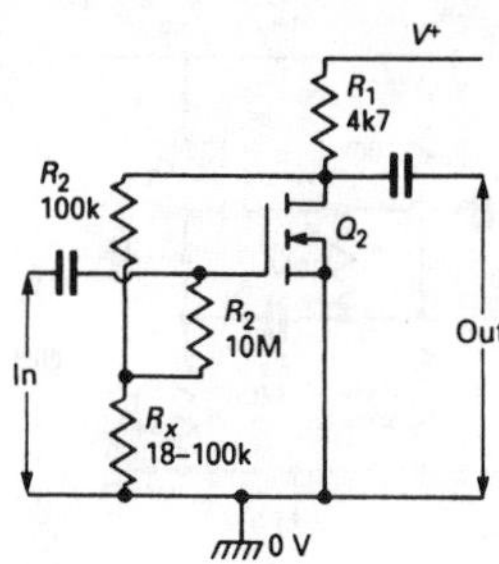

Figure 16.9. High input impedance version of the inverting amplifier.

Figure 16.10 shows how to use an n-channel MOSFET as a unity-gain non-inverting common-drain amplifier or source follower. The MOSFET gate is biased at half-supply volts by the R_2–R_3 divider, and the source terminal automatically takes up a quiescent value that is slightly more than V_{TH} below the gate value. The basic circuit has an input impedance equal to the paralleled values of R_2 and R_3 (=50k), but can be increased to greater than 10M by wiring R_4 as shown. Alternatively, the input impedance can be raised to several hundred megohms by bootstrapping R_4 via C_1 as shown in *Figure 16.11*.

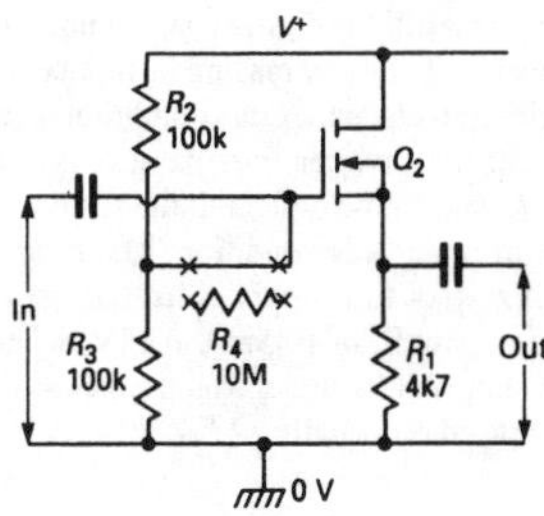

Figure 16.10. Methods of biasing n-channel 4007UB MOSFET as a unity-gain non-inverting amplifier or source follower.

Note from the above descriptions that the enhancement-mode MOSFET performs like a conventional bipolar transistor, except that it has an ultra-high input impedance and has a substantially larger input-offset voltage (the base-to-emitter offset of a bipolar is typically 600mV, while the gate-to-source offset of a MOSFET is typically 2 volts). Allowing for these differences, the enhancement mode MOSFET can thus be used as a direct replacement in many small-signal bipolar transistor circuits.

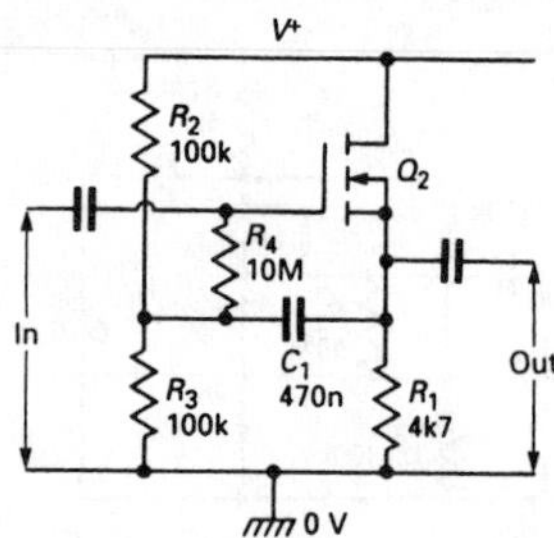

Figure 16.11. Bootstrapped source follower has ultra-high input impedance.

The CMOS inverter

A major application of enhancement-mode MOSFETs is in the basic CMOS inverting stage of *Figure 16.12(a)*, in which an n-channel and a p-channel pair of MOSFETs are wired in series but share common input and output terminals. This basic CMOS circuit is primarily meant for use in digital applications (as described in chapter 14), in which it consumes negligible quiescent current but can source or sink substantial output currents; *Figures 16.12(b)* and *16.12(c)* show the inverter's digital truth table and its circuit symbol. Note that Q_5 and Q_6 of the 4007UB IC are fixed-wired in the CMOS inverter configuration.

Although intended primarily for digital use, the basic CMOS inverter can be used as a linear amplifier by biasing its input to a value between the logic-0 and logic-1 levels; under this condition Q_1 and Q_2 are both biased partly on, and the inverter thus passes significant quiescent current. *Figure 16.13* shows the typical drain-current transfer characteristics of the circuit under this condition. The drain current is zero when the input is at zero or full supply volts, but rises to a maximum value (typically 0.5mA at 5V, or 10.5mA at 15V) when the input is at approximately half-supply volts, under which condition both MOSFETs of the inverter are biased on equally.

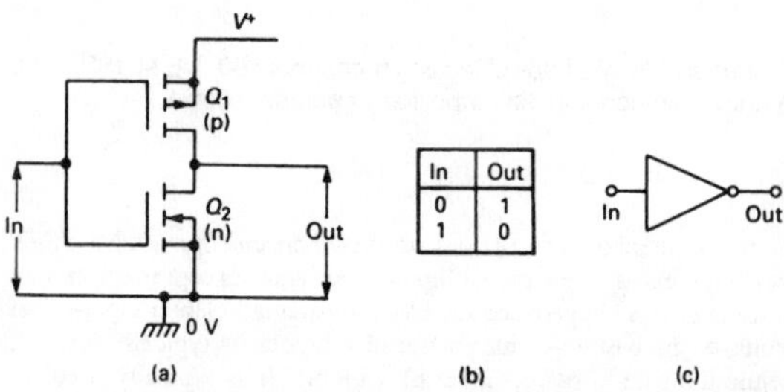

In	Out
0	1
1	0

Figure 16.12. Circuit *(a)*, truth table *(b)* and symbol *(c)* of the basic CMOS digital inverter.

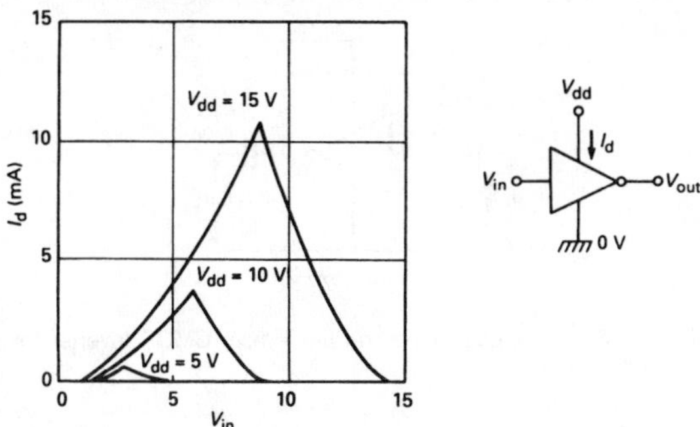

Figure 16.13. Drain-current transfer characteristics of the simple CMOS inverter.

Figure 16.14 shows the typical input-to-output voltage-transfer characteristics of the simple CMOS inverter at different supply voltage values. Note that the output voltage changes by only a small amount when the input voltage is shifted around the V_{DD} and 0V levels, but that when V_{in} is biased at roughly half-supply volts a small change of input voltage causes a large change of output voltage: typically, the inverter gives a voltage gain of about 30dB when used with a 15V supply, or 40dB at 5V.

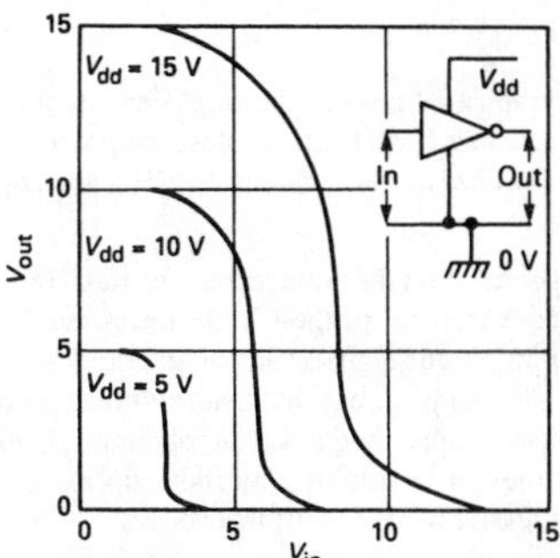

Figure 16.14. Typical input-to-output voltage transfer characteristics of the 4007UB simple CMOS inverter.

Figure 16.15 shows a practical linear CMOS inverting amplifier stage. It is biased by wiring 10M resistor R_1 between the input and output terminals, so that the output self-biases at approximately half-supply volts. *Figure 16.16* shows the typical voltage gain and frequency characteristics of this circuit when operated at three alternative supply

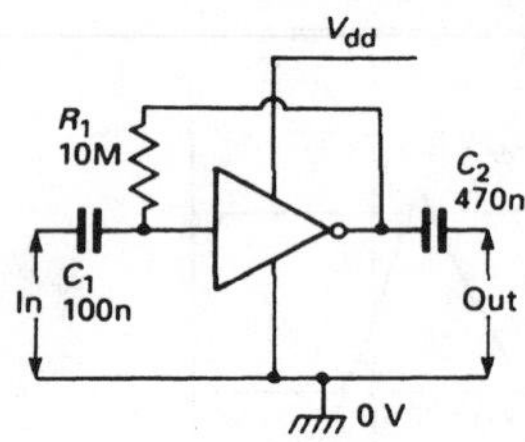

Figure 16.15. Method of biasing the simple CMOS inverter for linear operation.

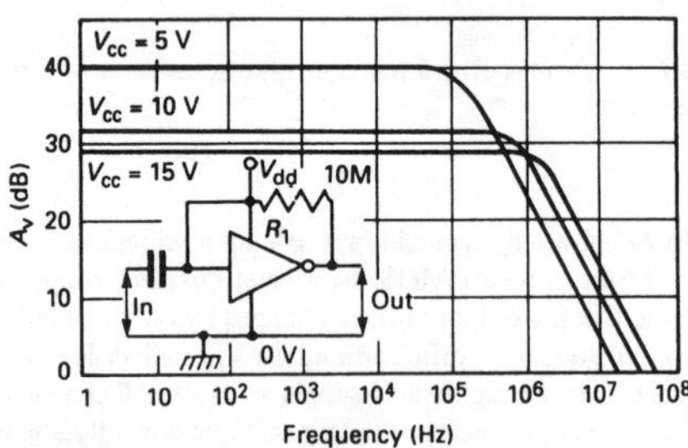

Figure 16.16. Typical A_V and frequency characteristics of the linear-mode basic CMOS amplifier.

rail values; this graph assumes that the amplifier output is feeding into the high impedance of a 10M/15pF oscilloscope probe, and under this condition the circuit has a bandwidth of 2.5MHz when operated from a 15 volt supply.

As would be expected from the voltage transfer graph of *Figure 16.14*, the distortion characteristics of the CMOS linear amplifier are quite good with small-amplitude signals (output amplitudes up to 3V peak-to-peak with a 15V supply), but the distortion then increases as the output approaches the upper and lower supply limits. Unlike a bipolar transistor circuit, the CMOS amplifier does not 'clip' excessive sinewave signals, but progressively rounds off their peaks.

Figure 16.17 shows the typical drain-current versus supply-voltage characteristics of the CMOS linear amplifier. The current typically varies from 0.5mA at 5V, to 12.5mA at 15V.

In many applications, the quiescent supply current of the 4007UB CMOS amplifier can usefully be reduced, at the cost of reduced amplifier bandwidth, by wiring external resistors in series with the source terminals of the two MOSFETs of the CMOS stage, as shown in the 'micropower' circuit of *Figure 16.18*. This diagram also lists the effects that different resistor values have on the drain current, voltage gain, and bandwidth of the amplifier when operated from a 15V supply and with its output loaded by a 10M/15pF oscilloscope probe.

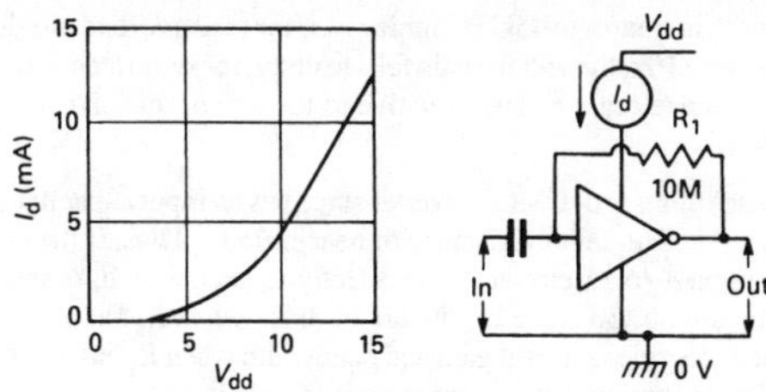

Figure 16.17. Typical I_D/V_{DD} characteristics of the linear-mode CMOS amplifier.

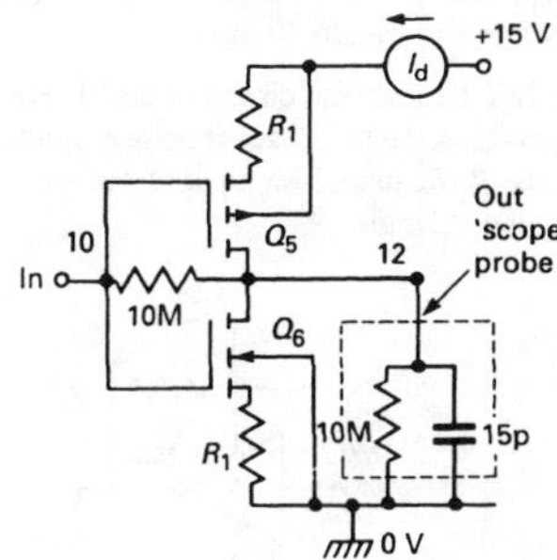

R_1	I_d (mA)	A_v (V_{in}/V_{out})	Upper 3 dB bandwidth
0	12.5 mA	20	2.7 MHz
100R	8.2 mA	20	1.5 MHz
560R	3.9 mA	25	300 kHz
1k0	2.5 mA	30	150 kHz
5k6	600 µA	40	25 kHz
10k	370 µA	40	15 kHz
100k	40 µA	30	2 kHz
1M0	4 µA	10	1 kHz

Figure 16.18. Micropower 4007UB CMOS linear amplifier, showing method of reducing I_D, with performance details.

Note that the additional resistors of the *Figure 16.18* circuit increase the output impedance of the amplifier (the output impedance is roughly equal to the R_1–A_V product), and this impedance and the external load resistance/capacitance has a great effect on the overall gain and bandwidth of the circuit. When using 10k values for R_1, for example, if the load capacitance is increased (from 15pF) to 50pF the bandwidth falls to about 4kHz, but if the capacitance is reduced to 5pF the bandwidth increases to 45kHz. Similarly, if the resistive load is reduced from 10M to 10k, the voltage gain falls to unity; for significant gain, the load resistance must be large relative to the output impedance of the amplifier.

The basic (unbiased) CMOS inverter stage has an input capacitance of about 5pF and an input resistance of near-infinity. Thus, if the output of the *Figure 16.18* circuit is fed directly to such a load, it shows a voltage gain of x30 and a bandwidth of 3kHz when R_1 has a value of 1M0; it even gives a useful gain and bandwidth when R_1 has a value of 10M, but consumes a quiescent current of only 0.4μA.

Practical CMOS

The CMOS linear amplifier can easily be used, in either its standard or micropower forms, to make a variety of fixed-gain amplifiers, mixers, integrators, active filters, and oscillators, etc. A selection of such circuits are shown in *Figures 16.19* to *16.23*.

Figure 16.19 shows the practical circuit of a x10 inverting amplifier. The CMOS stage is biased by feedback resistor R_2, and the voltage gain is set at x10 by the R_1/R_2 ratio. The input impedance of the circuit is 1M0, and equals the R_1 value.

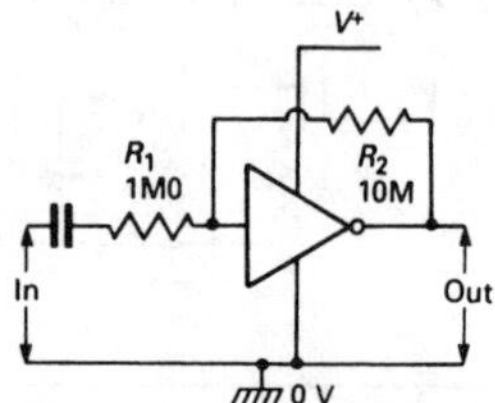

Figure 16.19. Linear CMOS amplifier wired as x10 inverting amplifier.

Figure 16.20 shows the above circuit modified for use as an audio 'mixer' or analogue voltage adder. The circuit has four input terminals, and the voltage gain between each input and the output is fixed at unity by the relative values of the 1M0 input resistor and the 1M0 feedback resistor. *Figure 16.21* shows the basic CMOS amplifier used as a simple integrator.

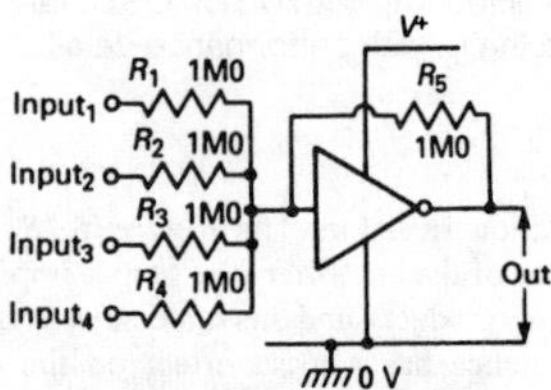

Figure 16.20. Linear CMOS amplifier wired as unity-gain four-input audio mixer.

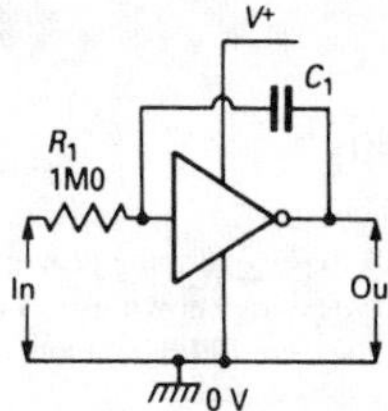

Figure 16.21. Linear CMOS amplifier wired as an integrator.

Figure 16.22 shows the linear CMOS amplifier used as a crystal oscillator. The amplifier is linearly biased via R_1 and provides 180° of phase shift at the crystal resonant frequency, thus enabling the circuit to oscillate. If the user wants the crystal to provide a frequency accuracy within 0.1% or so, R_X can be replaced by a short and C_1–C_2 can be omitted. For ultra-high accuracy, the correct values of R_X–C_1–C_2 must be individually determined (the diagram shows the typical range of values).

Finally, *Figure 16.23* shows a 'micropower' version of the CMOS crystal oscillator. In this case, R_X is actually incorporated in the amplifier. If desired, the output of this oscillator can be fed directly to the input of an additional CMOS inverter stage, for improved waveform shape/amplitude.

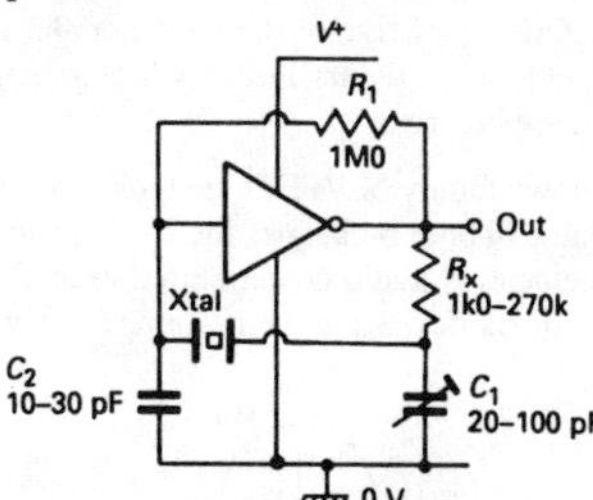

Figure 16.22. Linear CMOS amplifier wired as a crystal oscillator.

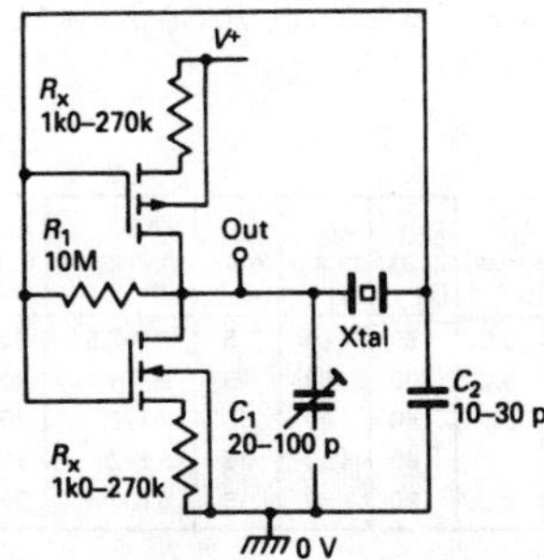

Figure 16.23. Micropower version of the crystal oscillator.

17 VMOS circuits

Chapter 14 explained the basic operating principles of those enhancement-mode power-FET devices known as a VFETs or VMOS. The present chapter concludes the 'FET' theme by showing practical VMOS applications.

A VMOS introduction

A VFET can, for most practical purposes, be simply regarded as a high-power version of a conventional enhancement-mode MOSFET. The specific form of VFET construction shown *Figure 14.7* (see Chapter 14) was pioneered by Siliconix in the mid-1970s, and the devices using this construction are marketed under the trade name of 'VMOS power FETs' (Vertically-structured Metal-Oxide Silicon power Field-Effect Transistors). This 'VMOS' name is normally associated with the V-shaped gate groove formed in the device's structure.

Siliconix VMOS power FETs are probably the best known type of VFETs. They are presently available as n-channel devices only, and usually incorporate an integral zener diode which gives the gate a high degree of protection against accidental damage; *Figure 17.1* shows the standard symbol used to represent such a device, and *Figure 17.2* lists the main characteristics of five of the most popular members of the VMOS family; note in particular the very high maximum operating frequencies of these devices.

Another well known family of VFETs are those produced by Hitachi. These are available in both n-channel and p-channel versions and are useful in complementary audio power amplifier applications. *Figure 17.3* lists details of six devices in their '7 amp, 100 watt' range.

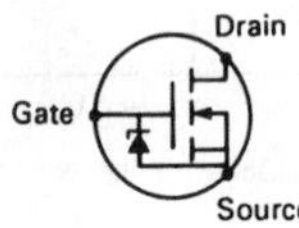

Figure 17.1. Symbol of Siliconix VMOS power FET with integral zener diode gate protection.

Device type no.	P_{tot} (max) (W)	I_d (max) (A)	V_{ds} (max) (V)	V_{dg} (max) (V)	V_{gs} (max) (V)	V_{th} (min–max) (V)	g_m (typ) (m℧)	C_{in} (max) (pF)	f_t (typ) (MHz)
VN10KM	1	0.5	60	60	5	0.3–2.5	200	48	–
VN1010	1	0.5	100	100	15	2 V max	200	48	–
VN46AF	12.5	2	40	40	15	0.8–2	250	50	600
VN66AF	12.5	2	60	60	15	0.8–2	250	50	600
VN88AF	12.5	2	80	80	15	0.8–2	250	50	600

Figure 17.2. Major parameters of five popular n-channel Siliconix VMOS power FETs.

Device type no.	P_{tot} (max) (W)	I_d (max) (A)	V_{ds} (max) (V)	V_{dg} (max) (V)	V_{gs} (max) (V)	V_{th} (min-max) (V)	g_m (typ) (m℧)	f_t (typ) (MHz)	Channel type
2SJ48	100	7	−120	−120	14	−0.8 to −1.5	1000	900	p
2SJ49	100	7	−140	−140	14	−0.8 to −1.5	1000	900	p
2SJ50	100	7	−160	−160	±14	−0.8 to −1.5	1000	900	p
2SK133	100	7	120	120	14	1 to 1.5	1000	600	n
2SK134	100	7	140	140	14	1 to 1.5	1000	600	n
2SK135	100	7	160	160	±14	1 to 1.5	1000	600	n

Figure 17.3. Major parameters of six popular high-power Hitachi VFETs.

The VN66AF

The best way to get to know VMOS is to actually 'play' with it, and the readily available Siliconix VN66AF is ideal for this purpose. It is normally housed in a TO202-style plastic-with-metal-tab package with the outline and pin connections shown in *Figure 17.4*.

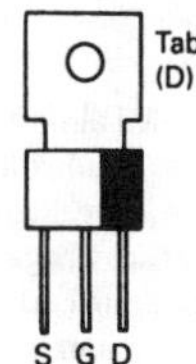

Figure 17.4. Outline and pin connections of the T0202-cased VN66AF VMOS power FET.

Figure 17.5 lists the major static and dynamic characteristics of the VN66AF. Points to note here are that the input (gate-to-source) signal must not exceed the unit's 15V zener rating, and that the device has a typical dynamic input capacitance of 50pF. This capacitance dictates the dynamic input impedance of the VN66AF; the static input impedance is of the order of a million megohms.

Figures 17.6 and *17.7* show the VN66AF's typical output and saturation characteristics. Note the following specific points from these graphs.

(1) The device passes negligible drain current until the gate voltage reaches a threshold value of about 1V; the drain current then increases non-linearly as the gate is varied up to about 4V, at which point the drain current value is about 400mA; the device has a square-law transfer characteristic below 400mA.

(2) The device has a highly linear transfer characteristic above 400mA (4V on the gate) and thus offers good results as a low-distortion class-A power amplifier.

	Parameter		Value
Static	Max drain-source voltage		60 V
	Max drain-gate voltage		60 V
	Max continuous drain current		2 A
	Max pulsed drain current		3 A
	Max continuous forward gate current		2 mA
	Max pulsed forward gate current		100 mA
	Max continuous reverse gate current		100 mA
	Max forward gate-source (zener) voltage		15 V
	Max reverse gate-source voltage		−0.3 V
	Max dissipation at 25°C case temperature		15 W
	Gate threshold voltage		0.8 V min, 1.2 V typical
	Zero-gate-voltage drain current at 25°C		10 μA max
	On-state drain current at $V_{gs}=10$ V		1.0 A min, 2.0 A typical
	Temperature operating and storage range		−40 to + 150°C
Dynamic	Forward transconductance (typical)		250 m℧
	Input capacitance (typical)		50 pF
	Reverse transfer capacitance (typical)		10 pF
	Common-source output capacitance (typical)		50 pF
	Typical switching times, 25 V supply, 23 R load 0–10 V gate drive from A 50R source	Turn-on delay	2 ns
		Rise time	2 ns
		Turn-off delay	2 ns
		Fall time	2 ns

Figure 17.5. Major static and dynamic characteristics of the VN66AF.

(3) The drain current is controlled almost entirely by the gate voltage and is almost independent of the drain voltage so long as the device is not saturated. A point not shown in the diagram is that, for a given value of gate voltage, the drain current has a negative temperature coefficient of about 0.7% per °C, so that the drain current decreases as temperature rises. This characteristic gives a fair degree of protection against thermal runaway.

(4) When the device is saturated (switched fully on) the drain-to-source path acts as an almost pure resistance with a value controlled by the gate voltage. The resistance is typically 2R0 when 10V is on the gate, and 10R when 2V is on the gate. The device's 'off' resistance is in the order of megohms. These features make the device highly suitable for use as a low-distortion high-speed analogue power switch.

Digital circuits

VMOS can be used in a wide variety of digital and analogue applications. It is delightfully easy to use in digital switching and amplifying applications; *Figure 17.8* shows the basic connections. The load is wired between the drain and the positive supply rail, and the digital input signal is fed directly to the gate terminal. Switch-off occurs when the input goes below the gate threshold value (typically about 1.2V). The drain ON current is determined by the peak amplitude of the gate signal, as shown in *Figure 17.6*, unless saturation occurs. In most digital applications the ON current should be chosen to ensure saturation.

The static input impedance of VMOS is virtually infinite, so zero drive power is needed to maintain the VN66AF in the ON or OFF state. Drive power is, however, needed to switch the device from one state to the other; this power is absorbed in charging or discharging the 50pF input capacitance of the VN66AF.

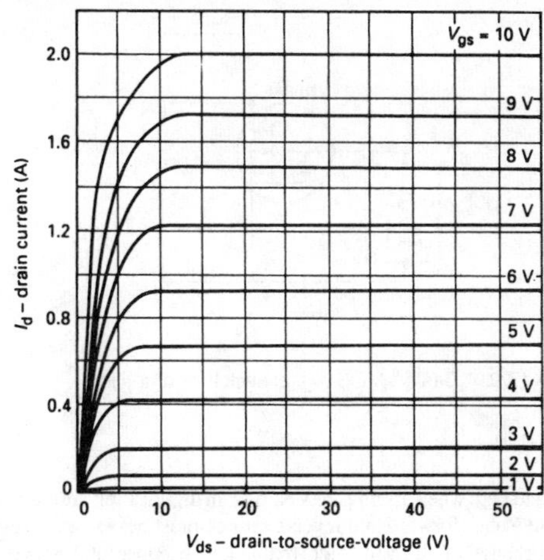

Figure 17.6. Typical output characteristics of the VN66AF.

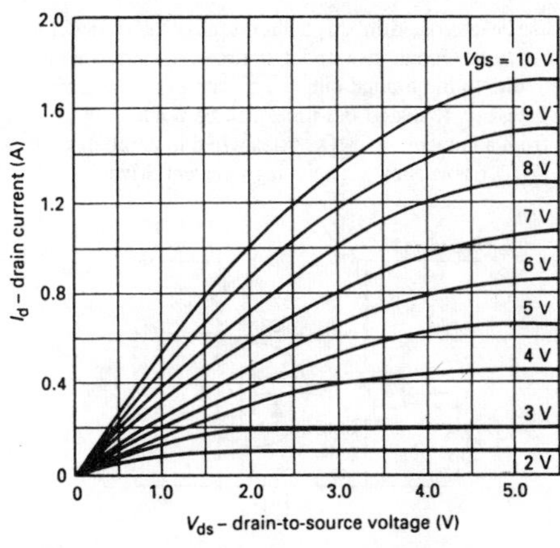

Figure 17.7. Typical saturation characteristics of the VN66AF.

The rise and fall times of the output of the *Figure 17.8* circuit are (assuming zero input rise and fall times) determined by the source impedance of the input signal, by the input capacitance and forward transconductance of the VMOS device, and by the value of R_L. If R_L is large compared to R_S, the VN66AF gives rise and fall times of roughly 0.11nS per ohm of R_S value. Thus, a 100R source impedance gives a 11nS rise or fall time. If R_L is not large compared to R_S, these times may be considerably changed.

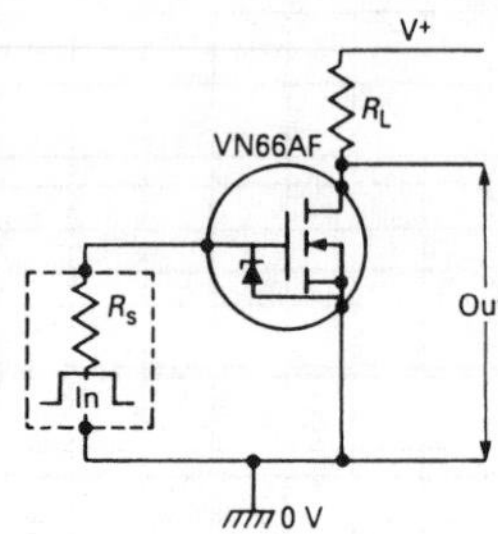

Figure 17.8. Basic VMOS digital switch or amplifier.

A point to note when driving the VN66AF in digital applications is that its input Zener forward and reverse ratings must never be exceeded. Also, because of the very high frequency response of VMOS, the device is prone to unwanted oscillations if its circuitry is poorly designed. Gate leads should be kept short, or be protected with a ferrite bead or a small resistor in series with the gate.

VMOS can be interfaced directly to the output of a CMOS IC, as shown in *Figure 17.9.* Output rise and fall times of about 60nS can be expected, due to the limited output currents available from a single CMOS gate, etc. Rise and fall times can be reduced by driving the VMOS from a number of CMOS gates wired in parallel, as shown in *Figure 17.10*, or by using a special high-current driver.

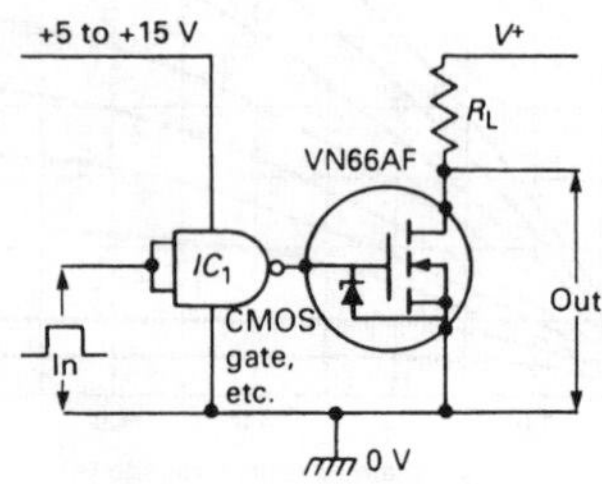

Figure 17.9. Methods of driving VMOS from CMOS.

VMOS can be interfaced to the output of TTL (either standard or LS types) by using a pull-up resistor on the TTL output, as shown in *Figure 17.11*. The 5 volt TTL output of this circuit is sufficient to drive 600mA through a single VN66AF. Higher output currents can be obtained either by wiring a level-shifter stage between the TTL output and the VMOS input, or by wiring a number of VMOS devices in parallel, as shown in *Figure 17.12*.

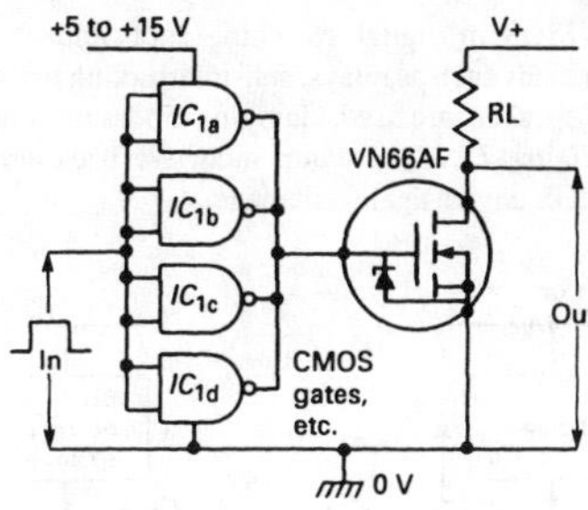

Figure 17.10. VMOS switching rise and fall times can be reduced by driving it from parallel-connected gates, etc. This circuit gives typical rise and fall times of 25nS.

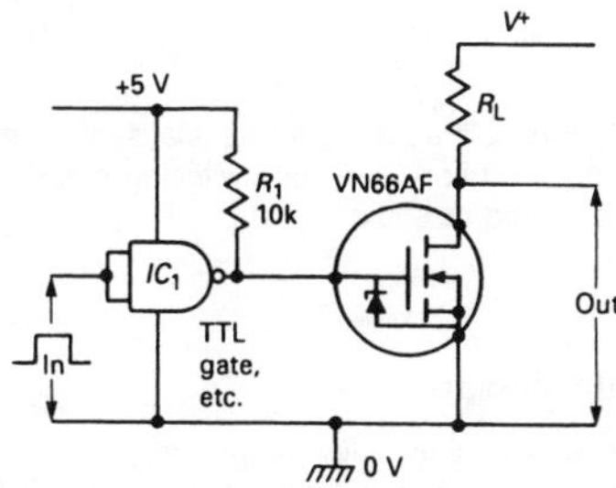

Figure 17.11. Method of driving VMOS from TTL.

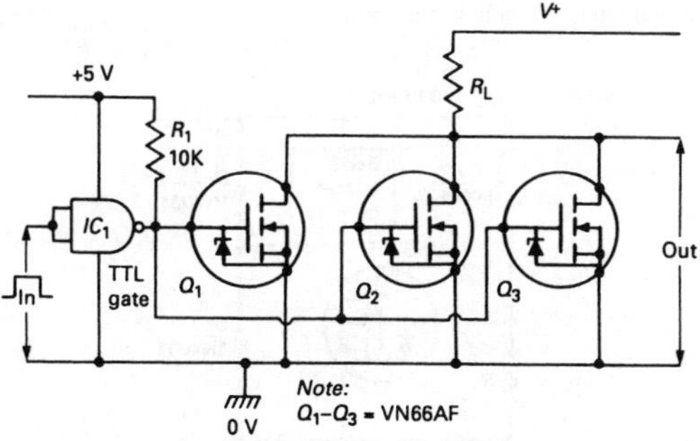

Figure 17.12. Method of boosting the output of *Figure 17.11* by driving three VN66AFs in parallel.

When using VMOS in digital switching applications, note that if inductive drain loads such as relays, self-interrupting bells or buzzers, or moving-coil speakers are used, clamping diodes must be connected as shown in *Figure 17.13*, to damp inductive back-emfs and thus protect the VMOS device against damage.

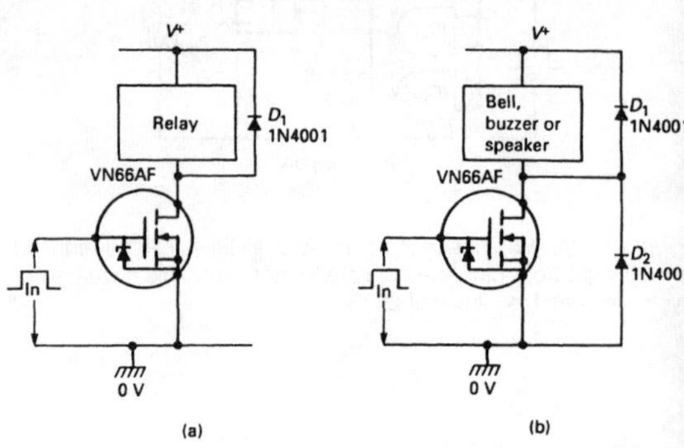

Figure 17.13. If inductive loads such as relays *(a)* or bells, buzzers or speakers *(b)* are used in digital switching circuits, protection diodes must be wired as shown.

Simple digital designs

Figures 17.14 to *17.19* show a few simple but useful digital applications of the VN66AF. The touch-activated power switch of *Figure 17.14* could not be simpler: when the 'contacts' are open, zero volts are on the gate of the VN66AF, so the device passes zero current. When a resistance (zero to tens of megohms) is placed across the contacts (by contact with skin resistance), a substantial gate voltage is developed by potential divider action and the VN66AF passes a high drain current, thus activating the bell, buzzer or relay.

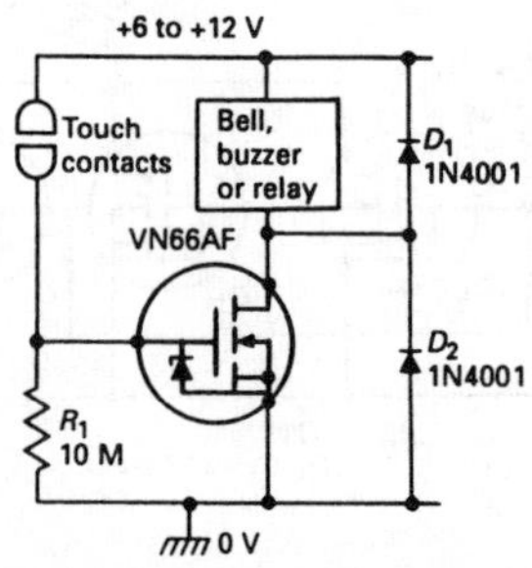

Figure 17.14. Touch-activated power switch.

The *Figure 17.15* circuit is similar to the above, but has two sets of touch contacts and gives a semi-latching relay action. When the ON contacts are touched, C_1 charges via the skin resistance and turns the relay on. The resulting C_1 charge then holds the relay on until the charge either leaks away naturally or is removed by briefly touching the OFF contacts.

The water-activated power switch of *Figure 17.16* is similar to that of *Figure 17.14*, except that it is activated by the relatively low resistance of water coming into contact with the two metal probes, rather than by skin contact.

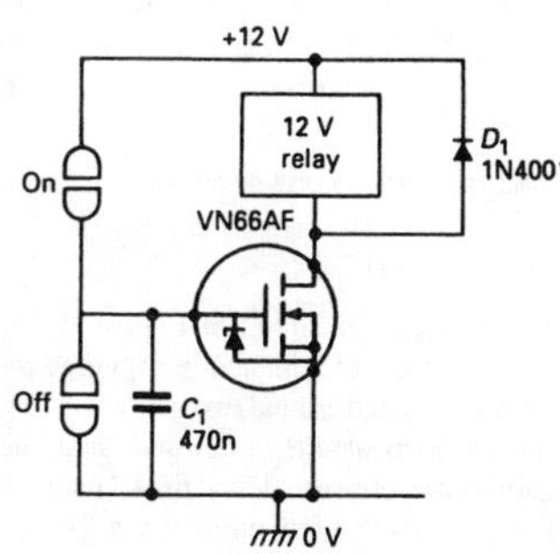

Figure 17.15. Semi-latching touch-activated relay switch.

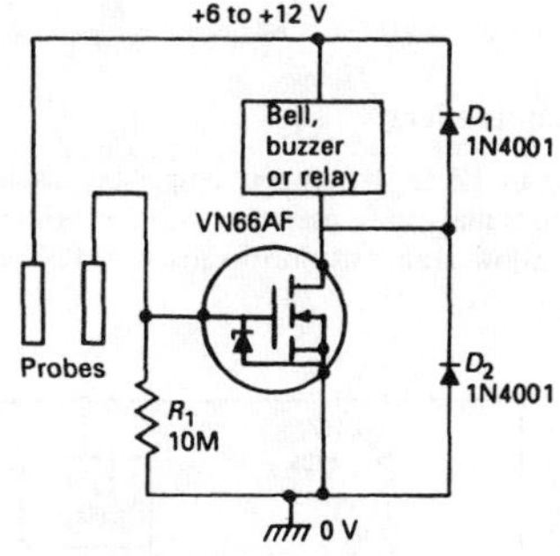

Figure 17.16. Water-activated power switch.

In the manually activated delayed-turn-off circuit of *Figure 17.17*, C_1 charges rapidly via R_1 when push-button switch PB_1 is closed, and discharges slowly via R_2 when PB_1 is open. The load thus activates as soon as PB_1 is closed, but does not deactivate until some tens of seconds after PB_1 is released.

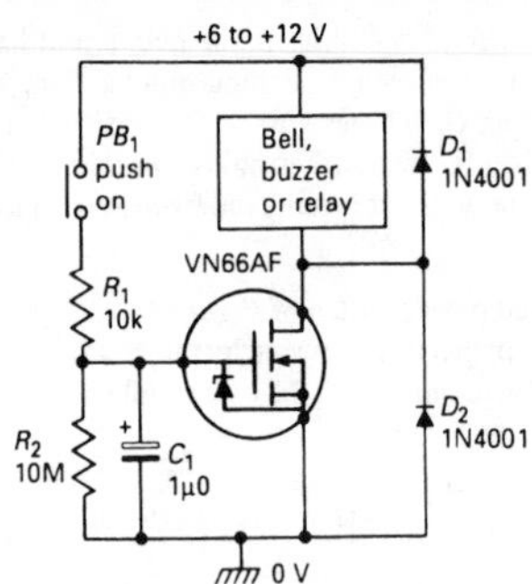

Figure 17.17. Delayed-turn-off power switch.

In the simple relay-output timer circuit of *Figure 17.18*, the VMOS device is driven by the output of a manually triggered monostable or one-shot multivibrator designed around two gates of a 4001 CMOS IC; the relay turns on as soon as PB_1 is closed, and then turns off automatically again some pre-set 'delay time' later. The delay is variable from a few seconds to a few minutes via RV_1.

Finally, *Figure 17.19* shows the practical circuit of an inexpensive but very impressive alarm-call generator that produces a 'dee-dah' sound like that of a British police car siren. The alarm can be turned on by closing PB_1 or by feeding a 'high' voltage to the R_1–R_2 junction. The circuit uses an 8R0 speaker and generates roughly 6 watts of output power.

D.C. lamp controllers

Figures 17.20 to *17.22* show three simple but useful D.C. lamp controller circuits that can be used to control the brilliance of any 12 volt lamp with a power rating of up to 6 watts. A VMOS power FET can,

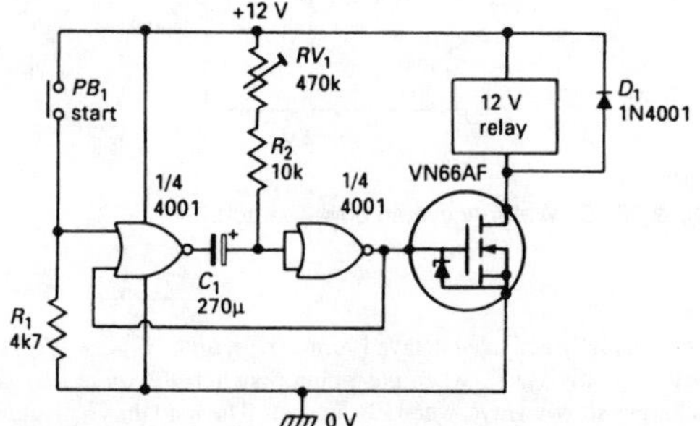

Figure 17.18. Simple relay-output timer circuit.

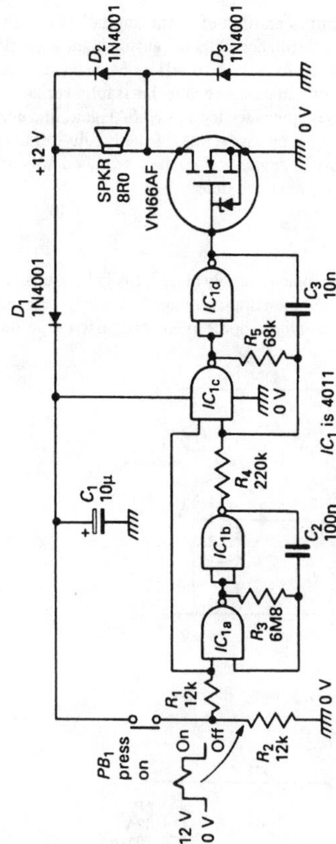

Figure 17.19. Warble-tone 6 watt alarm.

for many purposes, be regarded as a voltage controlled constant-current generator: thus, in *Figure 17.20*, the VMOS drain current (and thus the lamp brightness) is directly controlled by the variable voltage of RV_1's slider. The circuit thus functions as a manual lamp dimmer.

The *Figure 17.21* circuit is a simple modification of the above design, the action being such that the lamp turns on slowly when the switch is closed as C_1 charges up via R_3, and turns off slowly when the switch is opened as C_1 discharges via R_3.

The *Figure 17.22* circuit is an ultra-efficient 'digital' lamp dimmer which controls the lamp brilliance without causing significant power loss across the VMOS device. The two 4011 CMOS gates form an astable multivibrator with a mark/space ratio that is fully variable from 10:1 to 1:10 via RV_1; its output is fed to the VN66AF gate, and enables the mean lamp brightness to be varied from virtually fully-off to fully-on. In this circuit the VMOS device is alternately switched fully on and fully off, so power losses are negligible.

Linear circuits

VMOS power FETs can, when suitably biased, easily be used in either the common source or common drain (voltage follower) linear modes. The voltage gain in the common source mode is equal to the product of

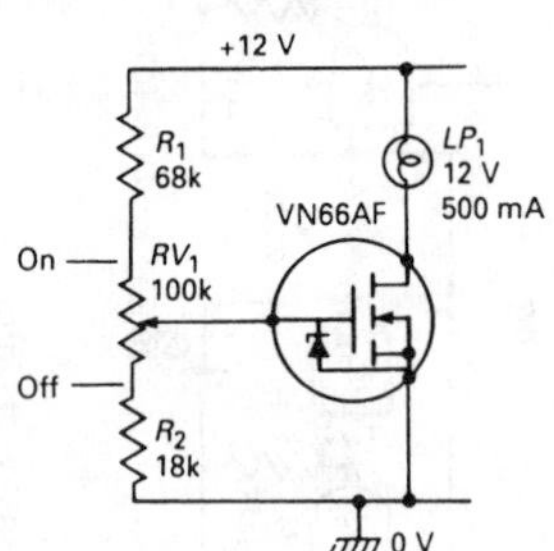

Figure 17.20. Simple D.C. lamp dimmer.

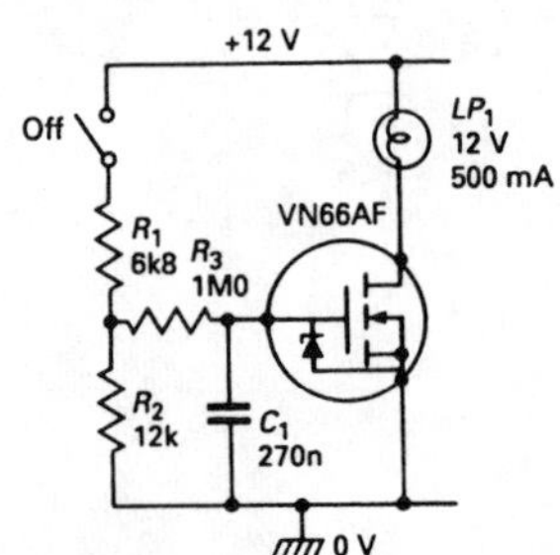

Figure 17.21. Soft-start lamp switch.

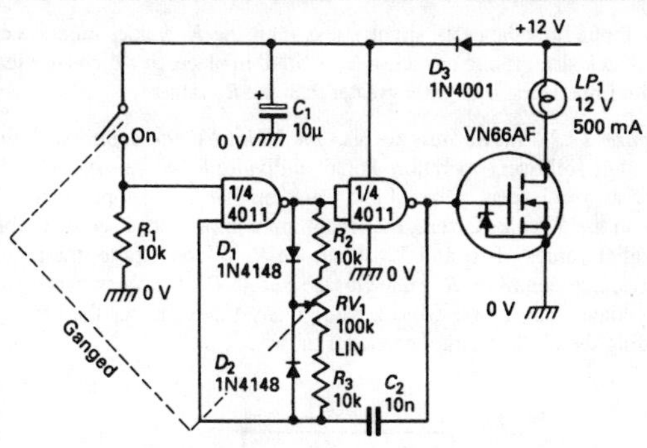

Figure 17.22. High-efficiency D.C. lamp dimmer.

R_L and the device's g_m or forward transconductance. In the case of the VN66AF, this gives a voltage gain of 0.25 per ohm of R_L value, i.e., a gain of x4 with a 16R load, or x25 with a 100R load. The voltage gain in the common drain mode is slightly less than unity.

A VMOS power FET can be biased into the linear common source mode by using the standard enhancement-mode MOSFET biasing technique shown in *Figure 17.23*, in which the R_1–R_2 potential divider is wired in the drain-to-gate negative feedback loop and sets the quiescent drain voltage at roughly half-supply value, so that maximal signal level swings can be accommodated before clipping occurs.

When, in the *Figure 17.23* circuit, R_3 has a value of zero ohms, the circuit exhibits an input impedance that, because of the a.c. negative feedback effects, is roughly equal to the parallel values of R_1 and R_2 divided by the circuit's voltage gain (R_L x g_m). If R_3 has a finite value,

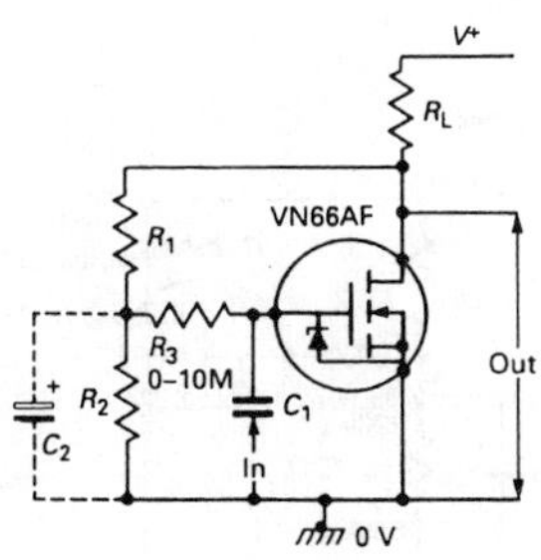

Figure 17.23. Biasing techniques for linear common source operation.

the input impedance is slightly less than the R_3 value, unless a.c. feedback-decoupling capacitor C_2 is fitted in place, in which case the input impedance is slightly greater than the R_3 value.

Figure 17.24 shows how to bias the VN66AF for common drain (voltage follower) operation. Potential divider R_1–R_2 sets the VMOS gate at a quiescent value slightly greater than half-supply voltage. When the R_3 value is zero, the circuit input impedance is equal to the parallel values of R_1 and R_2. When the R_3 value is finite, the input impedance equals the R_3 value plus the parallel R_1–R_2 values. The input impedance can be raised to a value many times greater than R_3 by adding the C_2 'bootstrap' capacitor to the circuit.

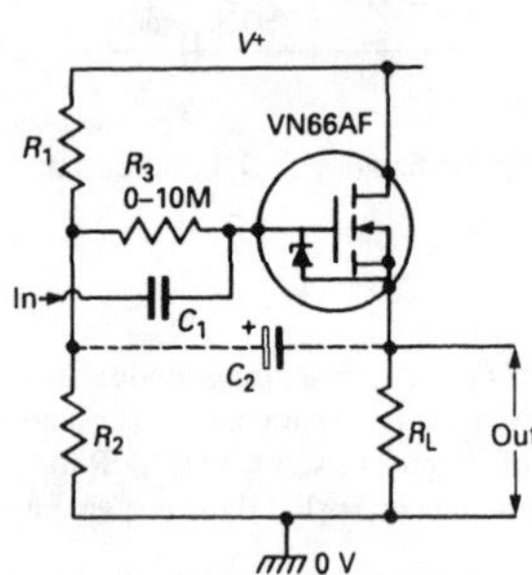

Figure 17.24. Biasing techniques for linear common drain (voltage follower) operation.

Figure 17.25 shows a practical example of a VMOS linear application. The circuit is wired as a class-A power amplifier which, because of the excellent linearity of the VN66AF, gives remarkably little distortion for so simple a design. The VN66AF must be mounted on a good heat sink in this application. When the design is used with a purely resistive 8R0 load, the amplifier bandwidth extends up to 10MHz.

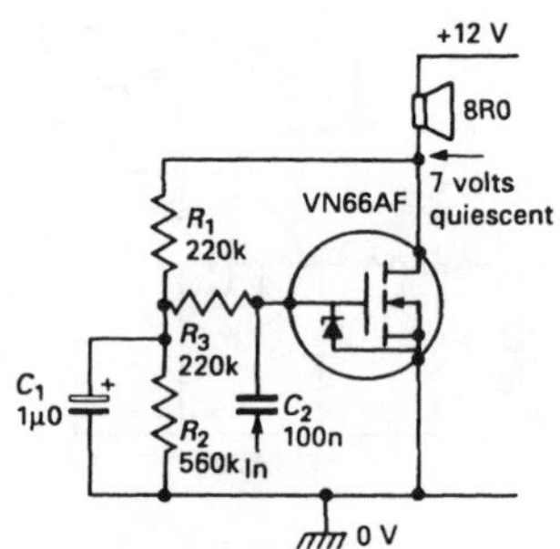

Figure 17.25. Simple class-A audio amplifier gives 1% THD at 1W.

Finally, *Figure 17.26* shows how a VN66AF can be used to make a simple but excellent 600mW radio control or CW transmitter output stage. The L_1–C_2 tank circuit and the L_2–C_3 antenna resonator component values must be chosen to suit the required operating frequency.

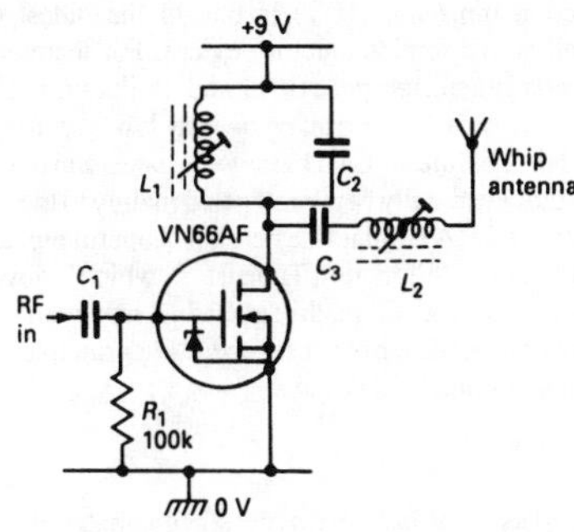

Figure 17.26. 600mW radio control or CW transmitter.

18 Modern UJT circuits

The Unijunction transistor (UJT) is one of the oldest (1952) and simplest of all active semiconductor devices. For many years it was widely used as a general-purpose timer and oscillator, but in the early 1970s many of these tasks were taken over by low-cost ICs such as the 555 timer and the ubiquitous CMOS range of gates, and the UJT slowly fell out of favour, eventually being relegated mainly to the high-energy pulse generating role. A similar fate befell its hopeful replacement, the PUT (Programmable Unijunction Transistor), which is now little used. Both of these devices are still readily available, however, and are quite versatile. This chapter describes their operating principles, and shows how to use them in practical circuits.

UJT basics

The UJT consists of a bar of n-type silicon material with a non-rectifying contact at either end (base 1 and base 2), and with a rectifying contact (emitter) alloyed into the bar part way along its length, to form the only junction within the device (hence the name 'unijunction'). *Figure 18.1* shows the symbol, construction, and equivalent circuit of the UJT. A simple 'inter-base' resistance (that of the silicon bar) appears between base 1 and base 2, and measures the same in either directions; it is given the symbol R_{BB} and has a typical value in the range 4k0 to 12k.

In use, base 2 is connected to a positive voltage and base 1 is taken to zero volts (see *Figure 18.1(c)*), and R_{BB} acts as a voltage divider with a division or 'intrinsic stand-off' ratio (η) that has a typical value

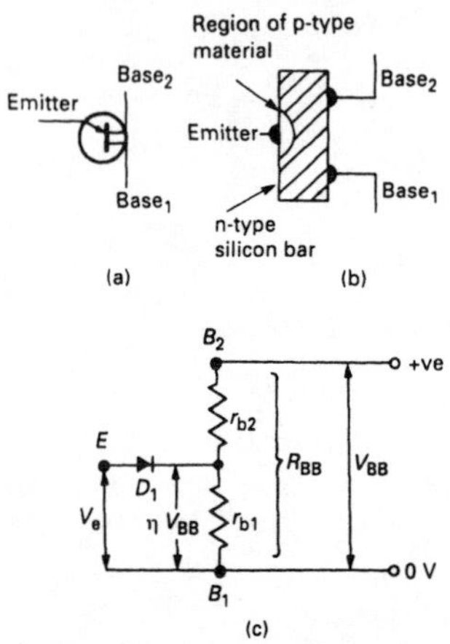

Figure 18.1. *(a)* UJT symbol, *(b)* UJT construction, *(c)* UJT equivalent circuit.

between 0.45 and 0.8. A 'stand-off' voltage of ηV_{BB} thus appears across the lower (r_{B1}) half of the bar under quiescent conditions. The UJT's emitter terminal is connected to this voltage via junction D_1. Normally the emitter input voltage (V_E) is less than ηV_{BB}, so D_1 is reverse biased and the emitter thus has a very high input impedance under this condition.

If V_E is steadily increased above ηV_{BB} a point is reached where D_1 becomes forward biased, and current starts to flow from the emitter to base 1. This current consists mainly of minority carriers injected into the silicon bar, and these drift to base 1 and lower the r_{B1} resistance. This decrease in r_{B1} lowers the ηV_{BB} voltage, so the emitter-to-base 1 current increases and makes the r_{B1} value fall even more. If V_E has a low source impedance, a regenerative action thus takes place, and the emitter input impedance falls sharply, typically to a value of about 20R. The UJT thus acts as a voltage-triggered switch that has a very high input (emitter) impedance when the UJT is off and a very low one when it is on. The precise point at which triggering occurs is called the 'peak-point' voltage, V_P, and is about 600mV above the ηV_{BB} value.

The UJT oscillator

The most basic UJT application is in the simple relaxation oscillator of *Figure 18.2*. Here, C_1 is fully discharged when the supply is initially connected, so the emitter is at ground potential and presents a very high impedance. C_1 immediately starts to charge exponentially towards V_{BB} via R_1, but when the emitter reaches V_P the UJT fires and rapidly discharges C_1 into the low impedance of the emitter; when the emitter current falls below a critical value the UJT switches off, and C_1 then starts to charge up again via R_1, and the whole process repeats *add infinitum*, generating a non-linear sawtooth waveform across C_1.

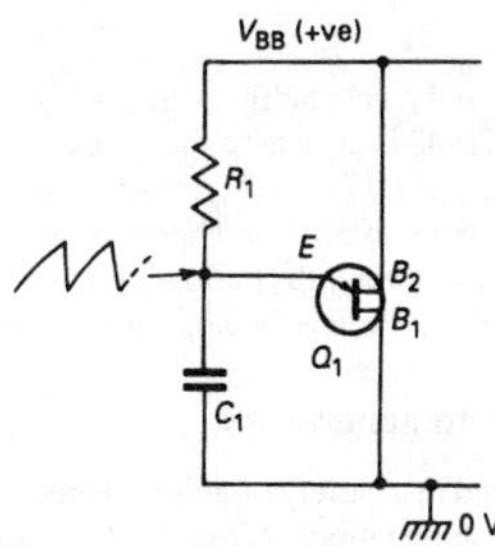

Figure 18.2. Basic UJT relaxation oscillator.

The UJT's switch-off action occurs in each cycle when the total emitter current (the C_1 discharge plus the R_1 current) falls to a 'valley-point' value, I_V (typically 4mA). A minimum 'peak-point emitter current', I_P, is needed to initiate the UJT switch-on action, and typically has a value of 5μA. Thus, R_1's maximum and minimum usable values are limited by the I_P and I_V characteristics.

The oscillation frequency of the *Figure 18.2* circuit is given approximately by $f = 1/(CR)$, and is almost independent of V_{BB} (typically, a 10% change in V_{BB} causes less than 1% change in f). The R_1 value can typically be varied from about 3k0 to 500k, enabling the circuit to span a 100:1 frequency range via a single variable resistor. The C_1 value can vary from a few hundred pF to hundreds of µF, enabling the circuit to be used over a very wide frequency range (from hundreds of kHz to below one cycle per minute).

In most practical UJT oscillators a resistor (R_3) is wired between base 1 and ground, as shown in *Figure 18.3*, either to control C_1's discharge time or (more usually) to give a brief high-energy positive output pulse from C_1's discharge. A resistor (R_2) may also be wired in series with base 2, either to enhance thermal stability or to enable a low-energy negative-going output pulse to be generated via C_1's discharge.

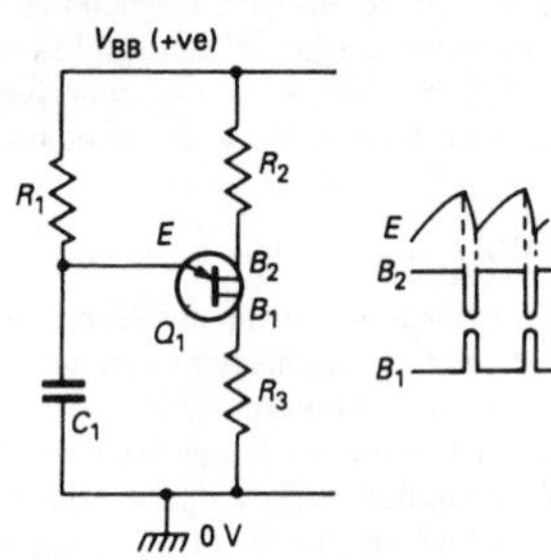

Figure 18.3. Alternative version of the UJT oscillator.

Practical UJTs

The two best known and most readily available UJT's are the 2N2646 and the TIS43. The TIS43 is the least expensive of the two, and is used as the basis of all practical UJT circuits shown in this chapter. It can use supplies up to a maximum of 30V, and has maximum I_P and I_V ratings of 5µA and 4mA respectively, thus allowing a wide range of timing resistor values to be used. *Figure 18.4* lists basic details of both UJTs.

Practical waveform generators

The TIS43 can be used in a variety of pulse, sawtooth, and rectangular waveform generator applications. *Figures 18.5* to *18.9* show a selection of practical circuits of these types.

Figure 18.5 is a wide-range pulse generator circuit, and generates a high-energy positive pulse across R_3 and a low-energy negative-going one across R_2. Both pulses are of similar form, but are in anti-phase. With the component values shown the pulse width is constant at about 30µS over the frequency range 25Hz to 3kHz (adjustable via RV_1). The pulse width and frequency range can be altered by changing the C_1 value; reducing it by a decade reduces the pulse width and raises the operating frequency by a factor of 10; C_1 can have any value from 100pF to 1000µF.

Parameter	2N2646	TIS43
Emitter reverse volts (max)	30 V	30 V
V_{BB} (max)	35 V	35 V
Peak emitter current (max)	2 A	1.5 A
Power dissipation (max)	300 mW	300 mW
Intrinsic stand-off ratio, η	0.56–0.75	0.55–0.82
R_{BB}	4k7–9k1	4k0–9k1
I_p (max)	5 μA	5 μA
I_V (max)	4 mA	4 mA
Outline	B_2, E, B_1 (To-18 case)	B_1, B_2, E (To-92 case)

Figure 18.4. 2N2646 and TIS43 UJT data.

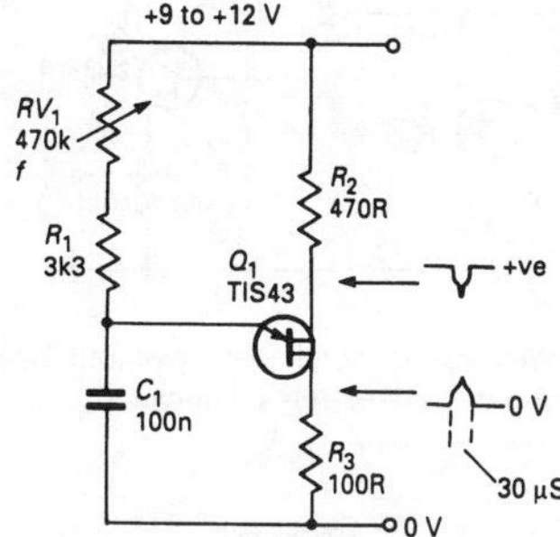

Figure 18.5. Wide-range pulse generator.

A non-linear sawtooth is generated across C_1 of the *Figure 18.5* circuit, but is at a high impedance level and is thus not readily available externally. Access can be gained to this sawtooth either by wiring a simple pnp emitter follower across the timing resistor network, as shown in *Figure 18.6*, or by wiring an npn Darlington emitter follower across C_1, as in *Figure 18.7*. Note that *Figure 18.6* gives a fixed-amplitude output that is referenced to the positive supply rail, and *Figure 18.7* gives a variable-amplitude output that is referenced to the zero volts line.

The UJT oscillator can be made to generate a linear sawtooth waveform by charging C_1 via a constant-current generator rather than via a simple resistance. *Figure 18.8* shows such a circuit. Q_1 and its associated network form the constant current generator, and the current magnitude (and thus the oscillation frequency) is variable via RV_1. C_1's linear

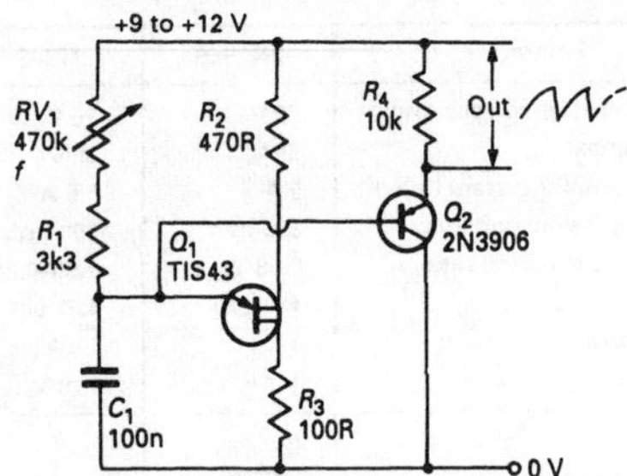

Figure 18.6. Wide-range non-linear sawtooth generator.

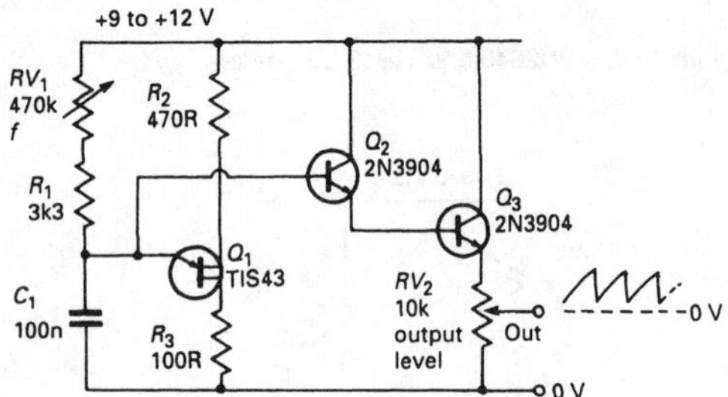

Figure 18.7. Wide-range non-linear sawtooth generator with variable-amplitude ground referenced output.

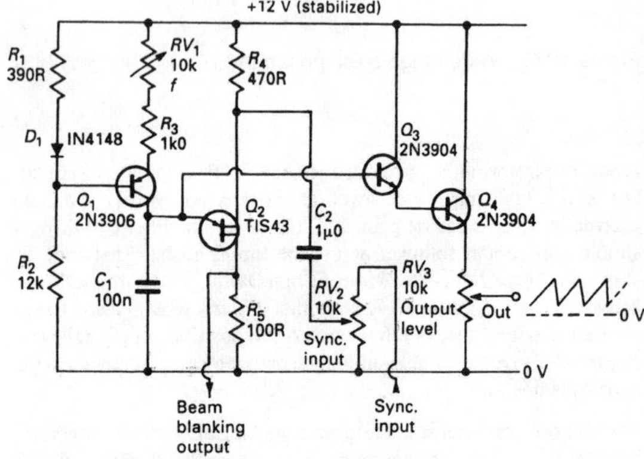

Figure 18.8. This linear sawtooth generator can be used as a simple oscilloscope timebase generator.

sawtooth waveform is made externally available via the Q_3–Q_4 Darlington emitter follower and its amplitude is variable via RV_3. With the component values shown the oscillation frequency is variable from 60Hz to 700Hz via RV_1. The circuit can be used as a simple oscilloscope timebase generator by taking its sawtooth output to the 'external timebase' socket of the oscilloscope and using the 'positive flyback pulses' from R_5 for beam blanking. The generator can be synchronised to any external signal that is fed to the *sync input* terminal.

Figure 18.9 shows how a UJT can be used to generate either a non-linear sawtooth or a rectangular waveform with an infinitely-variable mark-space ratio. The LF356 op-amp used here is a 'fast' device with a very high input impedance. When S_1 is in the sawtooth position this op-amp acts as a simple voltage follower, and C_1's sawtooth appears across output control RV_2. When S_1 is set to the rectangle position the op-amp is configured as a fast voltage comparator, with the sawtooth fed to its non-inverting input and a variable (via RV_3) d.c. reference voltage fed to its inverting input; this simple arrangement converts the sawtooth waveform into a rectangular output that has its mark-space ratio fully variable via RV_3.

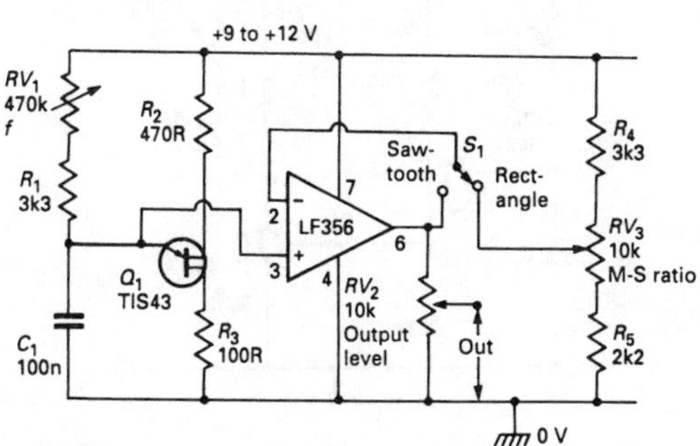

Figure 18.9. 25Hz to 3kHz generator produces a non-linear sawtooth or a rectangular waveform with fully variable M–S ratio.

Gadgets and novelties

Figures 18.10 to *18.14* show a variety of ways of using UJTs in handy gadgets and novelty circuits. *Figure 18.10* is a simple morse-code practice oscillator; it generates a fixed tone (adjustable from 300Hz to 3kHz) directly in a small speaker whenever the morse key is closed.

Figure 18.11 is a musicians metronome with a beat rate variable from 20 to 200 per minute via RV_1; the UJT's output pulses are fed to the speaker via Q_2, producing a distinct 'click' each time the UJT completes a timing cycle.

Figure 18.12 is a multi-tone signalling system that consumes zero quiescent current and generates a tone that is unique to each one of its three push-button operating switches; each switch connects the oscillator's supply via an isolating diode, but selects a unique value of tone-generating timing resistor.

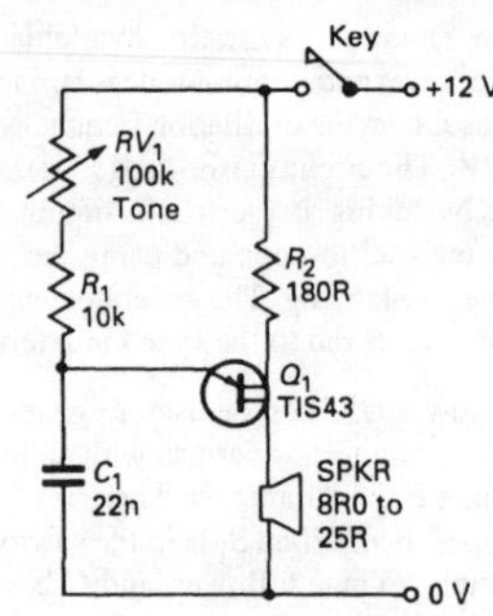

Figure 18.10. Simple code-practice oscillator; tone is variable from 300Hz to 3kHz.

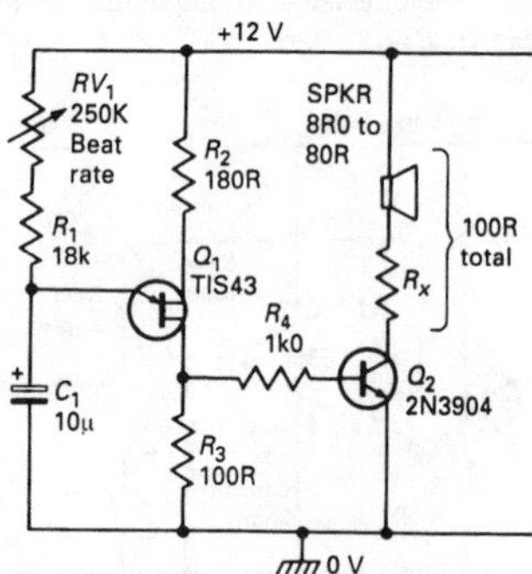

Figure 18.11. Metronome giving 20 to 200 beats per minute.

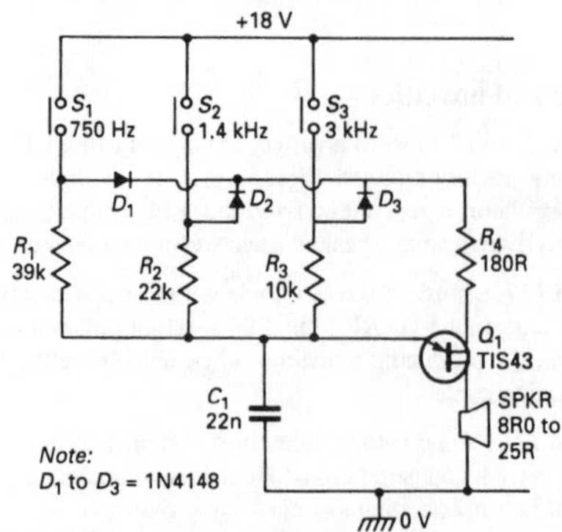

Figure 18.12. Simple multi-tone signalling system.

Figure 18.13 shows a simple rising-tone siren. When power is first applied C_1 is fully discharged, so the UJT operates at a frequency set only by R_3 and C_2; C_1 immediately starts to charge via R_1, however, and its voltage causes C_2's charge current to increase via D_1 and R_2, raising the UJT's frequency. Thus, the UJT's oscillation frequency slowly rises as C_1 charges up, as shown by the diagram's exponential graph, and generates a distinct 'rising' tone.

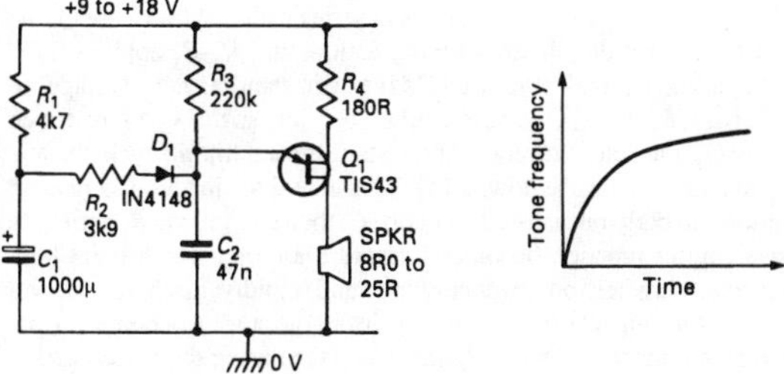

Figure 18.13. Simple rising-tone siren.

Figure 18.14 shows the UJT used as a light-sensitive oscillator, with an LDR acting as its main timing resistor. This LDR is a cadmium sulphide photocell; under dark conditions it acts as a very high resistance, so the operating frequency is low and is determined mainly by R_1; under bright condition the LDR resistance is very low, so the operating frequency is high and is determined mainly by R_2. At intermediate light levels the UJT frequency is set mainly by the LDR value and thus by the light level. This circuit can be used as a simple musical instrument that is played by the light of a torch or by shadows cast by the hand.

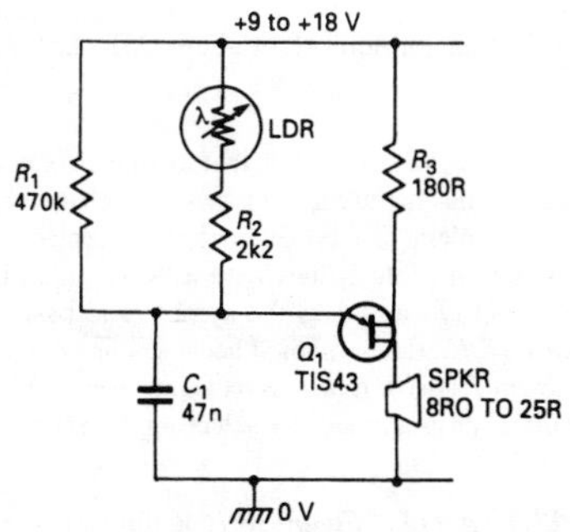

Figure 18.14. Light-sensitive oscillator.

A.C. power control circuits

The most important use of the UJT is in A.C. power control applications, where its high-energy time-delayed output pulses can be used to trigger TRIACs and thus control the power feed to A.C. lamps, heaters, or motors, etc. Circuits of this type are shown in Chapter 20.

PUTs and kindred devices

The action of a UJT oscillator can be simulated by the circuit of *Figure 18.15*, in which pnp transistor Q_1 is in series with npn transistor Q_2; R_1 and C_1 control the circuit's timing action, and R_2–R_3 apply a fixed voltage (the equivalent of a UJT's intrinsic standoff ratio voltage) to Q_1's base; R_5 shunts Q_2's base-emitter junction, so that Q_2 is not driven on by Q_1's leakage currents. At the start of each timing cycle the R_1–C_1 junction voltage is low, so Q_1's base-emitter junction is reverse biased and both transistors are cut off. C_1 then charges via R_1 until Q_1's base-emitter junction becomes forward biased, at which point both transistors switch on regeneratively and rapidly discharge C_1 via current-limiting resistor R_4, until the discharge current becomes so low that both transistors switch off again, and the timing sequence starts to repeat again.

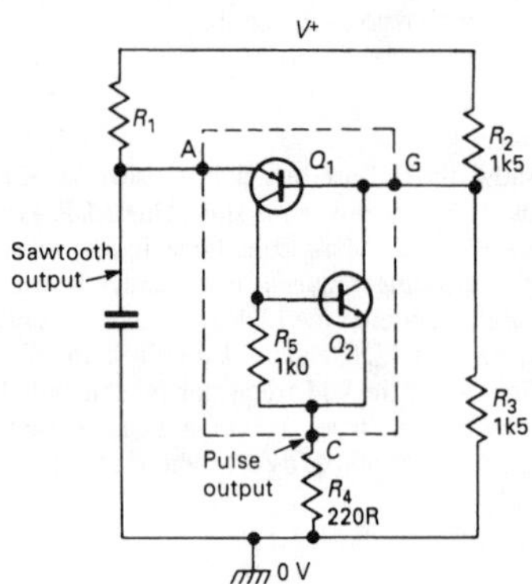

Figure 18.15. Transistor equivalent of the UJT oscillator.

A practical weakness of this circuit is that its transistors can easily be burnt out, since C_1's discharge current flows through their base-emitter junctions. This problem can be overcome by replacing the three components within the dotted lines with a PUT, which is the direct equivalent of Q_1–Q_2–R_5 and uses the symbol and basic application circuit of *Figure 18.16*; it is so named because it acts like a Programmable Unijunction Transistor, in which the intrinsic standoff ratio and R_{BB} values can be 'programmed' by selecting the external R_2 and R_3 values.

Note that the PUT symbol of *Figure 18.16* is similar to that of an SCR (see Chapter 19), except that the gate is related to the anode rather than the cathode; the PUT is in fact sometimes called an anode-controlled

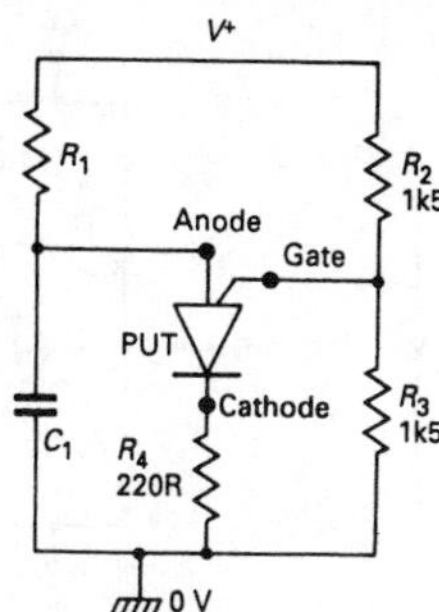

Figure 18.16. PUT symbol and basic oscillator circuit.

SCR, and is one of four closely related pnpn 'thyristor' devices; details of the other three members of the family are shown in *Figures 18.17* to *18.19*.

The SUS (*Figure 18.17*) or Silicon Unilateral Switch acts like a PUT with a built-in zener between its gate and cathode. The gate pin is normally left open, and the device acts as a voltage-triggered self-latching switch that turns on when the anode voltage rises high enough (above 8V) to make the zener start to break down via Q_1's base-emitter junction. Once the SUS has latched on, it can only be turned off again by reducing its anode current below the minimum holding value.

The SCS (*Figure 18.18*) or Silicon Controlled Switch has the same symbol as an ordinary SCR, but differs from it in one important respect; it acts as a self-latching switch that can be triggered on by applying a positive trigger signal to its gate, but can be turned off again either by reducing its anode current below its minimum holding value or (unlike an SCR) by briefly shorting or reverse biasing its gate-cathode junction.

Finally, the most versatile of all these devices is the thyristor tetrode which, as can be seen from *Figure 18.19*, can also be used as a PUT or SCS. This device has two gate terminals (G_C and G_A), and can be turned

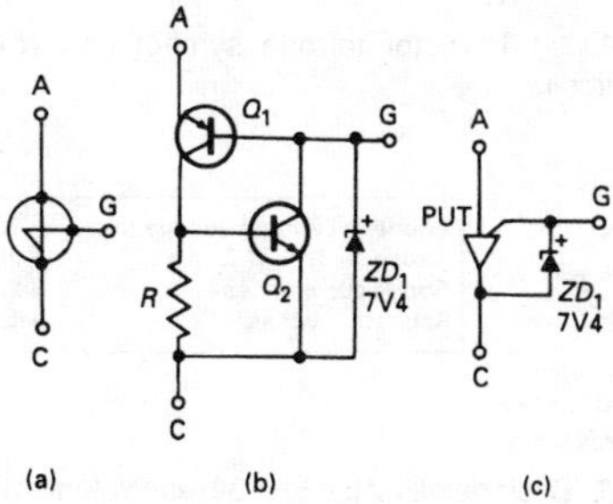

Figure 18.17. *(a)* SUS symbol and *(b)* and *(c)* equivalent circuits.

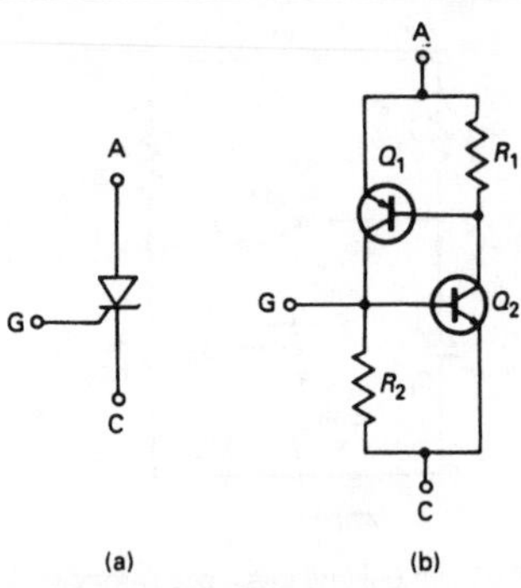

Figure 18.18. *(a)* SCS symbol and *(b)* transistor equivalent circuit.

on either by driving G_C positive to the cathode or by driving G_A negative to the anode, and can be gated off either by driving G_C negative to the cathode or by driving G_A positive to the anode.

The best known practical versions of the thyristor tetrode are the BRY39, the 2N6027, and the D13T1, which are virtually identical devices. *Figure 18.20* shows the basic details of the BRY39, which is housed in a TO-72 case. It can easily be used as a PUT or SCS.

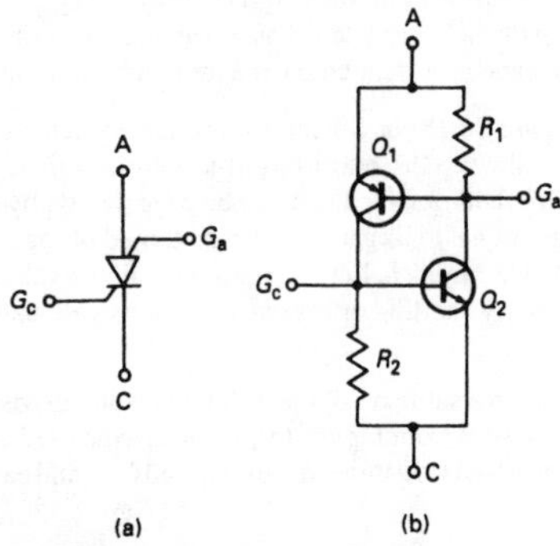

Figure 18.19. *(a)* Thyristor tetrode symbol and *(b)* transistor equivalent circuit.

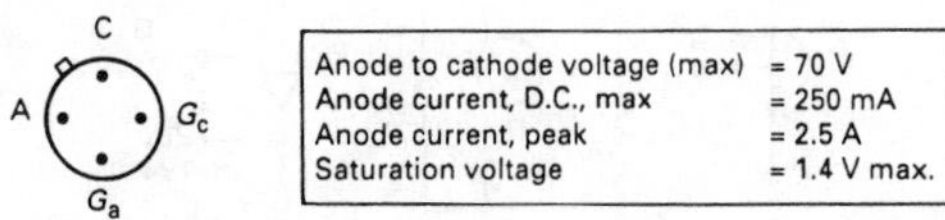

Anode to cathode voltage (max)	= 70 V
Anode current, D.C., max	= 250 mA
Anode current, peak	= 2.5 A
Saturation voltage	= 1.4 V max.

(G_a connected to case)
Pin connections

Figure 18.20. Basic details of the BRY39 (equivalent to the 2N6027 and D13T1) thyristor tetrode device, which can also be used as a PUT or SCS.

19 SCR circuits

An SCR (Silicon Controlled Rectifier) is a controllable medium- to high-power self-latching solid-state D.C. power switch. This chapter explains its basic operation and shows practical ways of using it.

SCR basics

An SCR is a four-layer pnpn silicon semiconductor device. It has three external terminals (anode, gate, and cathode) and uses the symbol of *Figure 19.1(a)* and has the transistor equivalent circuit of *Figure 19.1(b)*. *Figure 19.2* shows the basic way of using the SCR as a D.C. switch, with the anode positive relative to the cathode, and the SCR controlled via its gate. The basic characteristics of the SCR can be understood with the aid of these diagrams, as follows:-

(1) When power is first applied to the SCR by closing S_1 in *Figure 19.2*, the SCR is 'blocked' and acts (between anode and cathode) like an open switch. This action is implied by *Figure 19.1(b)*, i.e., Q_2's base

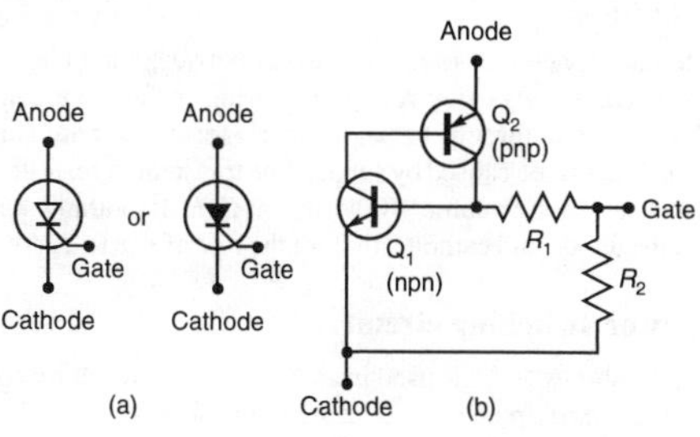

Figure 19.1. Alternative SCR symbols *(a)* and SCR equivalent circuit *(b)*.

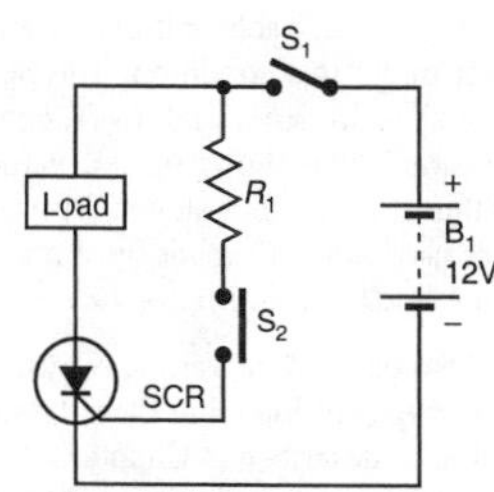

Figure 19.2. Basic ways of using an SCR as a D.C. switch

current is derived from Q_1 collector, and Q_1's base current is derived from either Q_2 collector or the gate terminal; in the latter case no base current is available, so both transistors are cut off, and only a small leakage current flows from anode to cathode.

(2) The SCR can be turned on and made to act like a forward-biased silicon rectifier by briefly applying gate current to it via S_2; the SCR quickly (in a few microseconds) self-latches into the on state under this condition, and stays on even when the gate drive is removed. This action is implied by *Figure 19.1(b)*; the initial gate current turns Q_1 on, and Q_1's collector current turns Q_2 on, and Q_2's collector current then holds Q_1 on even when the gate drive is removed: a 'saturation' potential of 1V or so is generated between the anode and cathode under the on condition.

(3) Only a brief pulse of gate current is needed to drive the SCR on. Once the SCR has self-latched, it can only be turned off again by briefly reducing its anode current to below a certain 'minimum holding current' value (typically a few milliamps); in A.C. applications turn-off occurs automatically at the zero-crossing point of each half-cycle.

(4) Considerable current gain is available between the gate and anode of the SCR, and low values of gate current (typically a few mA or less) can control high values of anode current (up to tens of amps). Most SCRs have anode ratings of hundreds of volts. The SCR gate characteristics are similar to those of a transistor base-emitter junction (see *Figure 19.1(b)*).

(5) Internal capacitance (a few pF) exists between the SCR's anode and gate, and a sharply rising voltage appearing on the anode can cause enough signal breakthrough to the gate to trigger the SCR on. This 'rate effect' turn-on can be caused by supply-line transients, etc. Rate-effect problems can be overcome by wiring a *C–R* smoothing network between the anode and cathode, to limit the rate of rise to a safe value.

A.C. power switching circuits

Figure 19.3 shows an SCR used in an A.C. power switching application; alternative component's values are shown for use with 120V or (in parentheses) 240VA.C. supplies. The A.C. power line signal is full-wave rectified via D_1–D_4 and applied to the SCR anode via lamp load LP_1. If S_1 is open, the SCR and lamp are off. If S_1 is closed, R_1–R_2 apply gate drive to the SCR, which turns on and self-latches just after the start of each half cycle and then turns off again automatically at the end of the half-cycle as its forward current falls below the minimum holding value. This process repeats in each half-cycle, and the lamp thus operates at full power under this condition. The SCR anode falls to about 1V when the SCR is on, so S_1 and R_1–R_2 consume little mean power. Note that the lamp load is shown placed on the D.C. side of the bridge rectifier, and this circuit is thus shown for use with D.C. loads; it can be modified for use with A.C. loads by simply placing the load on the A.C. side of the bridge, as in *Figure 19.4*.

Note that SCRs can also be used, in various ways, to apply *variable* A.C. power to various types of load, but these tasks are usually best carried out by TRIACs, as described in Chapter 20; two special types of SCR variable A.C. power control circuits are, however, shown at the end of this chapter.

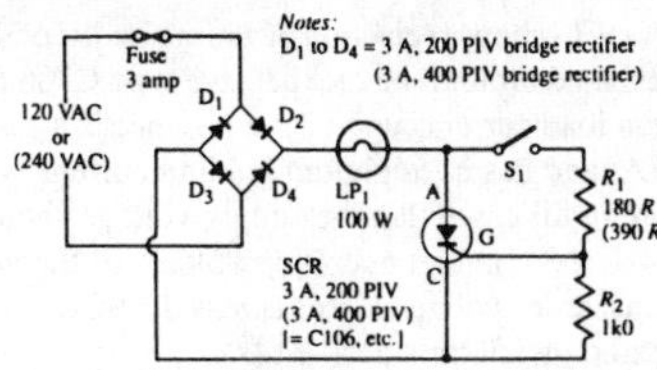

Figure 19.3. Full-wave on-off SCR circuit with D.C. power load.

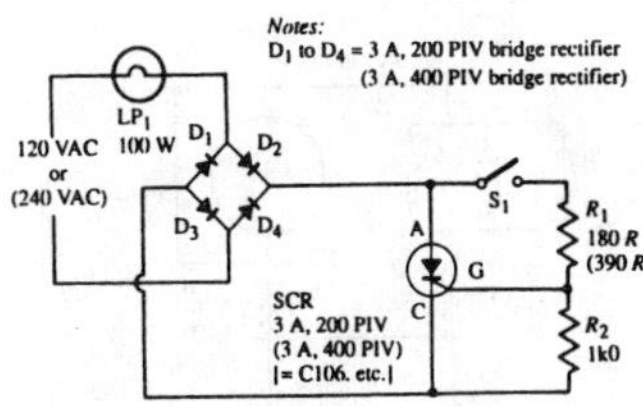

Figure 19.4. Full-wave on-off SCR circuit with A.C. power load.

Bell/buzzer alarm circuits

One useful application of the SCR is in D.C.-powered 'alarm' circuits that use self-interrupting loads such as bells or buzzers; these loads comprise a solenoid and a series switch, and give an action in which the solenoid first shoots forward via the closed switch, and in doing so forces the switch to open, thus making the solenoid fall back and re-close the switch, thus restarting the action, and so on. *Figure 19.5* shows such an alarm circuit; it effectively gives a non-latching load-driving action, since the SCR unlatches automatically each time the load self-interrupts. The circuit can be made fully self-latching, if desired, by shunting the load with resistor R_3, as shown, so that the SCR anode current does not fall below the SCR's minimum holding value as the load self-interrupts.

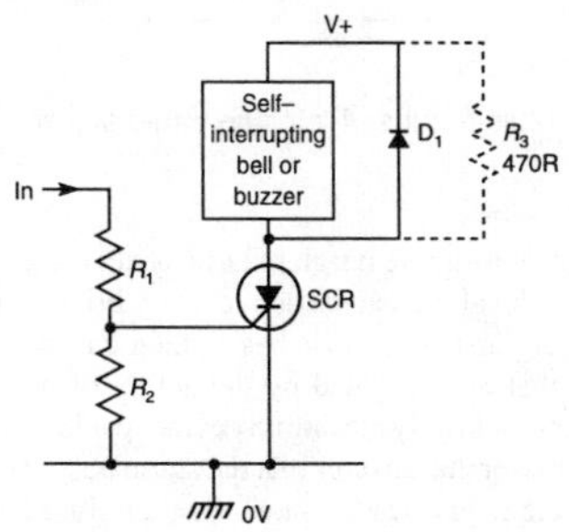

Figure 19.5. Basic SCR alarm circuit.

Figures 19.6 to *19.14* show a selection of alarm circuits of this type. All of these are designed around the inexpensive type C106 SCR, which can handle mean load currents up to 2.5 amps, needs a gate current of less than 200µA, and has a 'minimum holding current' value of less than 3mA. Note in all cases that the supply voltage should be about 1.5V greater than the nominal operating voltage of the alarm device used, to compensate for voltage losses across the SCR, and that diode D_1 is used to damp the alarm's back-emfs.

Figure 19.6 shows a simple non-latching multi-input alarm, in which the alarm activates when any of the S_1 to S_3 push-button input switches are closed, but stops operating as soon as the switch is released.

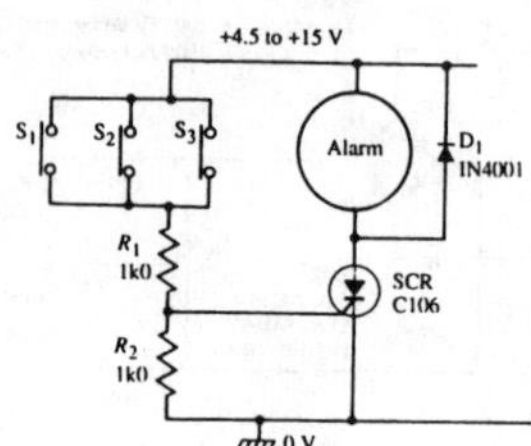

Figure 19.6. Multi-input non-latching alarm circuit.

Figure 19.7 shows the above circuit converted into a self-latching multi-input 'panic' alarm by wiring R_3 plus normally-closed reset switch S_4 in parallel with the alarm device. Once this circuit has latched, it can be unlatched again (reset) by briefly opening S_4.

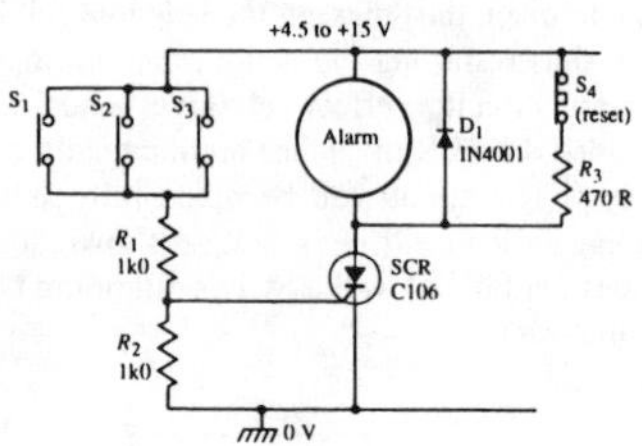

Figure 19.7. Multi-input self-latching panic alarm.

Figure 19.8 shows a simple burglar alarm system, complete with the 'panic' facility. The alarm can be activated by briefly opening any of the S_1 to S_3 'burglar alarm' switches (which can be reed-relays or microswitches that are activated by the action of opening doors or windows, etc.), or by briefly closing any of the 'panic' switches. C_1 acts as a noise-suppressor that ensures that the alarm only activates if the S_1 to S_3 switches are held open for more than a millisecond or so, thus enhancing the circuit's reliability. The circuit consumes a typical standby current of 0.5mA (via R_1) from a 6V supply.

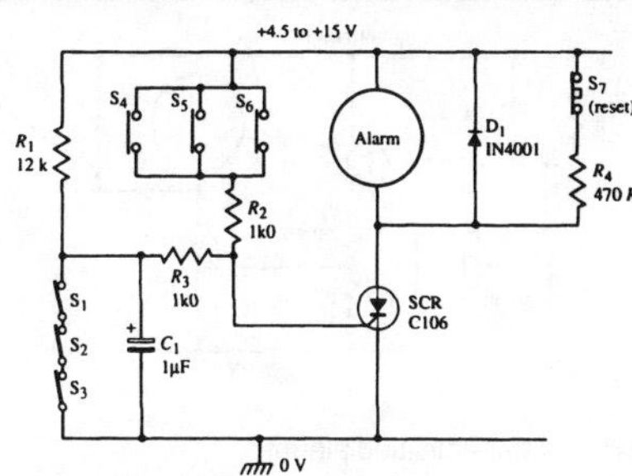

Figure 19.8. Simple burglar alarm system, with panic facility.

The standby current of the burglar alarm circuit can be reduced to a mere 1.4μA (at 6V) by modifying it as shown in *Figure 19.9*, where Q_1 and Q_2 are connected as a Darlington common emitter amplifier that inverts and boosts the R_1-derived 'burglar' signal and then passes it on to the gate of the SCR.

Water, light & heat alarms

The basic SCR-driven alarm circuit can be used to indicate the presence of excess water, light, or temperature levels by driving the SCR gate via suitable sensing circuitry; *Figures 19.10* to *19.14* show alarm circuits of this type.

The *Figure 19.10* 'water-activated' alarm uses Q_1 to activate the SCR when a resistance of less than about 220 kilohms appears a cross the two metal probes. It's operation as a water-activated alarm relies on the fact that the impurities in normal water (and many other liquids and vapours) make it act as a conductive medium with a moderately low electrical resistance, which thus causes the alarm to activate when water comes into contact with both probes simultaneously. C_1 suppresses unwanted a.c. signal pick-up, and R_2 limits Q_1's base current to a safe value. By suitably adjusting the placing of the two metal probes, this circuit can be used to sound an alarm when water rises above a pre-set level in a bath, tank, or cistern, etc.

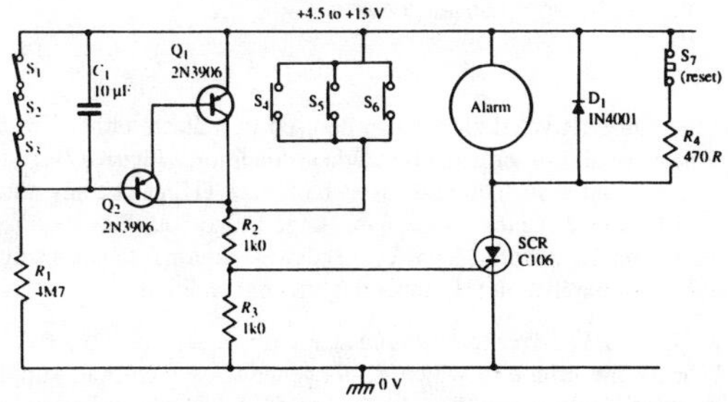

Figure 19.9. Improved burglar alarm circuit.

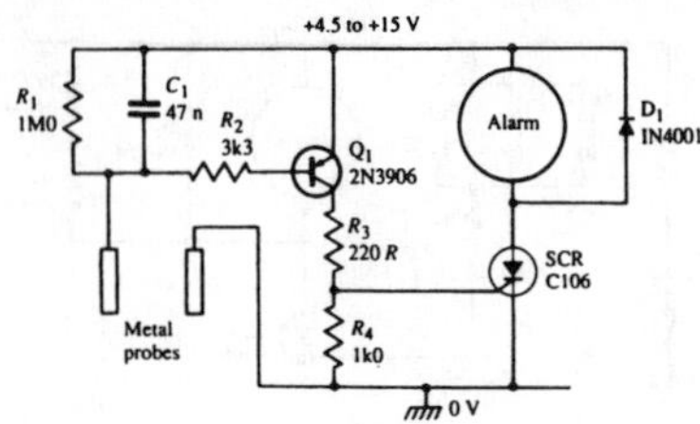

Figure 19.10. Water-activated alarm.

Figure 19.11 is a 'light-activated' circuit that can be used to sound an alarm when light enters a normally-dark area such as a drawer or wall safe, etc. The LDR and RV_1 form a light-sensitive potential divider that has its output buffered via Q_1 and fed to the SCR gate via R_1; this output is low under dark conditions (LDR resistance is high), but goes high under bright conditions (LDR resistance is low) and thus drives the SCR and alarm on; the light-triggering point can be pre-set via RV_1. Almost any small cadmium sulphide photocell can be used in the LDR position.

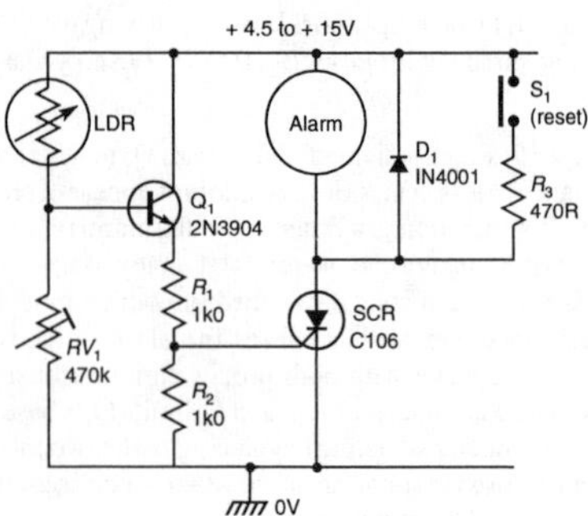

Figure 19.11. Light-activated alarm.

Temperature-activated alarms can be used to indicate either fire or overheat conditions, or frost or underheat conditions. *Figures 19.12* to *19.14* show three such circuits; in each of these TH_1 can be any n.t.c. thermistor that has a resistance in the range 1k0 to 20k at the required trigger temperature; pre-set pot RV_1 needs a maximum resistance value roughly double that of TH_1 under this trigger condition.

The *Figure 19.12* over-temperature alarm uses R_1–R_2 and TH_1–RV_1 as a Wheatstone bridge in which R_1–R_2 generates a fixed half-supply 'reference' voltage and TH_1–RV_1 generates a temperature-sensitive 'variable' voltage, and Q_1 is used as a bridge balance detector and SCR

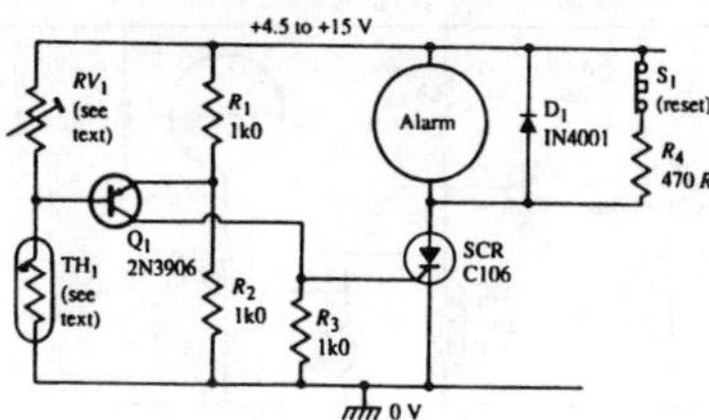

Figure 19.12. Simple over-temperature alarm.

gate driver. RV_1 is adjusted so that the 'reference' and 'variable' voltages are equal at a temperature just below the required trigger value, and under this condition Q_1 base and emitter are at equal voltages and Q_1 and the SCR are thus cut off. When the TH_1 temperature goes above this 'balance' value the TH_1–RV_1 voltage falls below the 'reference' value, so Q_1 becomes forward biased and drives the SCR on, thus sounding the alarm. The precise trigger point can be pre-set via RV_1. The circuit's action can be reversed, so that the alarm turns on when the temperature falls below a pre-set level, by simple transposing the TH_1 and RV_1 positions as shown in the frost or under-temperature alarm circuit of *Figure 19.13*.

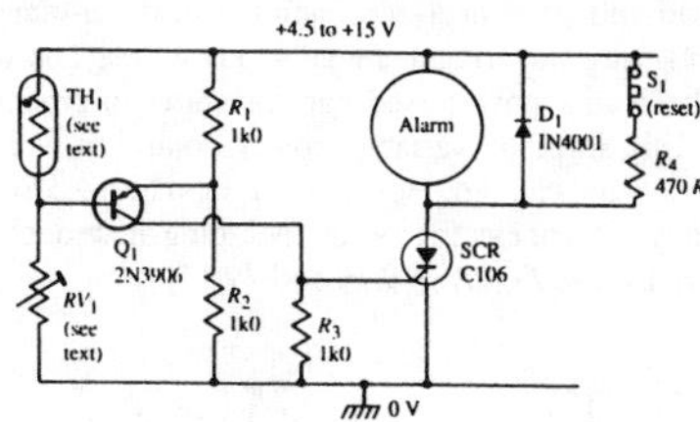

Figure 19.13. Simple frost or under-temperature alarm.

Note in these two circuits that if TH_1 and Q_1 are not mounted in the same environment, the precise trigger points are subject to slight variation with changes in Q_1 temperature, due to the temperature dependence of its base-emitter junction characteristics. These circuits are thus not suitable for use in precision applications, unless Q_1 and TH_1 operate at equal temperatures. This snag can be overcome by using a two-transistor differential detector in place of Q_1, as shown in the *Figure 19.14* over-temperature alarm, which can be made to act as a precision under-temperature alarm by simply transposing RV_1 and TH_1.

Variable A.C. power control

All SCR circuits shown so far give an on/off form of power control. SCRs (and TRIACs) can be used to give *variable* power control ofA.C. circuits in several ways. One of these is via the 'phase-delayed

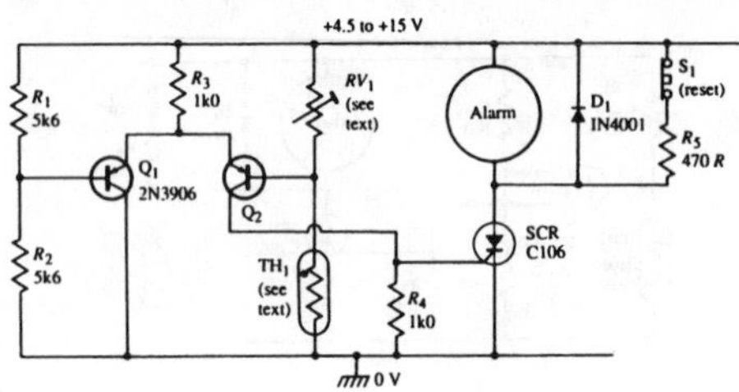

Figure 19.14. Precision over-temperature alarm.

switching' technique of *Figure 19.15*, in which power is fed to the load via a self-latching solid-state power switch that can be triggered (via a variable phase-delay network and a trigger-pulse generator) at any point in each power half-cycle, and automatically unlatches again at the end of each half-cycle. The diagram shows the load voltage waveforms that can be generated.

Thus, if the power switch is triggered near the start of each half cycle (with near-0° phase delay) the mean load voltage equals almost the full supply value, and the load consumes near-maximum power; if it is triggered near the end of each half cycle (with near-180° phase delay) the mean load voltage is near-zero, and the load consumes minimal power; by varying the trigger signal's phase-delay between these extremes, the load's power feed can be varied between zero and maximum. This form of variable power control is very efficient (typically 95%), and can (amongst other things) be used to control the speeds of many types of electric motor, including those of electric drills and model trains (see *Figures 19.16* and *19.17*).

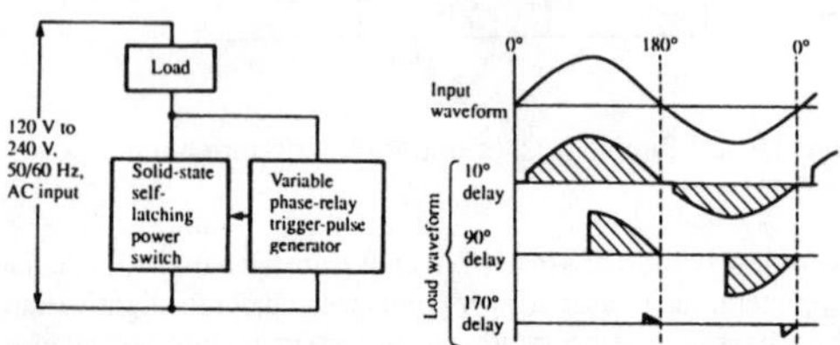

Figure 19.15. Variable phase-delay-switching A.C. power controller, with waveforms.

A drill-speed controller

Most electric drills are powered by series-wound 'universal' (A.C. or D.C.) electric motors. These motors generate a back-emf that is proportional to the motor speed, and the motor's *effective* applied voltage thus equals the true applied voltage minus the back-emf; this gives the motor a degree of speed self-regulation, since any increase in motor loading tends to reduce the speed and back-emf, thereby increasing the effective applied voltage and causing the motor speed to rise towards its original value, and so on.

The speed of an electric drill can be varied electronically by using the 'phase-delayed switching' technique. *Figure 19.16* shows a particularly effective yet simple variable speed-regulator circuit. This uses an SCR as the control element and feeds half-wave power to the motor (this causes only a 20% reduction in maximum available speed/power), but in the off half-cycles the back-emf of the motor is sensed by the SCR and used to give automatic adjustment of the next gating pulse, to give automatic speed regulation. The R_1–RV_1–D_1 network provides only 90° of phase adjustment, so all motor pulses have minimum durations of 90° and provide very high torque. At low speeds the circuit goes into a high-torque 'skip cycling' mode, in which power pulses are provided intermittently, to suit motor loading conditions.

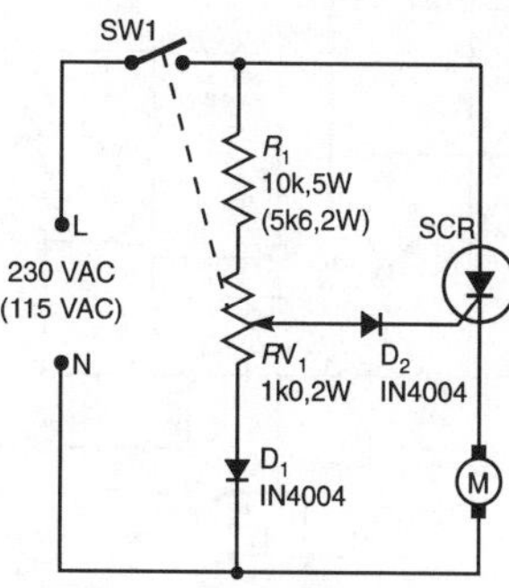

Figure 19.16. Electric drill speed-controller.

Model-train speed-controller

Figure 19.17 shows how the 'phase-delayed switching' technique can be used to make an excellent 12 volt model-train speed-controller that enables speed to be varied smoothly from zero to maximum. The maximum available output current is 1.5 amps, but the unit incorporates short-circuit sensing and protection circuitry that automatically limits the output current to a mean value of only 100mA if a short occurs on the track. The circuit operates as follows.

The circuit's power line voltage is stepped down via T_1 and full-wave (bridge) rectified via BR_1, to produce a raw (unsmoothed) D.C. supply that is fed to the model train (via the track rails) via the series-connected SCR and direction control switch SW_3. At the start of each raw D.C. half-cycle the SCR is off, so D.C. voltage is applied (via R_4 and ZD_1) to UJT Q_1 and its associated C_1–RV_1 (etc.) timing circuitry, and C_1 starts to charge up until eventually the UJT fires and triggers the SCR; as the SCR turns on it saturates, removing power from Q_1 (which thus resets) and feeding the rest of the power half-cycle to the model train via R_2/

R_3 and SW_3. This timing/switching process repeats in each raw D.C. half-cycle (i.e., at twice the power line frequency), giving a classic phase-triggered power control action that enables the train speed to be varied over a wide range via RV_1.

Note that the circuit's output current passes through R_2/R_3, which generate a proportional output voltage that is peak-detected and stored via D_1–C_2 and fed to Q_2 base via R_8–R_9. The overall action is such that, because of the voltage storing action of C_2, Q_2 turns on and disables the UJT's timing network (thus preventing the SCR from firing) for several half-cycles if the peak output current exceeds 1.5 amps. Thus, if a short occurs across the track the half-cycle output current is limited to a peak value of a few amps by the circuit's internal resistance, but the protection circuitry ensures that the SCR fires only once in (say) every fifteen half-cycles, thus limiting the *mean* output current to only 100mA or so.

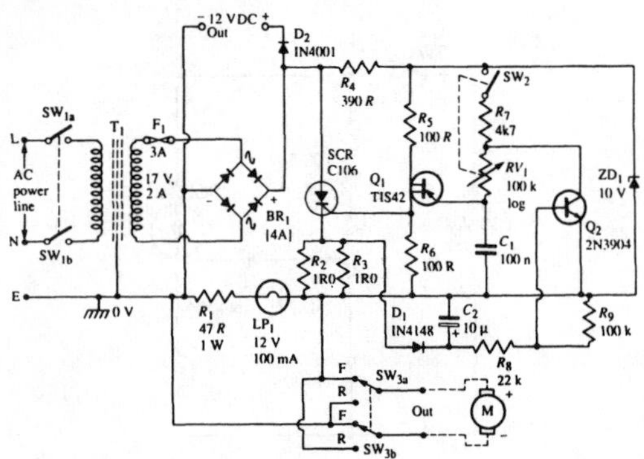

Figure 19.17. Model train speed-controller circuit with automatic short-circuit protection.

20 Triac circuits

A triac is a controllable medium- to high-power semi-latching solid-state A.C. power switch. This final chapter explains its basic operation and shows various ways of using it; most of the practical circuits show two sets of component values, for use with either 230V or (in parenthesis) 115V A.C. supplies; in each design, the user must use a triac with ratings to suit his own particular application.

Triac basics

A triac is a three-terminal (MT_1, gate, and MT_2) solid-state thyristor that uses the symbols of *Figure 20.1* and acts like a pair of SCRs wired in inverse parallel and controlled via a single gate terminal. It can conduct current in either direction between its MT_1 and MT_2 terminals and can thus be used to directly control A.C. power. It can be triggered by either positive or negative gate currents, irrespective of the polarity of the MT_2 current, and it thus has four possible triggering modes or 'quadrants', signified as follows.

I+ Mode = MT_2 current + ve, gate current + ve
I- Mode = MT_2 current + ve, gate current - ve
III+ Mode = MT_2 current - ve, gate current + ve
III- Mode = MT_2 current - ve, gate current - ve

The trigger current sensitivity is greatest when the MT_2 and gate currents are both of the same polarity (either both positive or both negative), and is usually about half as great when they are of opposite polarity.

Figure 20.2 shows a triac used as a simple A.C. power switch. When S_1 is open, the triac acts as an open switch and the lamp passes zero current. When S_1 is closed the triac is gated on via R_1 and self-latches shortly after the start of each half-cycle, thus switching full power to the lamp load.

Triac rate-effect

Triacs, like SCRs, are susceptible to 'rate-effect' problems. Internal capacitances inevitably exist between the main terminals and gate of a triac, and if a sharply rising voltage appears on either main terminal it

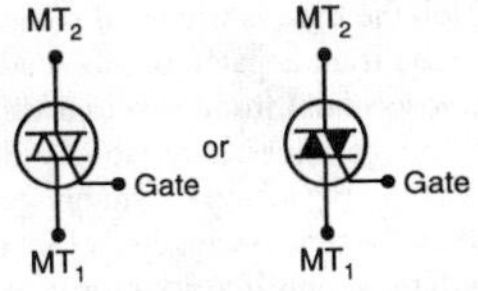

Figure 20.1. Triac symbols.

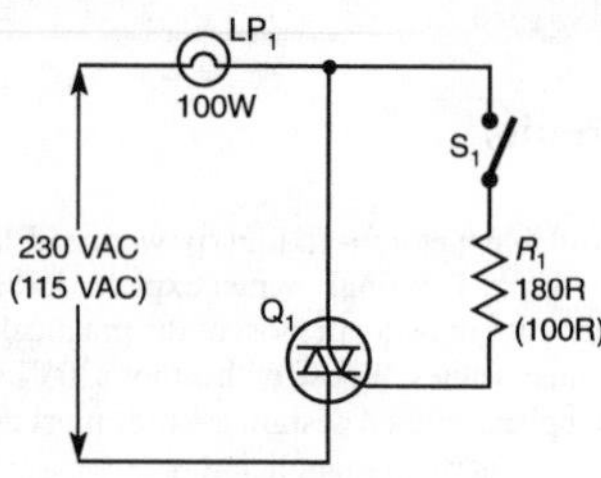

Figure 20.2. Simple A.C. power switch.

can cause enough break-through to the gate to trigger the triac on. This unwanted 'rate-effect' turn-on can be caused by supply line transients; the problem is particularly severe when driving inductive loads such as electric motors, in which load currents and voltages are out of phase. Rate-effect problems can usually be overcome by wiring an *R–C* 'snubber' network between MT_1 and MT_2, to limit the voltage rate-of-rise to a safe value as shown (for example) in the triac power switch circuit of *Figure 20.3*, where R_2–C_1 form the snubber network.

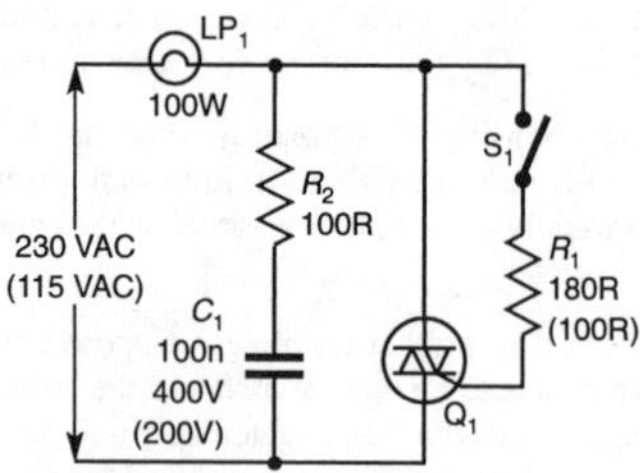

Figure 20.3. Simple power switch with C_1–R_2 snubber network to give rate-effect suppression.

RFI suppression

A triac can be used to give *variable*A.C. power control by using the 'phase-delayed switching' technique described in Chapter 19 (see *Figure 19.15*), in which the triac is triggered part-way through each half-cycle. Each time the triac is gated on, its load current switches sharply (in a few microseconds) from zero to a value set by its load resistance and supply voltage values; in resistively loaded circuits such as lamp dimmers this switching action inevitably generates a pulse of RFI, which is least when the triac is triggered close to the 0° and 180° 'zero crossing' points of the supply line waveform (at which the switch-on currents are minimal), and is greatest when the device is triggered 90° after the start of each half cycle (where the switch-on currents are at their greatest). The RFI pulses occur at twice the supply line frequency, and can be very annoying. In lamp dimmers, RFI can usually be eliminated by fitting the dimmer with a simple L–C filter network, as shown in *Figure 20.4*; the filter is fitted close to the triac, and greatly reduces the rate-of-rise of the A.C. power line currents.

Diacs and quadracs

A diac is a 2-terminal trigger device that can be used with voltages of either polarity and is often used in conjunction with a triac; *Figure 20.5* shows its circuit symbol. Its basic action is such that, when connected across a voltage source via a current-limiting load resistor, it acts like a high impedance until the applied voltage rises to about 35V, at which point it triggers and acts like a 30V zener diode, and 30V is developed across the diac and the remaining 5V appears across the load resistor. The diac remains in this state until its forward current falls below a minimum holding value (this occurs when the supply voltage is reduced below the 30V 'zener' value), at which point the diac turns off again.

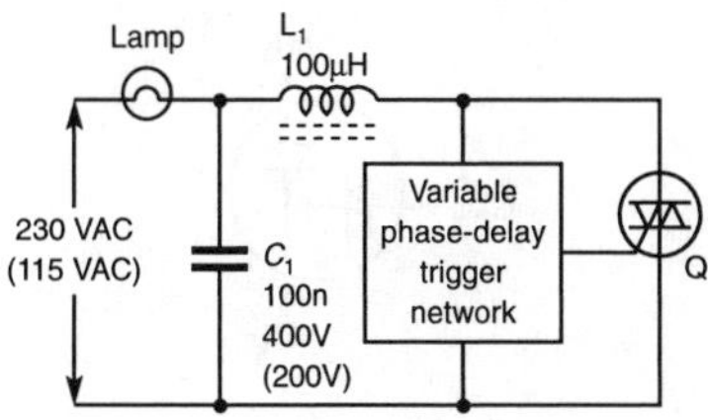

Figure 20.4. Basic lamp dimmer with RFI suppression via C_1–L_1.

Figure 20.5. Diac symbol.

The diac is most often used as a trigger device in phase-triggered triac variable power control applications, as in the basic lamp dimmer circuit of *Figure 20.6*. Here, in each power line half-cycle, the R_1–C_1 network applies a variable phase-delayed version of the half-cycle to the triac gate via the diac, and when the C_1 voltage rises to 35V the diac fires and delivers a 5V trigger pulse (from C_1) into the triac gate, thus turning the triac on and simultaneously applying power to the lamp load and removing the drive from the *R–C* network. The mean power to the load (integrated over a full half-cycle period) is thus fully variable from near-zero to maximum via R_1.

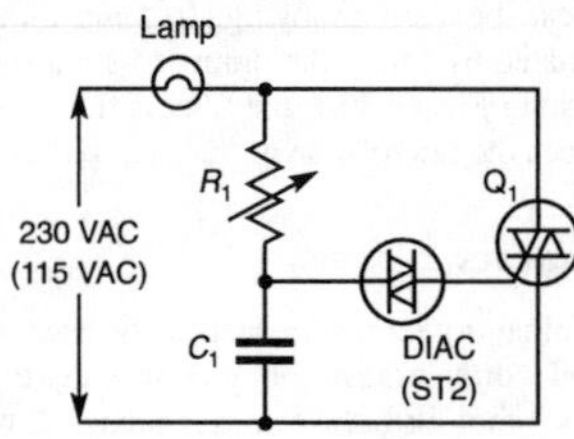

Figure 20.6. Basic diac-type variable phase-delay lamp dimmer circuit.

Some triacs are manufactured with a built-in diac in series with the triac gate; such devices are known as quadracs, and use the *Figure 20.7* circuit symbol.

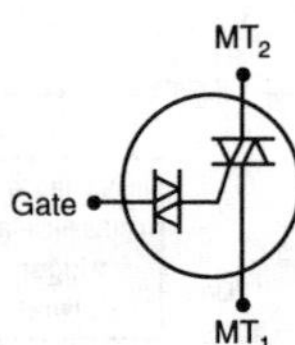

Figure 20.7. Quadrac symbol.

A.C. power switch variations

The simplest type of triac power switch is that of *Figure 20.2*, in which the triac is gated on via R_1 when S_1 is closed; only 1V or so is generated across the triac when it is on, so R_1 and S_1 consumes very little power; *Figure 20.3* shows the same circuit fitted with a 'snubber' network. There are many useful variations of these basic circuits. *Figure 20.8*, for example, shows a version that can be triggered via anA.C.-derived D.C. supply. C_1 charges (via R_1–D_1) to +10V on each positive a.c. power line half-cycle, and this charge triggers the triac when SW_1 is closed. Note that R_1 is subjected to almost the fullA.C. line voltage at all times, and thus needs a fairly high power rating, and that all parts of this circuit are 'live', making it difficult to interface to external control circuitry.

Figure 20.9 shows the above circuit modified to give 'isolated' interfacing to external control circuitry. SW_1 is simply replaced by transistor Q_2, which is driven from the phototransistor side of an optocoupler. The coupler's LED is driven via an external D.C. supply via R_4, and the triac turns on only when SW_1 is closed; SW_1 can be replaced by electronic switching circuitry if desired.

Figure 20.10 shows a variation in which the triac is A.C. triggered in each half-cycle via the A.C. impedance of C_1–R_1 and via back-to-back

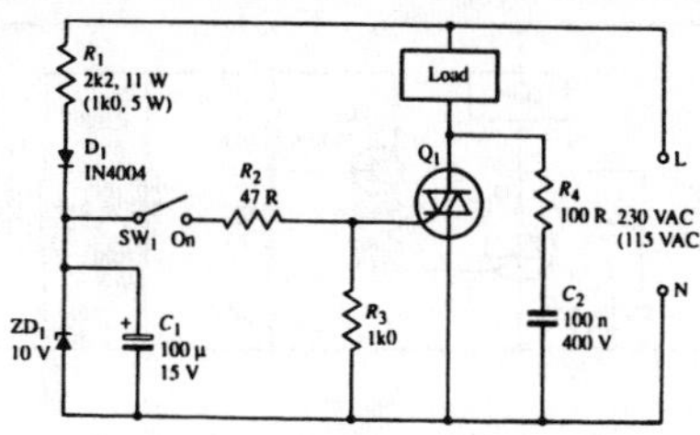

Figure 20.8. A.C. power switch with A.C.-derived D.C. triggering.

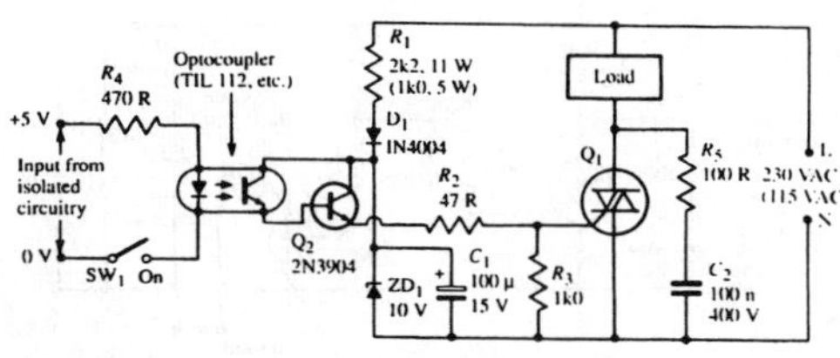

Figure 20.9. Isolated-input (optocoupled) A.C. power switch, D.C. triggered.

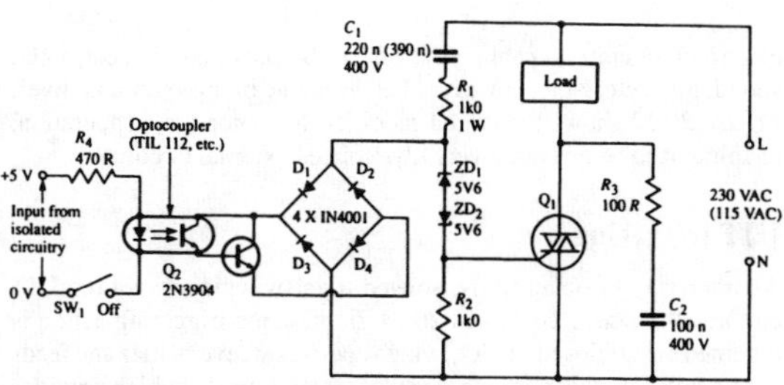

Figure 20.10. Isolated-input A.C. power switch, A.C. triggered.

zeners ZD_1–ZD_2, and C_1 dissipates near-zero power. Bridge rectifier D_1–D_4 is wired across the ZD_1–ZD_2–R_2 network and is loaded by Q_2; when Q_2 is off, the bridge is effectively open and the triac is gated on in each half-cycle, but when Q_2 is on a near-short appears across ZD_1–ZD_2–R_2, and the triac is off. Q_2 is driven via the optocoupler from the isolated external circuit, and the triac is on when SW_1 is open and off when SW_1 is closed.

Figures 20.11 and *20.12* show variations in which the triac is triggering via a transformer-derived D.C. supply and a transistor-aided switch. In *Figure 20.11* Q_1 and the triac are both driven on when SW_1 is closed, and are off when SW_1 is open. In practice, SW_1 can be replaced by

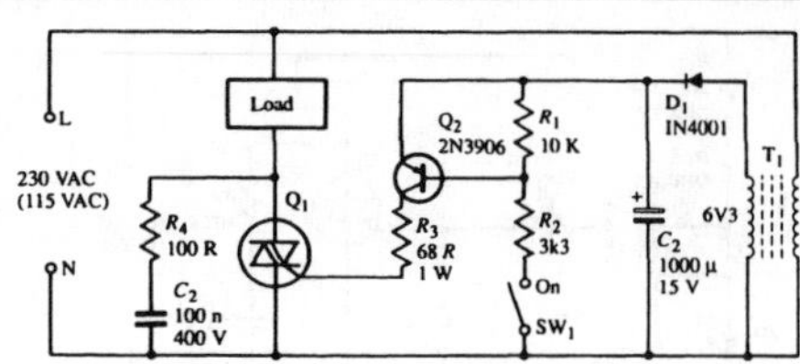

Figure 20.11. A.C. power switch with transistor-aided D.C.triggering.

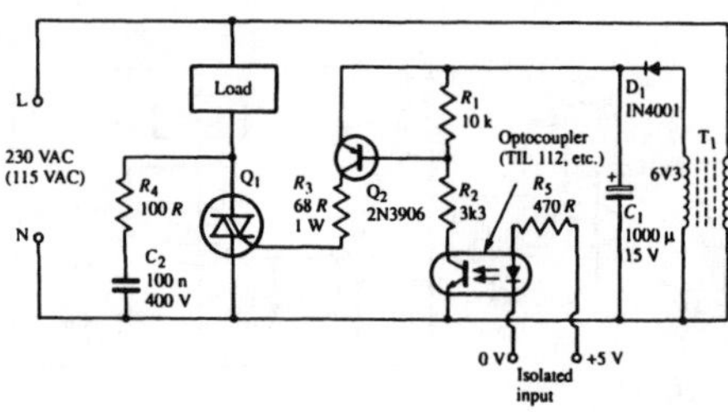

Figure 20.12. Isolated-input A.C. power switch with D.C. triggering.

electronic circuitry, enabling the triac to be activated via heat, light, sound, time, etc. Note, however, that the whole of this circuit is 'live'; *Figure 20.12* shows the circuit modified for optocoupler operation, enabling it to be activated via fully-isolated external circuitry.

UJT triggering

Another way to obtain fully isolated triac switching is via the UJT circuits of *Figures 20.13* and *20.14*. In these the triggering action is obtained via UJT oscillator Q_2, which operates at several kHz and feeds output pulses to the triac gate via pulse transformer T_1, which provides the desired 'isolation'. Because of its fairly high oscillating frequency, the UJT triggers the triac within a few degrees of the start of each A.C. power-line half-cycle when the oscillator is active.

In *Figure 20.13*, Q_3 is in series with the UJT's main timing resistor, so the UJT and triac turn on only when SW_1 is closed. In *Figure 20.14*, Q_3 is wired in parallel with the UJT's main timing capacitor, so the UJT and triac turn on only when SW_1 is open.

Optocoupled triacs

The gate junctions of a 'naked' triac are inherently photosensitive, and an optocoupled triac can thus be made by mounting a naked triac and LED close together in a single package. *Figure 20.15* shows the outline of a typical six-pin DIL version of such a device, in which the LED has

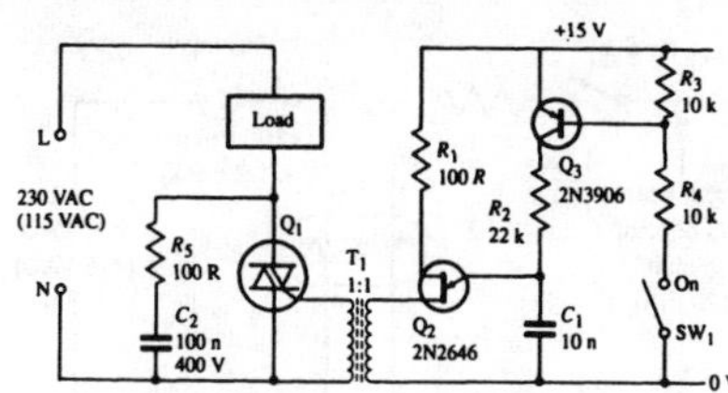

Figure 20.13. Isolated-input (transformer-coupled) A.C. power switch.

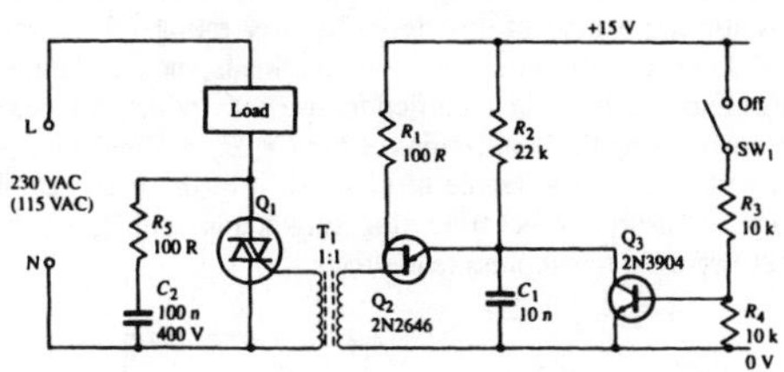

Figure 20.14. Isolated-input A.C. power switch.

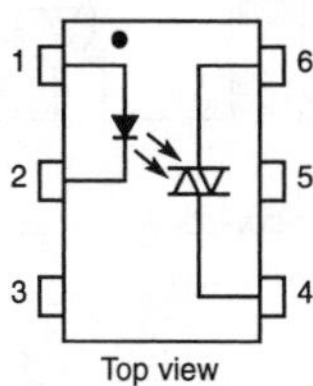

Figure 20.15. Typical optocoupled triac.

a maximum current rating of 50mA, the SCR has maximum ratings of 400V and 100mA r.m.s. (and a surge rating of 1.2A for 10μS), and the entire package has an isolating voltage rating of 1.5kV and a typical input current trigger sensitivity of 15mA.

Optocoupled triacs are easy to use and provide excellent electrical isolation between input and output. The input is used like a normal LED, and the output like a low-powered triac. *Figure 20.16* shows the device used to activate an A.C. line-powered filament lamp, which must have an r.m.s. rating below 100mA and a peak inrush current rating below 1.2A.

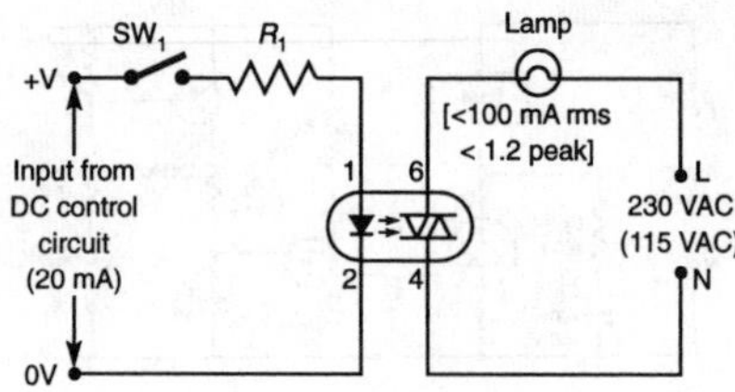

Figure 20.16. Low-power lamp control.

Figure 20.17 shows an optocoupled triac used to activate a slave triac, thereby driving a load of any desired power rating. This circuit is suitable for use only with non-inductive loads such as lamps and heating elements. It can be modified for use with inductive loads such as motors by using the connections of *Figure 20.18*. Here, the R_2–C_1–R_3 network provides a degree of phase-shift to the triac gate-drive network, to ensure correct triac triggering action, and R_4–C_2 form a snubber network, to suppress rate effects.

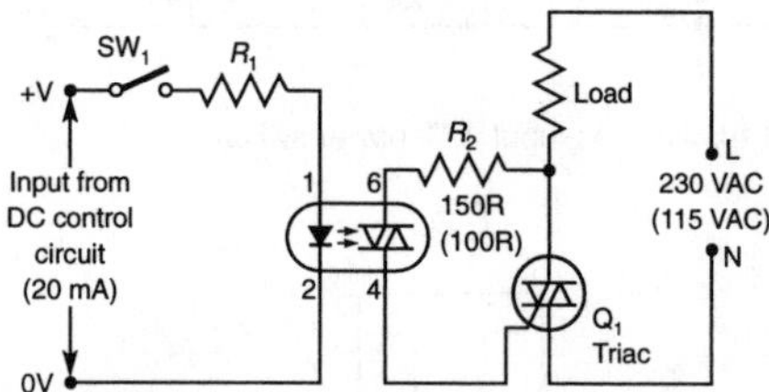

Figure 20.17. High-power control via a triac slave.

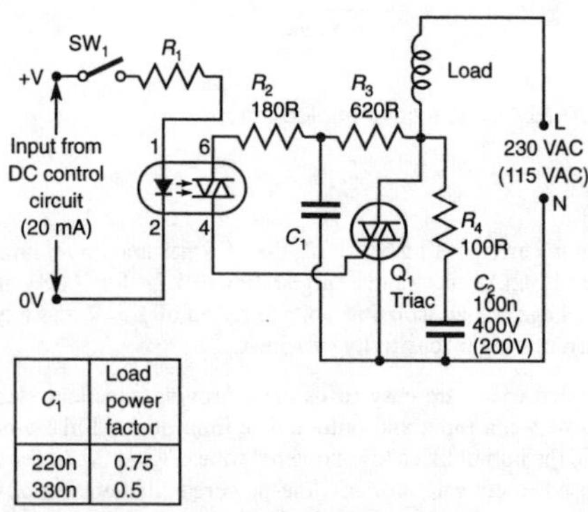

C_1	Load power factor
220n	0.75
330n	0.5

Figure 20.18. Driving an inductive load.

Synchronous power switching

A synchronous power switch is one in which the triac invariably turns on just after the start of each power half-cycle and then turns off again automatically at the end of it, thus generating minimal RFI. In all power switching circuits shown so far in this chapter the triac turns on at an arbitrary point in its initial 'switch-on' half-cycle, thus producing a potentially high *initial* burst of RFI, but then gives synchronous action on all subsequent half-cycles.

A truly synchronous circuit uses the switching system of *Figure 20.19*, in which the triac can only be gated on near the start or 'zero voltage' point of each half-cycle, and thus produces minimal switching RFI. This system is widely used to give on/off control of high-current loads such as electric heaters, etc.

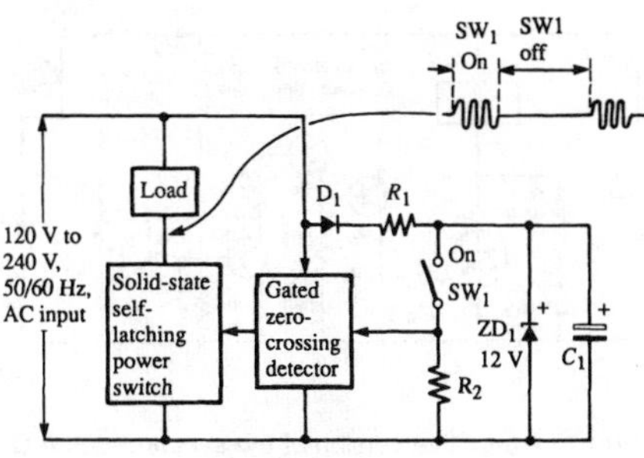

Figure 20.19. Synchronous zero-voltage A.C. power switching system.

Figure 20.20 shows a practical synchronous A.C. power switch. 10V D.C. is A.C.-derived via R_1–D_1–ZD_1 and C_1 and is switched to the triac gate via Q_5, which is controlled via SW_1 and 'zero-voltage' detector Q_2–Q_3–Q_4 and can supply gate current only when SW_1 is closed and Q_4 is off. In the zero-voltage detector, Q_2 or Q_3 are driven on whenever the A.C. line voltage is more than a few volts (set by RV_1) above or below zero, thereby driving Q_4 on via R_3 and inhibiting Q_5. Thus, gate current can only be fed to the triac when SW_1 is closed and the instantaneous A.C. line voltage is within a few volts of zero; this circuit thus generates minimal switching RFI. *Figure 20.21* shows the circuit modified so that the triac can only turn on when SW_1 is open. Note in both cases that only a narrow pulse of gate current is fed to the triac, and the mean gate current is thus only 1mA or so. SW_1 can be replaced by an electronic switch or optocoupler, if desired.

CA3059 circuits

Several dedicated synchronous 'zero-voltage' triac-gating ICs are available; the best known of these is the 14-pin CA3059, which has built-in A.C.-derived D.C. power supply circuitry, a 'zero-voltage'

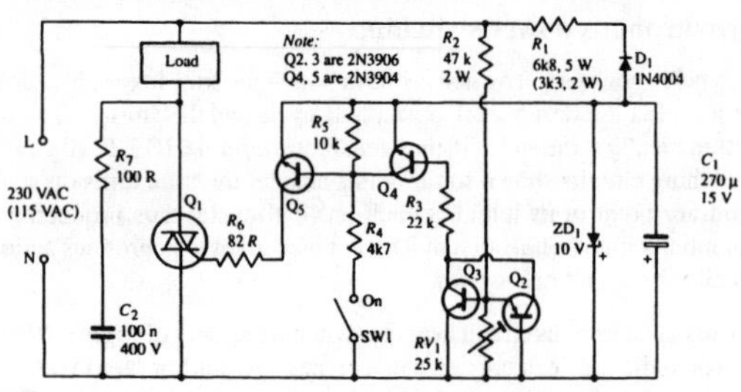

Figure 20.20. Synchronous A.C. power switch.

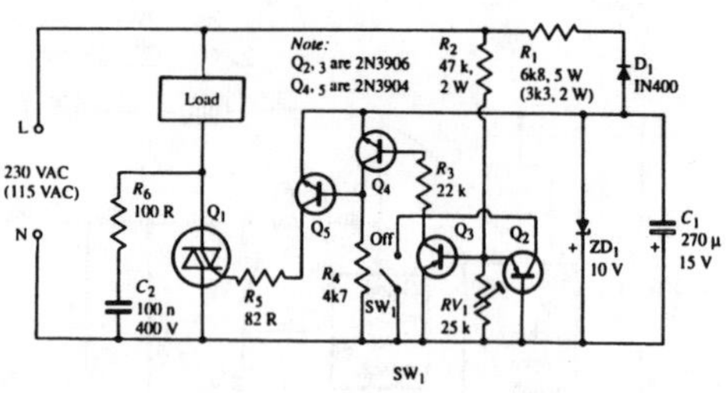

Figure 20.21. Alternative version of the synchronous A.C. power switch.

detector, triac gate drive circuitry, and a high-gain differential amplifier/gating network. *Figures 20.22* and *20.23* show basic ways of using it to give manually controlled synchronous on/off switching of a triac. The circuits use SW_1 to enable or disable the triac gate drive via the IC's internal differential amplifier (the drive is enabled only when pin 13 is biased above pin 9).

In *Figure 20.22*, pin 9 is biased at half-supply volts and pin 13 is biased via R_2–R_3 and SW_1, and the triac thus turns on only when SW_1 is closed. In *Figure 20.23*, pin 13 is biased at half-supply volts and pin 9 is biased via R_2–R_3 and SW_1, and the triac again turns on only when SW_1 is closed. In both circuits, SW_1 handles maximum potentials of 6V and maximum currents of about 1mA. C_2 is used to apply a slight phase delay to the pin 5 'zero-voltage detecting' terminal, and causes the gate pulses to occur after (rather than to 'straddle') the zero-voltage point.

Note in *Figure 20.23* that the triac can be turned on by pulling R_3 low or can be turned off by letting R_3 float. *Figures 20.24* and *20.25* show ways of using this fact to extend the versatility of the basic circuit. In *Figure 20.24* the triac can be turned on and off via Q_2, which in turn can be activated by on-board CMOS circuitry (such as one-shots, astables, etc.) that are powered from the 6V pin 2 supply. In *Figure 20.25* the circuit can be turned on and off by fully isolated external circuitry via an optocoupler.

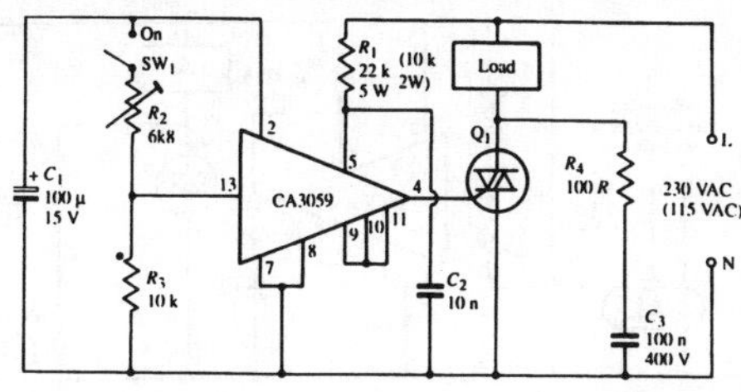

Figure 20.22. Direct-switching IC-gated zero-voltage A.C. switch.

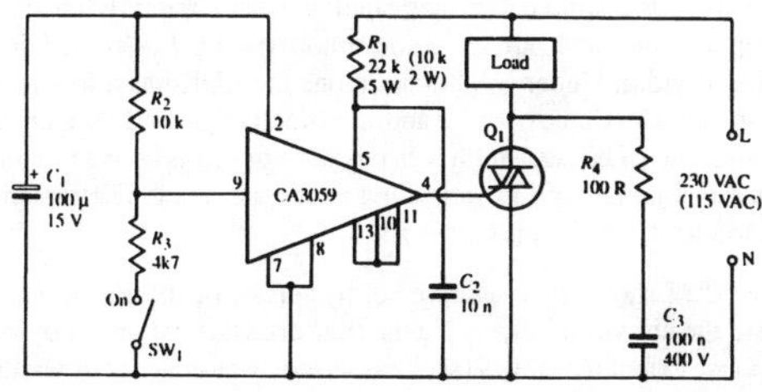

Figure 20.23. An alternative method of direct-switching the CA3059 IC.

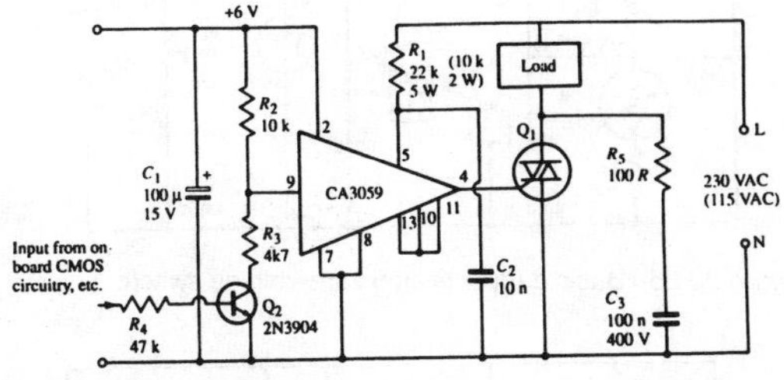

Figure 20.24. Method of transistor-switching the CA3059 via on-board CMOS circuitry, etc.

Figures 20.26 and *20.27* show ways of using the CA3059 to give light-sensitive 'dark-operated' synchronous A.C. power switch operation. In these designs the IC's built-in differential amplifier is used as a precision voltage comparator that turns the triac on or off when one of the comparator input voltages goes above or below the other.

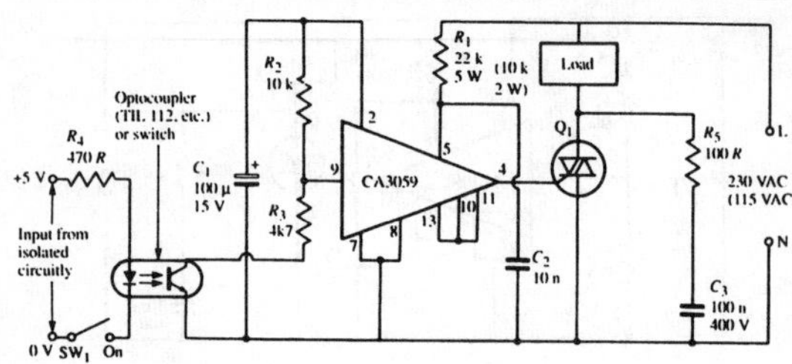

Figure 20.25. Method of remote-switching the CA3059 via an optocoupler.

Figure 20.26 is a simple dark-activated power switch. Pin 9 is tied to half-supply volts and pin 13 is controlled via the R_2–RV_1–LDR–R_3 potential divider. Under bright conditions the LDR has a low resistance, so pin 13 is below pin 9 and the triac is disabled. Under dark conditions the LDR has a high resistance, so pin 13 is above pin 9 and the triac is enabled and power is fed to the a.c. load. The threshold switching level can be preset via RV_1.

Figure 20.27 shows how a degree of hysteresis or 'backlash' can be added to the above circuit, so that the triac does not switch annoyingly in response to small changes (such as caused by passing shadows, etc.) in ambient light level. The hysteresis level is controlled via R_3, which can be selected to suit individual applications.

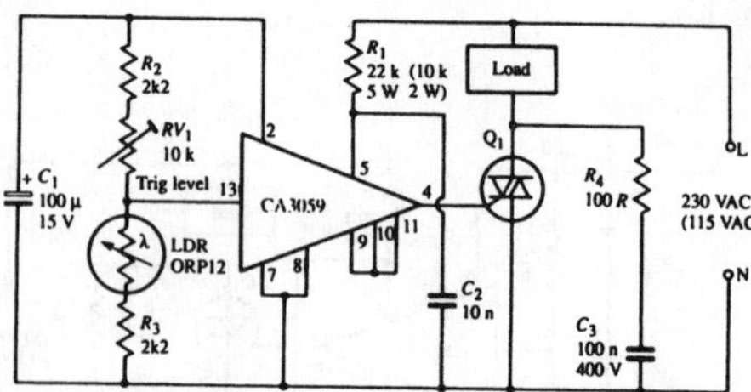

Figure 20.26. Basic dark-activated zero-voltage switch.

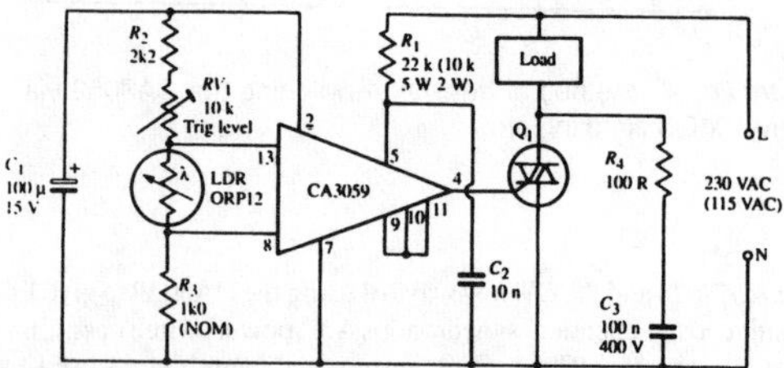

Figure 20.27. Dark-activated zero-voltage switch with hysteresis provided via R_3.

Electric heater controllers

Triacs can be used to give automatic room temperature control, via electric heaters, by using thermostats or thermistors as feedback control elements. Electric heaters consume high currents, and must be switched synchronously to avoid excessive RFI generation. The circuits of *Figures 20.22* of *20.23* can be used to give automatic on/off control by using a thermostat in the SW_1 position. Alternatively, thermistor control can be obtained by using the circuit of *Figure 20.28*, which is a simple modification of the *Figure 20.26* design except that n.t.c. thermistor TH_1 is used as the feedback control element.

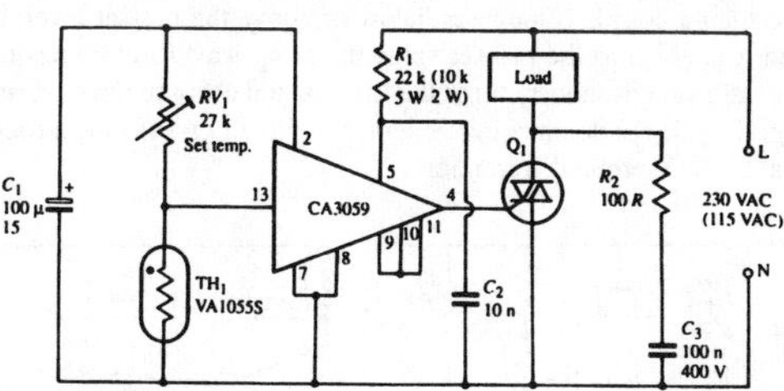

Figure 20.28. Heater controller with thermistor-regulated zero-voltage switching.

These automatic on/off control circuits can hold a room's temperature to within a degree or two of a pre-set value. Finer control can be obtained by using the synchronous 'burst-fire' or integral-cycle proportional power control system of *Figure 20.29*, which adjusts the mean heater power so that, when the room temperature is at the precise pre-set level, its heat output exactly balance the thermal losses of the room. In this system power bursts of complete (synchronous) half-cycles are fed to the heater at regular line-frequency-related intervals.

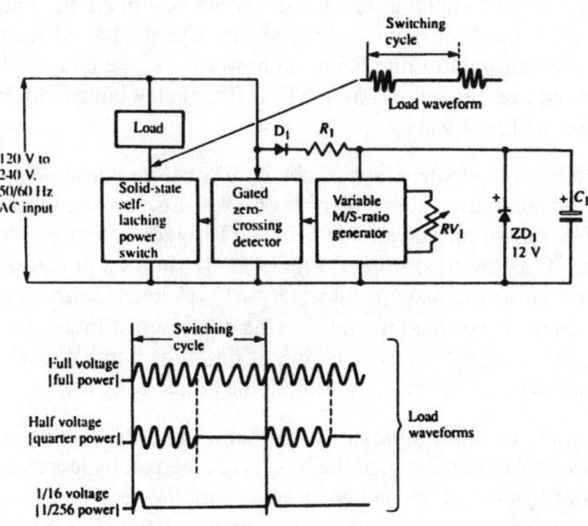

Figure 20.29. Burst-fire (integral-cycle) A.C. power controller.

Thus, if bursts are repeated at 8-cycle intervals, the mean load voltage equals the full supply line value if the bursts are of 8-cycle duration, or half-voltage (equals quarter power) at 4-cycle duration, or one sixteenth voltage (equals 1/256th power) at one half-cycle duration, etc.

Figure 20.30 shows a practical 'burst-fire' heater controller that can hold room temperatures to within ±0.5°C of a pre-set value. Here, a thermistor-controlled voltage is applied to the pin-13 side of the CA3059's comparator, and a repetitive 300mS ramp waveform, centred on half-supply volts, is applied to the pin-9 side of the comparator from CMOS astable IC_1. The action is such that the triac is synchronously gated fully on or is cut fully off if the ambient temperature is more than a couple of degrees below or above the pre-set level, but when it is close to the pre-set value the ramp waveform comes into effect and synchronously turns the triac on and off in the burst-fire or integral cycle mode once every 300mS, with an on/off ratio proportional to the thermal differential.

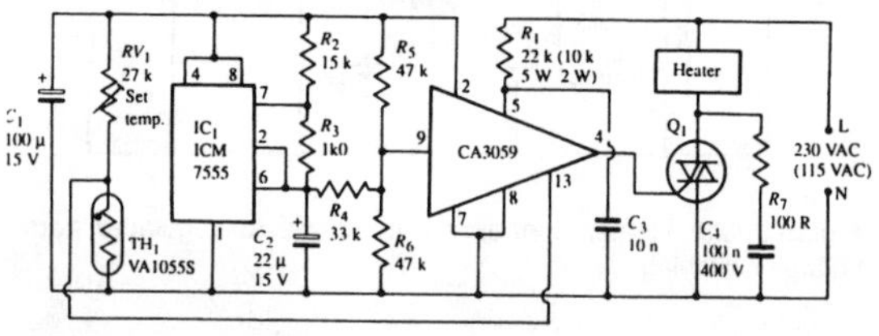

Figure 20.30. Heater controller giving integral-cycle precision temperature regulation.

A.C. lamp dimmer circuits

Triacs can be used to make very efficient lamp dimmers by using the 'phase-delayed switching' technique already described, in which the triac is gated on at some phase-delayed time after the start of each A.C. half-cycle, thus controlling the mean power fed to the lamp. All such circuits require the use of a simple L–C filter in the lamp feed line, to minimise RFI problems.

The three most popular ways of obtaining variable phase-delay triac triggering are to use either a diac plus *C–R* phase delay network, or to use a line-synchronised variable-delay UJT trigger, or to use a special-purpose IC as the triac trigger. *Figure 20.31* shows a practical diac-triggered lamp dimmer, in which R_1–RV_1–C_1 provide the variable phase-delay. This circuit is similar to the basic lamp dimmer circuit of *Figure 20.4*, except for the addition of on/off switch SW_1, which is ganged to RV_1 and enables the lamp to be turned fully off.

A weakness of this simple design is that it has considerable control hysteresis or backlash, e.g., if the lamp is dimmed off by increasing the RV_1 value to 470k, it will not go on again until RV_1 is reduced to about 400k, and then burns at a fairly high brightness level. This backlash is caused by the diac partially discharging C_1 each time the triac fires.

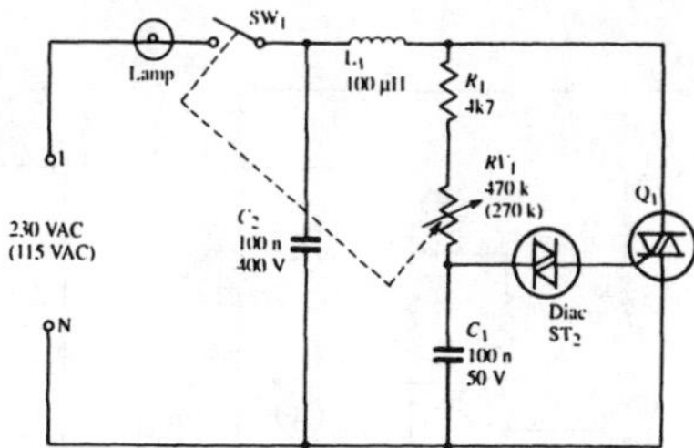

Figure 20.31. Practical circuit of a simple lamp dimmer.

Backlash can be greatly reduced by using the 'gate slaving' technique of *Figure 20.32*, in which the diac is triggered from C_2, which 'follows' the C_1 phase-delay voltage but protects C_1 from discharging when the Diac fires.

If absolutely zero backlash is needed, the UJT-triggered circuit of *Figure 20.33* can be used. The UJT is powered from a 12V d.c. supply derived from the A.C. line via R_1–D_1–ZD_1–C_1 and is line-synchronised via the Q_2–Q_3–Q_4 zero-voltage detector network, the action being such that Q_4 is turned on (applying power to the UJT) at all times other than when the A.C. line voltage is close to the zero-crossover point at the end and start of each A.C. half-cycle. Thus, shortly after the start of each half-cycle, power is applied to the UJT circuit via Q_4, and some time later (determined by R_5–RV_1–C_2) a trigger pulse is applied to the triac gate via Q_5. The UJT resets at the end of each half-cycle, and a new sequence then begins.

A smart lamp dimmer IC

Many modern lamp dimmers have their triac driven via a dedicated 'smart' IC that takes its commands via a touch-sensitive pad or push-button input switch. Siemens are leaders in this field, and their latest IC (introduced in 1990) is the SLB0586 (see *Figure 20.34*), which needs a 5.6V, 0.45mA, D.C. supply and incorporates touch conditioning circuitry and an 'options' control (pin-2) that allows the user to select a variety of dimming or switching modes.

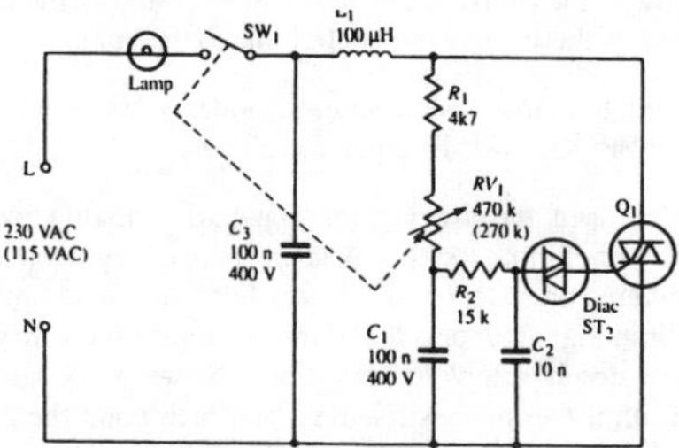

Figure 20.32. Improved lamp dimmer with gate slaving.

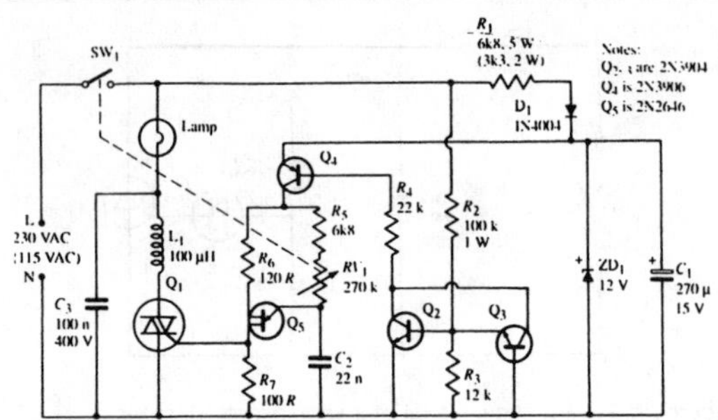

Figure 20.33. UJT-triggered zero-backlash lamp dimmer.

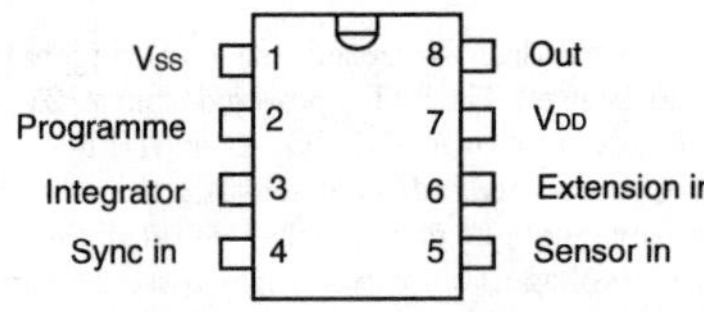

Figure 20.34. SLB0586 outline and pin notations.

Figure 20.35 shows the SLB0586 basic applications circuit, using a single touch-sensitive input control. The IC is powered (between pins 1 and 7) from a 5.6V D.C. supply derived from theA.C. line via R_2–C_2–ZD_1–D_1–C_3, and has its EXTENSION input disabled by shorting pin-6 to pin-7. The pin-5 SENSOR input works on the induced pick-up principle, in which the operators body picks up radiated A.C. power line signals that are detected via the conductive touch pad, which must be placed close to the IC to avoid false pick-up. The operator is protected from the power line voltage via R_7–R_8; for correct operation the A.C. power lines must be connected as shown, with the live or hot lead to pin-1 of the IC, and the neutral line to the lamp.

The SLB0586 has three basic operating modes, which can be selected via its pin-2 PROGRAMME input, as follows.

If pin-2 is left open, the circuit gives 'standard' dimming operation in which a very brief input touch makes the lamp change state (from off to a remembered on state, or *vice versa*), but a sustained (greater than 400mS) 'dimming' input puts the IC into a ramping mode in which the lamp power slowly ramps up and down (between 3% and 97% of maximum) until the input is released, at which point the prevailing power level is held and 'remembered'; the ramp direction reversed on alternate 'dimming' touches.

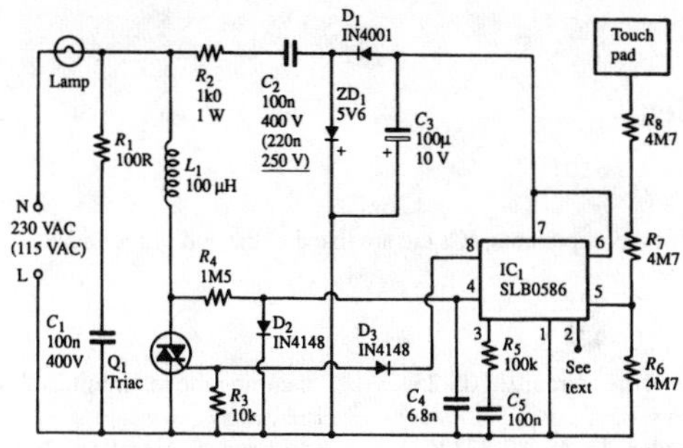

Figure 20.35. Basic SLB0586 lamp dimmer circuit, with touch-sensitive control

If pin-2 is shorted to pin-7, the lamp goes to maximum brightness when switched on, and in dimming operations the lamp starts at minimum brightness and then slowly ramps up and down until the sensor is released; the ramping direction does not reverse on successive dimming operations.

If pin-2 is shorted to pin-1, the lamp operation is like that just described, except that the ramping direction reverses on successive dimming operations.

Index

Devices, components, ICs etc are listed at the end of the index

Devices, components, ICs etc. by type number